Materials Science and Technologies

nova
science publishers
www.novapublishers.com

Materials Science and Technologies

Ferrite Materials and Technologies
Ravi Panwar, PhD (Editor)
Dharmendra Singh (Editor)
2023. ISBN: 979-8-89113-086-9 (Hardcover)

Perovskite Solar Cells: From Materials Science to Device Engineering
Thembinkosi Donald Malevu, PhD (Editor)
2023. ISBN: 979-8-89113-130-9 (Softcover)
2023. ISBN: 979-8-89113-188-0 (eBook)

Physics and Mechanics of New Materials and Their Applications, 2021 – 2022
Ivan A. Parinov, DrSc (Editor)
Shun-Hsyung Chang, PhD (Editor)
Arkady N. Soloviev, DrSc (Editor)
2023. ISBN: 979-8-88697-542-0 (eBook)

Recycling of Discarded Carpets for Structural Polymer Composites
Rajesh Kumar Verma, PhD (Editor)
Jogendra Kumar, PhD (Editor)
Shivi Kesarwani (Editor)
2023. ISBN: 979-8-88697-538-3 (Softcover)
2023. ISBN: 979-8-88697-588-8 (eBook)

Magnesium Alloys: Advances in Research and Applications
Catalin Iulian Pruncu, PhD (Editor)
Kavian Omar Cooke, PhD (Editor)
2022. ISBN: 978-1-68507-975-8 (Softcover)
2022. ISBN: 979-8-88697-237-5 (eBook)

More information about this series can be found at
https://novapublishers.com/product-category/series/materials-science-and-technologies/

James R. Bush
Editor

Biocomposites

Advances in Research and Applications

nova
science publishers
www.novapublishers.com

NOTICE TO THE READER

Library of Congress Cataloging-in-Publication Data

ISBN: 979-8-89113-267-2

Published by Nova Science Publishers, Inc. † New York

Contents

Preface

This book contains six chapters that detail advances in the research and applications of biocomposites. Chapter One discusses calcium orthophosphate (CaPO4) biocomposites with natural and synthetic polymers suitable for biomedical applications. Chapter Two provides a summary of the recent advances in the development of boron biomaterials that respond to specific biomedical applications. Chapter Three reveals the current findings on biocomposite materials used in the field of orthodontics. The goal of Chapter Four is to determine the effects of multi-wall carbon nanotubes and calcined Sepia Officinalis cuttlefish bone hydroxyapatite as reinforcing material. Chapter Five provides information on various literature that reported on drug release studies of naturally-synthesized hydroxyapatite composites. Lastly, Chapter Six discusses the advantages of dental bioceramics, their compositions, biological evolutions, and how their physical properties relate to clinical findings.

Chapter 1

Calcium Orthophosphate ($CaPO_4$) Biocomposites with Natural and Synthetic Polymers

Sergey V. Dorozhkin*
Faculty of Physics, M. V. Lomonosov Moscow State University, Moscow, Russia

Abstract

The state-of-the-art on calcium orthophosphate ($CaPO_4$) biocomposites with natural and synthetic polymers suitable for biomedical applications is presented. Since these types of biomaterials offer many significant and exciting possibilities for hard tissue regeneration, this subject belongs to a rapidly expanding area of biomedical research. Through the successful combinations of the desired properties of matrix materials with those of fillers (in such systems, both $CaPO_4$ and polymers might play either role), innovative bone graft biomaterials can be designed. Various types of $CaPO_4$ biocomposites with polymers that are either already in use or being investigated for biomedical applications are extensively discussed. Many different formulations in terms of the material constituents, fabrication technologies, structural and bioactive properties, as well as both *in vitro* and *in vivo* characteristics have been already proposed. Among them, the nano-structurally controlled biocomposites, those containing nanodimensional compounds, biomimetically fabricated formulations with collagen, chitin and/or gelatin seem to be the most promising candidates for clinical applications. The specific advantages of using $CaPO_4$-based biocomposites with polymers are highlighted.

* Corresponding Author's Email: sedorozhkin@yandex.ru.

In: Biocomposites
Editor: James R. Bush
ISBN: 979-8-89113-267-2

Keywords: calcium orthophosphates, hydroxyapatite, biocomposites, bone grafts, biomaterials

Abbreviations

BMP	bone morphogenetic protein
BSA	bovine serum albumin
CMC	carboxymethylcellulose
DNA	desoxynucleic acid
EVOH	a copolymer of ethylene and vinyl alcohol
IBS	injectable bone substitute
HDPE	high-density polyethylene
HPMC	hydroxypropylmethylcellulose
PA	polyamide
PAA	polyacrylic acid
PBT	polybutyleneterephthalate
PCL	poly(ε-caprolactone)
PDLA	poly(D-lactic acid)
PDLLA	poly(D,L-lactic acid)
PE	polyethylene
PEEK	polyetheretherketone
PEG	polyethylene glycol
PGA	polyglycolic acid
PHB	polyhydroxybutyrate
PHBHV	poly(hydroxybutyrate-*co*-hydroxyvalerate)
PHEMA	polyhydroxyethyl methacrylate
PLA	polylactic acid
PLGA	poly(lactic-*co*-glycolic) acid
PLGC	*co*-polyester lactide-*co*-glycolide-*co*-ε-caprolactone
PLLA	poly(L-lactic acid)
PMMA	polymethylmethacrylate
PP	polypropylene
PPF	poly(propylene-*co*-fumarate)
PSZ	partially stabilized zirconia
PTMC	poly(trimethylene carbonate)
PU	polyurethane
PVA	polyvinyl alcohol
PVAP	polyvinyl alcohol phosphate
SEVA-C	a blend of EVOH with starch

1. Introduction

The fracture of bones due to various traumas or natural aging is a typical type of a tissue failure. An operative treatment frequently requires implantation of a temporary or a permanent prosthesis, which still is a challenge for orthopedic surgeons, especially in the cases of large bone defects. A fast aging of the population and serious drawbacks of natural bone grafts make the situation even worse; therefore, there is a high clinical demand for bone substitutes. Unfortunately, a medical application of xenografts (e.g., bovine bone) is generally associated with potential viral infections. In addition, xenografts have a low osteogenicity, an increased immunogenicity and, usually, resorb more rapidly than autogenous bone. Similar limitations are also valid for human allografts (i.e., tissue transplantation between individuals of the same species but of non-identical genetic composition), where the concerns about potential risks of transmitting tumor cells, a variety of bacterial and viral infections, as well as immunological and blood group incompatibility are even stronger. Moreover, harvesting and conservation of allografts (exogenous bones) are additional limiting factors [1-3]. Autografts (endogenous bones) are still the "golden standard" among any substitution materials because they are osteogenic, osteoinductive, osteoconductive, completely biocompatible, non-toxic and do not cause any immunological problems (non-allergic). They contain viable osteogenic cells, bone matrix proteins and support bone growth. Usually, autografts are well accepted by the body and rapidly integrated into the surrounding bone tissues. Due to these reasons, they are used routinely for a long period with good clinical results [2-5]; however, it is fair to say on complication cases, those frequently happened in the past [6]. Unfortunately, a limited number of donor sites restrict the quantity of autografts harvested from the iliac crest or other locations of the patient's own body. In addition, their medical application is always associated with additional traumas and scars resulting from the extraction of a donor tissue during a superfluous surgical operation, which requires further healing at the donation site and can involve long-term postoperative pain. Thus, any types of biologically derived transplants appear to be imperfect solutions, mainly due to a restricted quantity of donor tissues, donor site morbidity, as well as potential risks of an immunological incompatibility and disease transfer [7-9]. In this light, fabricated materials (alloplastic or synthetic bone grafts) stand out as a reasonable option because they are easily available, might be processed and modified to suit the specific needs of a given application. What is more, there are no concerns about potential infections, immunological incompatibility, sterility and donor site morbidity. Therefore, investigations on artificial materials for bone tissue repair appear to be one of the key subjects in the field of biomaterials research for clinical applications [10, 11].

Currently, there are several classes of synthetic bone grafting biomaterials for *in vivo* applications. The examples include natural coral, coral-derived materials, bovine porous demineralized bone, human demineralized bone matrix, bioactive glasses, glass-ceramics and $CaPO_4$ [12, 13]. Among them, porous bioceramics made of $CaPO_4$ appear to be very prominent due to both the excellent biocompatibility and bonding ability to living bone in the body. This is directly related to the fact that the inorganic material of mammalian calcified tissues, i.e., of bone and teeth, consists of $CaPO_4$ [14, 15]. Due to this reason, other artificial materials are normally encapsulated by fibrous tissue, when implanted in body defects, while $CaPO_4$ are not. Many types of $CaPO_4$-based bioceramics with different chemical composition are already on the market [16, 17]. Unfortunately, as for any ceramic material, $CaPO_4$ bioceramics alone lack the mechanical and elastic properties of the calcified tissues. Namely, scaffolds made of $CaPO_4$ only suffer from a low elasticity, a high brittleness, a poor tensile strength, a low mechanical reliability and fracture toughness, which leads to various concerns about their mechanical performance after implantation. Besides, in many cases, it is difficult to form $CaPO_4$ bioceramics into the desired shapes [16, 17].

The superior strength and partial elasticity of biological calcified tissues (e.g., bones) are due to the presence of bioorganic polymers (mainly, collagen type I fibers) rather than to a natural ceramic (mainly, a poorly crystalline ion-substituted calcium-deficient hydroxyapatite (CDHA), often referred to as "biological apatite") phase [18-21]. The elastic collagen fibers are aligned in bone along the main stress directions. The biochemical composition of bones is given in Table 1 [22]. A decalcified bone becomes very flexible being easily twisted, whereas a bone without collagen is very brittle; thus, the inorganic nano-sized crystals of biological apatite provide with the hardness and stiffness, while the bioorganic fibers are responsible for the elasticity and toughness. In bones, both types of materials integrate each other into a nanometric scale in such a way that the crystallite size, fibers orientation, short-range order between the components, *etc.* determine its nanostructure and therefore the function and mechanical properties of the entire composite. From the mechanical point of view, bone is a tough material at low strain rates but fractures more like a brittle material at high strain rates; generally, it is rather weak in tension and shear, particularly along the longitudinal plane.

Besides, bone is an anisotropic material because its properties are directionally dependent [18-21].

Table 1. The biochemical composition* of bones [22]

Inorganic phases	wt. %	Bioorganic phases	wt. %
$CaPO_4$ (biological apatite)	~ 60	collagen type I	~ 20
water	~ 9	non-collagenous proteins: osteocalcin, osteonectin, osteopontin, thrombospondin, morphogenetic proteins, sialoprotein, serum proteins	~ 3
carbonates	~ 4	other traces: polysaccharides, lipids, cytokines	balance
citrates	~ 0.9	primary bone cells: osteoblasts, osteocytes, osteoclasts	balance
sodium	~ 0.7		
magnesium	~ 0.5		
other traces: Cl^-, F^-, K^+ Sr^{2+}, Pb^{2+}, Zn^{2+}, Cu^{2+}, Fe^{2+}	balance		

* The composition is varied from species to species and from bone to bone.

It remains a great challenge to design the ideal bone graft that emulates nature's own structures or functions. Certainly, the successful design requires an appreciation of the bones' structure. According to expectations, the ideal bone graft should be benign, available in a variety of forms and sizes, all with sufficient mechanical properties for use in load-bearing sites, form a chemical bond at the bone/implant interface, as well as be osteogenic, osteoinductive, osteoconductive, biocompatible, completely biodegradable at the expense of bone growth and moldable to fill and restore bone defects [23, 24]. Further, it should resemble the chemical composition of bones (thus, the presence of $CaPO_4$ is mandatory), exhibit contiguous porosity to encourage invasion by the live host tissue, as well as possess both viscoelastic and semi-brittle behavior, as bones do [25-27]. Moreover, the degradation kinetics of the ideal implant should be adjusted to the healing rate of the human tissue with absence of any chemical or biological irritation and/or toxicity caused by substances, which are released due to corrosion or degradation. Ideally, the combined mechanical strength of the implant and the ingrowing bone should remain constant throughout the regenerative process. Furthermore, the substitution implant material should not disturb significantly the stress environment of the surrounding living tissue [28]. Finally, there is an opinion, that in the case of a serious trauma, bone should fracture rather than the implant [23]. A good sterilizability, storability and processability, as well as a relatively low cost are also of a great importance to permit a clinical application. Unfortunately, no artificial biomaterial is yet available, which embodies all these requirements and unlikely it will appear in the nearest future. To date, most of the available biomaterials appear to be either predominantly osteogenic or osteoinductive or else purely osteoconductive [1].

Mimicking the structure of calcified tissues and addressing the limitations of the individual materials, development of organic-inorganic composite biomaterials provides excellent possibilities for improving the conventional bone implants. In this sense, suitable biocomposites of tailored physical, biological and mechanical properties with the predictable degradation behavior can be prepared combining biologically relevant $CaPO_4$ with bioresorbable polymers [29]. As a rule, the general behavior of such biocomposites is dependent on nature, structure and relative contents of the constitutive components, although other parameters such as the preparation conditions also determine the properties of the final materials. Currently, $CaPO_4$ is incorporated as either a filler or a coating (or both) either into or onto a biodegradable polymer matrix, in the form of particles or fibers, are increasingly considered for using as bone tissue engineering scaffolds due to their improved physical, biologic and mechanical properties [30-33]. In addition, such biocomposites could fulfill general requirements to the next generation of biomaterials, those should combine the bioactive and bioresorbable properties to activate *in vivo* mechanisms of tissue regeneration, stimulating the body to heal itself and leading to replacement of the implants by the regenerating tissue. Thus, through the successful combinations of ductile polymer matrixes with hard and bioactive particulate bioceramic fillers, optimal materials can be designed and, ideally, this approach could lead to a superior construction to be used as either implants or posterior dental restorative material [29, 34].

A lint-reinforced plaster was the first composite used in clinical orthopedics as an external immobilizer (bandage) in the treatment of bone fracture by Mathijsen in 1852 [35], followed by Dreesman in 1892 [36]. Great progress in the clinical application of various types of composite materials has been achieved since then. Based on both the experience and the newly gained knowledge, various composite materials with tailored mechanical and biological performance can be manufactured and used to meet various clinical requirements [37]. However, this review presents only a brief history and advances in the field of $CaPO_4$-based biocomposites suitable for biomedical application. Most of the reviewed literature is restricted to the recent publications; a limited number of papers published in the XX-th century have been cited. Various aspects of the material constituents, fabrication technologies, structural and bioactive properties, as well as phase interaction have been considered and discussed in details. Finally, several critical issues and scientific challenges that are needed for further advancement are outlined.

2. General Information and Knowledge

According to Wikipedia, the free encyclopedia, "A *composite material* (also called a *composition material* or shortened to *composite*, which is the common name) is a material which is produced from two or more constituent materials. These constituent materials have notably dissimilar chemical or physical properties and are merged to create a material with properties unlike the individual elements. Within the finished structure, the individual elements remain separate and distinct, distinguishing composites from mixtures and solid solutions" [38]. Thus, composites are always heterogeneous. Furthermore, the phases of any composite retain their identities and properties, and are bonded, which is why an interface is maintained between them. This provides improved specific or synergistic characteristics that are not obtainable by any of the original phases alone [39]. Following the point of view of some predecessors, we also consider that "for the purpose of this review, composites are defined as those having a distinct phase distributed through their bulk, as opposed to modular or coated components" [40, p. 1329]. For this reason, with a few important exceptions, the structures obtained by soaking of various materials in supersaturated solutions containing ions of calcium and orthophosphate (e.g., Refs. [41-44]), those obtained by coating of various materials by $CaPO_4$ (reviewed in Refs. [45-47]), as well as $CaPO_4$ coated by other compounds [48-51] have not been considered; however, composite coatings have been considered. Occasionally, porous $CaPO_4$ scaffolds filled by cells inside the pores [52-54], as well as $CaPO_4$ impregnated by biologically active substances [55-57] are also defined as composites; nevertheless, such structures have not been considered either.

In any composite, there are two major categories of constituent materials: a matrix (or a continuous phase) and (a) dispersed phase(s). To create a composite, at least one portion of each type is required. General information on the major fabrication and processing techniques might be found elsewhere [40, 58]. The continuous phase is responsible for filling the volume, as well as it surrounds and supports the dispersed material(s) by maintaining their relative positions. The dispersed phase(s) is(are) usually responsible for enhancing one or more properties of the matrix. Most of the composites target an enhancement of mechanical properties of the matrix, such as stiffness and strength; however, other properties, such as erosion stability, transport properties (electrical or thermal), radiopacity, density or biocompatibility might also be of a great interest. This synergism produces the properties, which are unavailable from the individual constituent materials [58, 59]. What is more, by controlling the volume fractions and local and global arrangement of the dispersed phase, the properties and design of composites can be varied and tailored to suit the necessary conditions. For example, in the case of ceramics, the dispersed phase serves to impede crack growth. In this case, it acts as reinforcement. Several methods, including deflecting crack tips, forming bridges across crack faces, absorbing energy during pullout and causing a redistribution of stresses in regions, adjacent to crack tips, can be used to accomplish this [60]. Other factors to be considered in composites are the volume percentage of the dispersed phase(s), its(their) dimensions, shape and orientation, a reinforcement/matrix interfacial state, as well as a homogeneity of the overall composite. For example, higher volume fractions of reinforcement phases tend to improve the mechanical properties of the composites, while continuous and aligned fibers best prevent crack propagation with the added property of anisotropic behavior. From a structural point of view, composites are anisotropic in nature: their mechanical properties are different in different directions. Furthermore, the uniform distribution of the dispersed phase is also desirable, as it imparts consistent properties to the composite [38, 58, 59].

To prepare any type of a composite, at least, two different materials must be mixed. Thus, a phase miscibility phenomenon appears to be of the paramount importance [61, 62]. Furthermore, the interfacial strength among the phases is a very important factor because a lack of adhesion among the phases will result in an early failure at the interface and thus in a decrease in the mechanical properties, especially the tensile strength. From a chemical point of view, one can distinguish several types of the interactions among the composite components: materials with strong (covalent, coordination, ionic) interactions; those with weak interactions (van der Waals forces, hydrogen bonds, hydrophilic-hydrophobic balance) or without chemical interactions among the components [63]. Wetting is also important in bonding or adherence of the materials. It depends on the hydrophilicity or polarity of the filler(s) and the available polar groups of the matrix.

Regarding *bio*composites, they are defined by 2 complementary interpretations. First, biocomposites are nontoxic composites able to interact well with the human body *in vivo* and, ideally, contain one or more component(s) that stimulate(s) the healing process and uptake of the implant [64]. Second, biocomposites are biomaterials of a composite nature, while the latest definition for biomaterials is this: "biomaterial is a material designed to take a form that can direct, through interactions with living systems, the course of any therapeutic or diagnostic procedure" [65]. In any case, a biological compatibility for biocomposites appears to be much more important than any other type of compatibility [37, 66, 67]. Interestingly that according to Scopus database, the earliest paper with the term "biocomposite" in the title was published in 1987 [68] and the one containing a combination of terms "biocomposite" and HA in the title was published in 1991 [69]. However, without a prefix "bio," the earliest paper with the combination of terms "composite" and HA in the title was published in 1976 [70]. Currently (August 2023), according to Scopus, there are 6513 publications with the combination of terms "composite" and "apatite" in their titles and 1743 publications with the combination of terms "composite" and "calcium phosphate" in their titles; thus, this is a hot topic of research.

Table 2. General respective properties from the bioorganic and inorganic domains, to be combined in biocomposites [24]

inorganic	bioorganic
hardness, brittleness	elasticity, plasticity
high density	low density
thermal stability	permeability
hydrophilicity	hydrophobicity
high refractive index	selective complexation
mixed valence slate (red-ox)	chemical reactivity
strength	bioactivity

The most common properties from the bioorganic and inorganic domains to be combined in biocomposites have been summarized in Table 2 [24]. For general advantages of the modern $CaPO_4$-based biocomposites over $CaPO_4$ bioceramics and bioresorbable polymers individually, the interested readers are advised to get through "Composite materials strategy" section of Ref. [29].

3. Two Major Constituents

3.1. $CaPO_4$

$CaPO_4$ were first mentioned in 1769 as the major constituents of bones and have been investigated since then [71, 72]. The main driving force behind the use of $CaPO_4$ as bone substitute materials is their chemical similarity to the mineral component of mammalian bones and teeth [14-16]. As a result, in addition to being non-toxic, they are biocompatible, not recognized as foreign materials in the body and, most importantly, both exhibit bioactive behavior and integrate into living tissue by the same processes active in remodeling healthy bone. This leads to an intimate physicochemical bond between the implants and bone, termed osteointegration. More to the point, $CaPO_4$ are also known to support osteoblast adhesion and proliferation. Even so, the major limitations to use $CaPO_4$ as load-bearing biomaterials are their mechanical properties; namely, they are brittle with poor fatigue resistance [23]. The poor mechanical behavior is even more evident for highly porous ceramics and scaffolds because porosity greater than 100 μm is considered as the requirement for proper vascularization and bone cell colonization [73, 74]. That is why, in biomedical applications $CaPO_4$ are used primarily as fillers and coatings [16].

The complete list of known $CaPO_4$, including their standard abbreviations and the major properties, is given in Table 3, while the detailed information on $CaPO_4$, might be found in special books and monographs [16, 75-78].

Table 3. Existing $CaPO_4$ and their major properties [16]

Ca/P molar ratio	Compound	Formula	Solubility at 25°C, -log(K_s)	Solubility at 25°C, g/L	pH stability range in aqueous solutions at 25°C
0.5	Monocalcium phosphate monohydrate (MCPM)	$Ca(H_2PO_4)_2 \cdot H_2O$	1.14	~ 18	0.0 – 2.0
0.5	Monocalcium phosphate anhydrous (MCPA or MCP)	$Ca(H_2PO_4)_2$	1.14	~ 17	[c]
1.0	Dicalcium phosphate dihydrate (DCPD), mineral brushite	$CaHPO_4 \cdot 2H_2O$	6.59	~ 0.088	2.0 – 6.0
1.0	Dicalcium phosphate anhydrous (DCPA or DCP), mineral monetite	$CaHPO_4$	6.90	~ 0.048	[c]
1.33	Octacalcium phosphate (OCP)	$Ca_8(HPO_4)_2(PO_4)_4 \cdot 5H_2O$	96.6	~ 0.0081	5.5 – 7.0
1.5	α-Tricalcium phosphate (α-TCP)	α-$Ca_3(PO_4)_2$	25.5	~ 0.0025	[a]
1.5	β-Tricalcium phosphate (β-TCP)	β-$Ca_3(PO_4)_2$	28.9	~ 0.0005	[a]
1.2 – 2.2	Amorphous calcium phosphates (ACP)	$Ca_xH_y(PO_4)_z \cdot nH_2O$, n – 3 – 4.5; 15 – 20% H_2O	[b]	[b]	~ 5 – 12 [d]
1.5 – 1.67	Calcium-deficient hydroxyapatite (CDHA or Ca-def HA)[e]	$Ca_{10-x}(HPO_4)_x(PO_4)_{6-x}(OH)_{2-x}$ $(0<x<1)$	~ 85	~ 0.0094	6.5 – 9.5
1.67	Hydroxyapatite (HA, HAp or OHAp)	$Ca_{10}(PO_4)_6(OH)_2$	116.8	~ 0.0003	9.5 – 12
1.67	Fluorapatite (FA or FAp)	$Ca_{10}(PO_4)_6F_2$	120.0	~ 0.0002	7 – 12
1.67	Oxyapatite (OA, OAp or OXA)[f], mineral voelckerite	$Ca_{10}(PO_4)_6O$	~ 69	~ 0.087	[a]
2.0	Tetracalcium phosphate (TTCP or TetCP), mineral hilgenstockite	$Ca_4(PO_4)_2O$	38 – 44	~ 0.0007	[a]

[a] These compounds cannot be precipitated from aqueous solutions.

[b] Cannot be measured precisely. However, the following values were found: 25.7±0.1 (pH = 7.40), 29.9±0.1 (pH = 6.00), 32.7±0.1 (pH = 5.28). The comparative extent of dissolution in acidic buffer is: ACP >> α-TCP >> β-TCP > CDHA >> HA > FA.

[c] Stable at temperatures above 100°C.

[d] Always metastable.

[e] Occasionally, it is called "precipitated HA (PHA)."

[f] Existence of OA remains questionable.

3.2. Polymers

Polymers are a class of organic materials composed of large molecules, often containing many thousands of small units, or monomers, joined altogether by covalent bonds to form one giant chain, thus creating very ductile materials. In this respect, polymers are comparable with major functional components of the biological environment: lipids, proteins and polysaccharides. They differ from each other in chemical composition, molecular weight, polydispersity, crystallinity, hydrophobicity, solubility and thermal transitions. Besides, their properties can be fine-tuned over a wide range by varying the type of polymer, chain length, as well as by copolymerization or blending of two or more polymers [79, 80]. Opposite to ceramics, polymers exhibit substantial viscoelastic properties and easily can be fabricated into complex structures, such as sponge-like sheets, gels or complex structures with intricate porous networks and channels [81]. Being X-ray transparent and non-magnetic polymeric materials are fully compatible with the modern diagnostic methods such as computed tomography and magnetic resonance imaging. Unfortunately, most of them are unable to meet the strict demands of the *in vivo* physiological environment. Namely, the main requirements to polymers suitable for biomedical applications are that they must be biocompatible, not eliciting an excessive or chronic inflammatory response upon implantation and, for those that degrade, that they breakdown into non-toxic products only. Unfortunately, polymers, for the most part, lack rigidity, ductility and ultimate mechanical properties required in load bearing applications. Thus, despite their good biocompatibility, many of the polymeric materials are mainly used for soft tissue replacements (such as skin, blood vessel, cartilage, ligament replacement, *etc.*). Moreover, the sterilization processes (autoclave, ethylene oxide and ^{60}Co irradiation) may affect the polymer properties [82].

There is a variety of biocompatible polymers suitable for biomedical applications [83-85]. For example, polyacrylates, poly(acrylonitrile-*co*-vinylchloride) and polylysine have been investigated for cell encapsulation and immunoisolation [86, 87]. Polyorthoesters and PCL have been investigated as drug delivery devices, the latter for long-term sustained release because of their slow degradation rates [88]. PCL is an aliphatic linear polyester, bioresorbable and biocompatible polymer, having appropriate resorption period and releases nontoxic byproducts upon degradation. Therefore, it is generally used in pharmaceutical products and wound dressings [89, 90]. PU is in use in engineering of both hard and soft tissues, as well as in nanomedicine [91]. Polymers considered for orthopedic purposes include polyanhydrides, which have also been investigated as delivery devices (due to their rapid and well-defined surface erosion), for bone augmentation or replacement since they can be photopolymerized *in situ* [88, 92, 93]. To overcome their poor mechanical properties, they have been co-polymerized with imides or formulated to be crosslinkable *in situ* [93]. Other polymers, such as polyphosphazenes, can have their properties (e.g., degradation rate) easily modified by varying the nature of their side groups and have been shown to support osteoblast adhesion, which makes them candidate materials for skeletal tissue regeneration [93]. PPF has emerged as a good bone replacement material, exhibiting good mechanical properties (comparable to trabecular bone), possessing the capability to crosslink *in vivo* through the C=C bond and being hydrolytically degradable. It has also been examined as a material for drug delivery devices [88, 92-95]. Polycarbonates have been suggested as suitable materials to make scaffolds for bone replacement and have been modified with tyrosine-derived amino acids to render them biodegradable [88, 96]. Polydioxanone has been also tested for biomedical applications [97]. PMMA is widely used in orthopedics, as a bone cement for implant fixation, as well as to repair certain fractures and bone defects, for example, osteoporotic vertebral bodies [98, 99]. However, PMMA sets by a polymerization of toxic monomers, which also evolves significant amounts of heat that damages tissues. Moreover, it is neither degradable nor bioactive, does not bond chemically to bones and might generate particulate debris leading to an inflammatory foreign body response [92, 100]. Several other non-degradable polymers applied in orthopedic surgery include PE in its different modifications such as a low-density PE, HDPE and ultrahigh molecular weight PE (used as the articular surface of total hip replacement implants [101, 102]), polyethylene terepthalate and PP, which are applied to repair knee ligaments [103]. Polyactive™, a block copolymer of PEG and PBT, was also considered for biomedical application [104-106]. Cellulose [107, 108] and its esters [109, 110] are also popular. Finally yet importantly, polyethylene oxide, PHB and blends thereof have also been tested for biomedical applications [29].

Table 4. Major properties of several FDA approved biodegradable polymers [111]

polymer	thermal properties*, °C	tensile modulus, GPa	degradation time, months
polyglycolic acid (PGA)	$t_g = 35 - 40$ $t_m = 225 - 230$	7.06	6 – 12 (strength loss within 3 weeks)
L-polylactic acid (LPLA)	$t_g = 60 - 65$ $t_m = 173 - 178$	2.7	> 24
D,L-polylactic acid (DLPLA)	$t_g = 55 - 60$ amorphous	1.9	12 – 16
85/15 D,L-polylactic-*co*-glycolic acid (85/15 DLPLGA)	$t_g = 50 - 55$ amorphous	2.0	5 – 6
75/25 D,L-polylactic-*co*-glycolic acid (75/25 DLPLGA)	$t_g = 50 - 55$ amorphous	2.0	4 – 5
65/35 D,L-polylactic-*co*-glycolic acid (65/35 DLPLGA)	$t_g = 45 - 50$ amorphous	2.0	3 – 4
50/50 D,L-polylactic-*co*-glycolic acid (50/50 DLPLGA)	$t_g = 45 - 50$ amorphous	2.0	1 – 2
poly(ε-caprolactone) (PCL)	$t_g = (-60) - (-65)$ $t_m = 58 - 63$	0.4	> 24

*t_g – glass transition temperature; t_m – melting point.

Nonetheless, the most popular synthetic polymers used in medicine are the linear aliphatic poly(α-hydroxyesters) such as PLA, PGA and a series of their copolymers with tunable properties (for example, a degree of crystallinity, a degradation rate, etc., depending on their specific composition) – PLGA (Table 4). These materials are fully biobased and biodegradable thermoplastic polyesters, which can be produced from different renewable resources such as starch, corn and sugarcane. Currently they are the second most consumed bioplastics in the world due to their use as both commodity and specialty applications. These polyesters have been extensively studied; they appear to be the only synthetic and biodegradable polymers with an extensive FDA approval history [29, 93, 111, 112]. They are biocompatible, mostly non-inflammatory, as well as degrade *in vivo* through hydrolysis and possible enzymatic action into products that are removed from the body by regular metabolic pathways [88, 93, 113]. Besides, they might be used for drug delivery purposes [114]. Poly(α-hydroxyesters) have been investigated as scaffolds for replacement and regeneration of a variety of tissues, cell carriers, controlled delivery devices for drugs or proteins (e.g., growth factors), membranes or films, screws, pins and plates for orthopedic applications [88, 93, 115, 116]. Additionally, the degradation rate of PLGA can be adjusted by varying the amounts of the two component monomers (Table 4), which in orthopedic applications can be exploited to create materials that degrade in concert with bone ingrowth [117]. Furthermore, PLGA is known to support osteoblast migration and proliferation [93, 118], which is a necessity for bone tissue regeneration. Unfortunately, such polymers on their own, though they reduce the effect of stress-shielding, are too weak to be used in load bearing situations and are only recommended in certain clinical indications, such as ankle and elbow fractures [113]. In addition, they exhibit bulk degradation, leading to both a loss in mechanical properties and lowering of the local solution pH that accelerates further degradation in an autocatalytic manner. As the body is unable to cope with the vast amounts of implant degradation products, this might lead to an inflammatory foreign body response. Finally, poly(α-hydroxyesters) do not possess the bioactive and osteoconductive properties [93, 119].

Several classifications of the biomedically relevant polymers are possible. For example, some authors distinguish between synthetic polymers like PE, PMMA, PLA, PGA, PCL, etc. and polymers of biological origin, which comprise polysaccharides (starch, alginate, chitin/chitosan [120, 121], gellan gum, cellulose, hyaluronic acid and its derivatives), proteins (soy, collagen, gelatin, fibrin, silk) and a variety of biofibers, such as lignocellulosic natural fibers [122, 123]. Among them, natural polymers often possess highly organized structures. In addition, they may contain an extracellular substance, called ligand, which is necessary to bind with cell receptors. However, they always contain various impurities, which should be removed prior use. As synthetic polymers can be produced under the controlled conditions, they in general exhibit predictable and reproducible mechanical and physical properties such as tensile strength, elastic modulus and degradation rate. In addition, synthetic polymers offer more possibilities for chemical modifications and molecular alterations that facilitate tailoring of the system's properties to the specific application requirements. Control of impurities is a further advantage of synthetic polymers. Other authors differentiate between resorbable or biodegradable (e.g., poly(α-hydroxyesters), polysaccharides and proteins) and non-resorbable (e.g., PE, PP, PMMA and cellulose) polymers [123]. Furthermore, polymeric materials can be broadly classified as thermoplastics and thermosets. For example, HDPE and PEEK are the examples of thermoplastics, while polydimethylsiloxane and PMMA are the examples of thermosets [82]. The list of synthetic biodegradable polymers used for biomedical application as scaffold materials is available as Table 1 in Ref. [123], while further details on polymers suitable for biomedical applications are available in literature [82, 116, 124-129] where the interested readers are referred to. Good reviews on the synthesis of various biodegradable polymers [130], as well as on the experimental trends in polymer composites [131] are available elsewhere.

4. Preparation Techniques

In general, preparation of any composites includes wetting, mixing or saturating the reinforcement with the matrix, followed by compacting. The matrix is then induced to bind together (with heat or a chemical reaction) into a rigid structure. To undergo a melding event, commonly, these operations are done in an open or closed forming mold. Since different material types require diverse processing conditions (such as temperature and solvent type), the order and ways of introducing the constituents alters considerably. Depending on the materials, composite fabrication is achieved by a wide variety of methods, including fiber placement, filament winding, lanxide process, tufting, z-pinning, various types of casting and molding, braiding (onto a former), filament winding and slip forming [38]. Nevertheless, prior blending, one needs to consider the granulometric dimensions of the mixing components. Regarding $CaPO_4$ and metal oxide ceramics in most of the cases, they represent fine-grade powders, which are ready to be mixed; however, this is not always the case for other ingredients. For example, polymers are usually supplied as pellets, granules or coarse powders, normally requiring an additional process of milling, which for polymers with a low melting point, such as PCL, may benefit from a cryogenic milling. Similar is valid for glasses, carbon and metals, which also may require a size reduction step.

Regarding $CaPO_4$-based biocomposites, different ways were also realized to bring matrix and dispersed components together into biocomposites, like mechanical blending or mixing, ball milling, compounding, compaction, casting, dispersion of $CaPO_4$ particles or whiskers into a polymer-solvent solution followed by solvent evaporation *e.g.*, by freeze-drying, dispersion of $CaPO_4$ particles or whiskers into a liquid monomer, followed by polymerization, a melt extrusion of $CaPO_4$/polymer mixtures, co-precipitation or co-deposition, as well as a rapid prototyping, selective laser sintering and 3D printing [22, 37, 132-134]. For example, three methods for preparing a homogeneous blend of HA with PLLA were compared [132]. A dry process, consisting in mixing ceramic powder and polymer pellets before a compression-molding step, was used. The second technique was based on the dispersion of ceramic fillers into a polymer-solvent solution. The third method was a melt extrusion of a ceramic/polymer powder mixture. Mixing dry powders led to a ceramic particle network around the polymer pellets, whereas the solvent and melt methods also produced a homogeneous dispersion of HA in the matrix. The main drawback of the solvent casting method is the risk of potentially toxic organic solvent residues. The melt extrusion method was shown to be a good way to prepare homogeneous ceramic/polymer blends [132].

Besides, there is *in situ* formation, which involves either synthesizing the reinforcement inside a preformed matrix material or synthesizing the matrix material around the reinforcement [37, 135-138]. This is one of the most attractive routes, since it avoids extensive particle agglomeration. For example, several papers have reported *in situ* formation technique to produce various composites of $CaPO_4$ with graphene. Other examples comprise a baiomimetic synthesis [137, 138] and preparation of nano-sized HA/PA biocomposites [139, 140]. In certain cases, a mechano-chemical route [141, 142], emulsions [143-148], freeze-drying [149, 150] and freeze-thawing techniques [151] or gel-templated mineralization [152] might be applied to produce $CaPO_4$-based biocomposites. Namely, just to prepare graphene/HA nanocomposites, the following techniques were used: *in situ* preparation, spark plasms synthesis, hydrothermal fabrication, biomimetic mineralization, hot isostatic pressing, electrochemical deposition and ball milling method [138]. Various fabrication procedures are well described elsewhere [22, 37, 132], where the interested readers are referred.

It is also possible to introduce porosity into $CaPO_4$-based biocomposites, which is advantageous for most applications as bone substitution material. The porosity facilitates migration of osteoblasts from surrounding bones to the implant site. Various material processing strategies to prepare composite scaffolds with interconnected porosity comprise thermally induced phase separation, solvent casting and particle leaching, solid freeform fabrication techniques, microsphere sintering and coating [123, 153-156]. A supercritical gas foaming technique might be used as well [132, 157-159]. Finally, various types of 3D printing techniques are also used. The majority of used porosity forming techniques were taked from $CaPO_4$-based bioceramics [160]. Namely, according to the topical review, a fused deposition modeling appears to be the most common technique for fabrication of porous scaffolds from $CaPO_4$-based biocomposites with polymers (64%), followed by a low-temperature deposition manufacturing (19%), selective laser sintering (12%) and stereolithography (5%) [161].

5. Biocomposites of $CaPO_4$ with Polymers

5.1. Biocomposites with Synthetic Polymers

It is logical to assume that the proper biocomposite of a $CaPO_4$ (for instance, CDHA) with a bioorganic polymer (for instance, collagen) would yield the physical, chemical and mechanical properties more or less similar to those of human bones. Therefore, $CaPO_4$/polymer biocomposites have been investigated extensively for bone regeneration. In such formulations, hard and stiff $CaPO_4$ bioceramics provide building blocks that are essential for both the mechanical strength and biomineralization, while flexible and soft polymeric components comprise those that both have shown a good biocompatibility and are routinely used in surgical applications. Thus, the formation of $CaPO_4$/polymer biocomposites capitalizes the advantages of both material types and minimizes their shortcomings. In general, since polymers have a low modulus (2 – 7 GPa, as the maximum) as compared to that of bones (3 – 30 GPa), to mimic the latter, $CaPO_4$ bioceramics need to be loaded at a high weight % ratio. Besides, general knowledge on composite mechanics suggests that any high aspect ratio particles, such as whiskers or fibers, significantly improve the modulus at a lower loading. Thus, some attempts have been already performed to prepare biocomposites containing whisker-like [162-168] or needle-like [169-172] $CaPO_4$, as well as $CaPO_4$ fibers [173].

The history of implantable $CaPO_4$/polymer formulations started in 1981 (however, a more general topic "ceramic-plastic material as a bone substitute" is, at least, 18 years older [174]) from the pioneering study by Prof. William Bonfield and colleagues at Queen Mary College, University of London, performed on HA/PE blends [175, 176]. That initial study introduced a bone-analogue concept, when proposed biocomposites comprised a polymer ductile matrix of PE and a ceramic stiff phase of HA, and was substantially extended and developed in further investigations by that research group [66, 177-187]. More recent studies included investigations on the influence of surface topography of HA/PE composites on cell proliferation and attachment [188-191]. The material is composed of a particular combination of HA particles at a volume loading of ~ 40% uniformly dispensed in a HDPE matrix. The idea was to mimic bones by using a polymeric matrix that can develop a considerable anisotropic character through adequate orientation techniques reinforced with a bone-like bioceramics that assures both a mechanical reinforcement and a bioactive character of the composite. Following FDA approval in 1994, in 1995 this material has become commercially available under the trade-name HAPEX™ (Smith and Nephew Richards, Bartlett, TN, USA), and to date it has been implanted in over 300 000 patients with the successful results. It remains the only clinically successful bioactive composite, which was a major step in the implant field [192]. The major production stages of HAPEX™ include blending, compounding and centrifugal milling. A bulk material or device is then created from this powder by compression and injection molding [37]. Besides, HA/HDPE biocomposites might be prepared by a hot rolling technique that facilitated uniform dispersion and blending of the reinforcements in the matrix [193]. In addition, PP might be used instead of PE [194-197].

A mechanical interlock between the both phases of HAPEX™ is formed by shrinkage of HDPE onto the HA particles during cooling [66, 67, 198]. Both HA particle size and their distribution in the HDPE matrix were recognized as important parameters affecting the mechanical behavior of HAPEX™. Namely, smaller HA particles were found to lead to stiffer composites due to general increasing of interfaces between the polymer and the ceramics; furthermore, rigidity of HAPEX™ was found to be proportional to HA volume fraction [182]. Furthermore, coupling agents, *e.g.*, 3-trimethoxysiyl propylmethacrylate for HA and acrylic acid for HDPE might be used to improve bonding (by both chemical adhesion and mechanical coupling) between HA and HDPE [199, 200]. Obviously, other types of $CaPO_4$ might be used instead of HA in biocomposites with PE [201]. Furthermore, attempts were performed to improve the mechanical properties of HAPEX™ by incorporating other ceramic phases into the polymer matrix, such as PSZ [202] and alumina [203]. For example, a partial replacement of HA filler particles by PSZ particles was found to lead to an increase in the strength and fracture toughness of HA/HDPE biocomposites. The compressive stress, set up by the volume expansion associated with tetragonal to monoclinic phase transformation of PSZ, inhibits or retards the crack propagation within the composite. This results in an enhanced fracture toughness of the HA/ZrO_2/HDPE biocomposite [202].

Various studies revealed that HAPEX™ attached directly to bones by chemical bonding (a bioactive fixation), rather than forming fibrous encapsulation (a morphological fixation). Initial clinical applications of HAPEX™ came in orbital reconstruction [204] but since 1995, the main uses of this composite have been in the shafts of middle ear implants for the treatment of conductive hearing loss [205, 206]. In both applications, HAPEX™ offers the advantage of *in situ* shaping, so a surgeon can make final alterations to optimize the fit of the prosthesis to the bone of a patient and subsequent activity requires only limited mechanical loading with virtually no risk of failure from insufficient tensile strength [66, 67]. As compared to cortical bones, HA/PE composites have a superior fracture toughness for HA concentrations below ~ 40% and similar fracture toughness in the 45 – 50% range. Their Young's modulus is in the range of 1 – 8 GPa, which is quite close to that of bone. The examination of the fracture surfaces revealed that only mechanical bond occurs between HA and PE. Unfortunately, the HA/PE composites are not biodegradable, the available surface area of HA is low and the presence of bioinert PE decreases the ability to bond to bones. Furthermore, HAPEX™ has been designed with a maximized density to increase its strength but the resulting lack of porosity limits the ingrowth of osteoblasts when the implant is placed into the body [23]. Further details on HAPEX™ are available elsewhere [66, 67]. Except of HAPEX™, other types of HA/PE biocomposites are also known [207-215].

Both linear and branched PE was used as a matrix and the biocomposites with the former were found to give a higher modulus [208]. The reinforcing mechanisms in $CaPO_4$/polymer formulations have yet to be convincingly disclosed. Generally, if a poor filler choice is made, the polymeric matrix might be affected by the filler through reduction of molecular weight during composite processing, formation of an immobilized shell of polymer around the particles (transcrystallization, surface-induced crystallization or epitaxial growth) and changes in conformation of the polymer due to particle surfaces and inter-particle spacing [66, 67]. On the other hand, the reinforcing effect of $CaPO_4$ particles might depend on the molding technique employed: a higher orientation of the polymeric matrix was found to result in a higher mechanical performance of the composite [213, 214].

Many other blends of $CaPO_4$ with various polymers are possible, including rather unusual formulations with dendrimers [216]. Even light-curable $CaPO_4$/polymer formulations are known [217]. The list of the appropriate $CaPO_4$ is shown in Table 3 (except of MCPM and MCPA – both are too acidic and, therefore, are not biocompatible [16]; nevertheless, to overcome this drawback, they might be mixed with basic compounds, such as HA, TTCP, $CaCO_3$, CaO, etc.) many biomedically suitable polymers have been listed above. The combination of $CaPO_4$ and polymers into biocomposites has a twofold purpose. The desirable mechanical properties of polymers compensate for a poor mechanical behavior of $CaPO_4$ bioceramics, while in turn the desirable bioactive properties of $CaPO_4$ improve those of polymers, expanding the possible uses of each material within the body [218-

221]. Namely, polymers have been added to $CaPO_4$ to improve their mechanical strength [218], while $CaPO_4$ fillers have been blended with polymers to improve their compressive strength and modulus, in addition to increasing their osteoconductive properties [119, 222-235]. In 1990-s, it was established that with increasing of $CaPO_4$ content, both Young's modulus and bioactivity of the biocomposites generally increased, while the ductility decreased [23]. However, the later investigations revealed that the mechanical properties of $CaPO_4$/polymer biocomposites were not so straightforward: the strength was found to decrease with increasing the $CaPO_4$ content in such biocomposites [226]. Nevertheless, biocompatibility of such biocomposites is enhanced because $CaPO_4$ fillers induce an increased initial flash spread of serum proteins compared to the more hydrophobic polymer surfaces [227]. What's more, experimental results of these biocomposites indicated favorable cell-material interactions with increased cell activities as compared to each polymer alone [220]. Furthermore, such formulations can provide a sustained release of calcium and orthophosphate ions into the milieus, which is important for mineralized tissue regeneration [219]. Indeed, a combination of two different materials draws on the advantages of each one to create a superior biocomposite with respect to the materials on their own. However, there are cases, when $CaPO_4$ inclusion into highly biocompatible, mechanically stable and 3D printable polymer improves neither *in vitro* nor *in vivo* osteogenesis [228].

The interfacial bonding between the phases is an important issue of any $CaPO_4$/polymer biocomposite. Four types of mutual arrangements of nanodimensional particles to polymer chains have been classified by Kickelbick (Figure 1): (1) inorganic particles embedded in inorganic polymer, (2) incorporation of particles by bonding to the polymer backbone, (3) an interpenetrating network with chemical bonds, (4) an inorganic-organic hybrid polymer [229]. If adhesion among the phases is poor, the mechanical properties of a biocomposite suffer. To solve the problem, various approaches have been already introduced. For example, a diisocyanate coupling agent was used to bind PEG/PBT (Polyactive™) block copolymers to HA filler particles. Using surface-modified HA particles as a filler in a PEG/PBT matrix significantly improved the elastic modulus and strength of the polymer as compared to the polymers filled with ungrafted HA [223, 230]. Another group used processing conditions to achieve a better adhesion of the filler to the matrix by pressing blends of varying PLLA and HA content at different temperatures and pressures [231]. The researchers found that maximum compressive strength was achieved at ~ 15 wt.% of PLLA. By using blends with 20 wt.% of PLLA, the authors also established that increasing the pressing temperature and pressure improved the mechanical properties. The former was explained by decrease in viscosity of the PLLA associated with a temperature increase, hence leading to improved wettability of HA particles. The latter was explained by increased compaction and penetration of pores at higher pressure, in conjunction with a greater fluidity of the polymer at higher temperatures. The combination of high pressures and temperatures was found to decrease porosity and guarantee a close apposition of a polymer to the particles, thereby improving the compressive strength [218] and fracture energy [232] of $CaPO_4$-containing biocomposites. Interestingly, that to improve an interfacial bonding between PLLA and HA by stereocomplexation, PDLA was initially grafted onto the surface of HA particles and then the PDLA-grafted HA particles were incorporated into a PLLA matrix [233]. In addition, compatibly to the chemical nature of both organic and $CaPO_4$ constituents, various crosslinking methodologies of polymers have been established to keep constant the mechanical properties of the biocomposites [234].

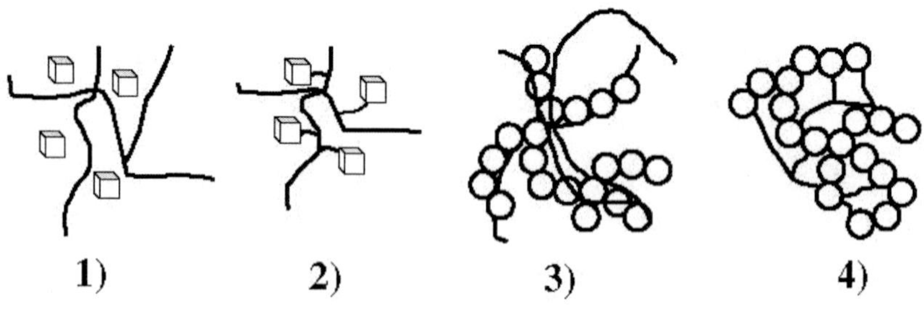

Figure 1. Four types of mutual arrangements of nano-sized particles to a polymer chain: (1) inorganic particles embedded in an inorganic polymer, (2) incorporation of particles by bonding to the polymer backbone, (3) interpenetrating network with chemical bonds, (4) inorganic-organic hybrid polymer. Reprinted from Ref. [229] with permission.

5.1.1. Apatite-Based Formulations

A biological apatite is known to be the major inorganic phase of mammalian calcified tissues [14, 15]. Consequently, CDHA, HA, carbonateapatite and, occasionally, FA (all of them may be both pure and with dopants) have been applied to prepare biocomposites with other compounds, usually, with the aim to improve the bioactivity. For example, polysulfone composed with HA can be used as a starting material for long-term implants [235-237]. Retrieved *in vivo*, HA/polysulfone biocomposite coated samples from rabbit distal femurs demonstrated direct bone apposition to the coatings, as compared to the fibrous encapsulation that occurred when uncoated samples were used [235]. The resorption time of such biocomposites is a very important factor, which depends on polymer's microstructure and the presence of modifying phases [236].

Various apatite-containing biocomposites with PVA [151, 238-243], PVAP [244] and several other polymeric components [245-256] have been already developed. Namely, PVA/CDHA biocomposite blocks were prepared by precipitation of CDHA in aqueous solutions of PVA [151]. An artificial cornea consisted of a porous nano-sized HA/PVA hydrogel skirt and a transparent center of PVA hydrogel has been prepared as well. The results displayed a good biocompatibility and interlocking between artificial cornea and host tissues [238, 239]. PVAP has been chosen as a polymer matrix, because its phosphate groups can act as a coupling/anchoring agent, which has a higher affinity toward the HA surface [244]. Greish and Brown developed HA/Ca poly(vinyl phosphonate) biocomposites [248-250]. A template-driven nucleation and mineral growth process for the high-affinity integration of CDHA with PHEMA hydrogel scaffold has been developed as well [254].

PEEK [162, 163, 257-262] and high impact polystyrene [263, 264] were also applied to create biocomposites with HA having a potential for clinical use in load bearing applications. The study on reinforcing PEEK with thermally sprayed HA particles revealed that the mechanical properties increased monotonically with the reinforcement concentration, with a maximum value in the study of ~ 40% volume fraction of HA particles [257-259]. The reported ranges of stiffness within 2.8 – 16.0 GPa and strength within 45.5 – 69 MPa exceeded the lower values for human bone (7 – 30 GPa and 50 – 150 MPa, respectively) [258]. Modeling of the mechanical behavior of HA/PEEK biocomposites is available elsewhere [260].

Biodegradable poly(α-hydroxyesters) are well established in clinical medicine. Currently, they provide with a good choice when a suitable polymeric filler material is sought. For example, HA/PLGA formulations were developed which appeared to possess a cellular-compatibility suitable for bone tissue regeneration [265-273]. Zhang and Ma seeded highly porous PLLA foams with HA particles in order to improve the osteoconductivity of polymer scaffolds for bone tissue engineering [222]. They pointed out that hydration of the foams prior to incubation in simulated body fluid increased the amount of carbonated CDHA material due to an increase of COOH and OH groups on the polymer surface, which apparently acted as nucleation sites for apatite. The mechanical properties of PLA/$CaPO_4$ biocomposites fabricated using different techniques, as well as the results of *in vitro* and *in vivo* experiments with them are available in literature [269].

On their own, poly(α-hydroxyesters), such as PGA and PLA, are known to degrade to acidic products (glycolic and lactic acids, respectively) that both catalyze polymer degradation and cause inflammatory reactions of the surrounding tissues [274]. Thus, in biocomposites of poly(α-hydroxyesters) with $CaPO_4$, the presence of slightly basic compounds (HA, TTCP) to some extent neutralizes the acid molecules, provides with a weak pH-buffering effect at the polymer surface and, therefore, partially compensates these drawbacks [119, 269, 275-277]. However, additives of even more basic chemicals (e.g., CaO, $CaCO_3$) might be necessary [123, 277-279]. Extensive cell culture experiments on pH-stabilized composites of PGA and carbonateapatite were reported, which afterwards were supported by extensive *in vitro* pH-studies [280]. A consequent development of this approach has led to designing of functionally graded composite skull implants consisting of polylactides, carbonateapatite and $CaCO_3$ [281, 282]. Besides the pH-buffering effect, inclusion of $CaPO_4$ was found to modify both surface and bulk properties of the biodegradable poly(α-hydroxyesters) by increasing the hydrophilicity and water absorption of the polymer matrix, thus altering the scaffold degradation kinetics. For example, polymer biocomposites filled with HA particles was found to hydrolyze homogeneously due to water penetrating interfacial regions [283].

Biocomposites of poly(α-hydroxyesters) with $CaPO_4$ are prepared mainly by incorporating the inorganic phase into a polymeric solution, followed by drying under vacuum. The resulting solid biocomposites might be shaped using different processing techniques. One can also prepare these biocomposites by mixing HA particles with L-lactide prior the polymerization [275] or by a combination of slip-casting technique and hot pressing [284]; however, other production techniques are known [269, 271, 285]. Addition of a surfactant (surface active agent) might be useful to keep the suspension homogeneity [286]. Furthermore, HA/PLA [144, 145] and HA/PLGA [146] microspheres might be prepared by a microemulsion technique. More complex formulations, such as carbonated-FA/PLA [287] and PLGA/carbon nanotubes/HA [288], are also known. An interesting list of references, assigned to the different ways of preparing HA/poly(α-hydroxyesters) biodegradable composites, is available in publications by Durucan and Brown [289-291]. The authors prepared CDHA/PLA and CDHA/PLGA biocomposites by solvent casting technique with a subsequent hydrolysis of α-TCP to CDHA in aqueous solutions. The presence of both polymers was found to inhibit α-TCP hydrolysis, if compared with that of single-phase α-TCP; what is more, the inhibiting effect of PLA exceeded that of PLGA [289-291]. The physical interactions between $CaPO_4$ and poly(α-hydroxyesters) might be easily seen in Figure 2 [291]. Another set of good pictures might be found in Ref. [51]. Nevertheless, it should not be forgotten that typically non-melting-based routes lead to development of composites with lower mechanical performance and many times require the use of toxic solvents and intensive hand labor [125].

Figure 2. SEM micrographs of a) α-TCP compact; b) α-TCP/PLGA biocomposite (bars = 5 μm). Reprinted from Ref. [291] with permission.

The mechanical properties of poly(α-hydroxyesters) could be substantially improved by addition of $CaPO_4$ [292, 293]. Namely, CDHA/PLLA biocomposites of very high mechanical properties were developed [119] and fixation tools (screws and plates) made of these composites were manufactured and tested. These fixation tools revealed an easy handling and shaping according to the implant site geometry, total resorbability, good ability to bond directly to the bone tissue without interposed fibrous tissue, osteoconductivity, biocompatibility and high stiffness retainable for the period necessary to achieve bone union [283, 285].

The initial bending strength of ~ 280 MPa exceeded that of cortical bone (120 – 210 MPa), while the modulus was as high as 12 GPa [119]. The strength could be maintained above 200 MPa up to 25 weeks in phosphate-buffered saline solution. These biocomposites could be obtained by precipitation from PLLA/dichloromethane solutions, in which small particles of $CaPO_4$ were distributed [118, 294]. Porous scaffolds of PDLLA + HA [158, 295, 296], PLGA + HA [297], as well as more complex formulations, such as PLA/ethyl cellulose/HA [298] and PLA/Fe_3O_4/BCP (HA+ β-TCP) [299], were produced as well. Upon implantation into rabbit femora, a newly formed bone was observed and biodegradation was significantly enhanced if compared to single-phase HA bioceramics. This might be due to a local release of lactic acid, which in turn dissolves HA. In other studies, PLA and PGA fibers were combined with porous HA scaffolds. Such reinforcement did not hinder bone ingrowth into the implants, which supported further development of such biocomposites as bone graft substitutes [29, 269, 270].

Blends (named as SEVA-C) of EVOH with starch filled with 10 – 30 wt.% HA have been fabricated to yield biocomposites with modulus up to ~ 7 GPa with a 30% HA loading [300-304]. The incorporation of bioactive fillers such as HA into SEVA-C aimed to assure the bioactive behavior of the composite and to provide the necessary stiffness within the typical range of human cortical bone properties. These biocomposites exhibited a strong *in vitro* bioactivity that was supported by the polymer's water-uptake capability [305]. However, the reinforcement of SEVA-C by HA particles was found to affect the rheological behavior of the blend. A degradation model of these biocomposites has been developed [306].

Higher homologues poly(3-hydroxybutyrate), 3-PHB, and poly(3-hydroxyvalerate) show almost no biodegradation. Nevertheless, biocomposites of these polymers with $CaPO_4$ showed a good biocompatibility both *in vitro* and *in vivo* [307-311]. Both bioactivity and mechanical properties of these biocomposites can be tailored by varying the volume percentage of $CaPO_4$. Similarly, biocomposites of PHBHV with both HA and amorphous carbonated apatite (almost ACP) appeared to have a promising potential for repair and replacement of damaged bones [312-315].

Along this line, PCL is used as a slowly biodegradable but well biocompatible polymer. PCL/HA and PCL/CDHA biocomposites have been already discussed as suitable materials for substitution, regeneration and repair of bone tissues [153, 316-325]. For example, biocomposites were obtained by infiltration of ε-caprolactone monomer into porous apatite blocks and *in situ* polymerization [317]. The composites were found to be biodegradable and might be applied as cancellous or trabecular bone replacement material or for a cartilage regeneration. Both the mechanical performance and biocompatibility in osteoblast cell culture of PCL were shown to be strongly increased when HA was added [326]. Several preparation techniques of PCL/HA biocomposites are known [153, 320]. For example, to make biocomposite fibers of PCL with nanodimensional HA, the desired amount of nanodimensional HA powder was dispersed in a solvent using magnetic stirrer followed by ultrasonication for 30 min. Then, PCL was dissolved in this suspension, followed by the solvent evaporation [327]. The opposite preparation order is also possible: PCL was initially dissolved in chloroform at room temperature (7 – 10% weight/volume), then HA (~ 10 μm particle size) was suspended in the solution, sonicated for 60 s, followed by the solvent evaporation [328] or salt-leaching [329]. The mechanical properties obtained by this technique were about one-third that of trabecular bone. In a comparative study, PCL and biological apatite were mixed in the ratio 19:1 in an extruder [330]. At the end of the preparation, the mixture was cooled in an atmosphere of nitrogen. The authors observed that the presence of biological apatite improved the modulus while concurrently increasing the hydrophilicity of the polymeric substrate. Besides, an increase in apatite concentration was found to increase both the modulus and yield stress of the composite, which indicated to good interfacial interactions between the biological apatite and PCL. It was also observed that the presence of biological apatite stimulated osteoblasts attachment to the biomaterial and cell proliferation [330]. In another study, a PCL/HA biocomposite was prepared by blending in melt form at 120°C until the torque reached equilibrium in the rheometer that was attached to the blender [331]. Then the sample was compression molded and cut into specimens of appropriate size for testing. It was observed that the composite containing 20 wt. % HA had the highest strength [331]. However, a direct grafting of PCL on the surface of HA particles seems to be the most interesting preparation technique [316]. In another study, HA porous scaffolds were coated by a PCL/HA composite coating [332]. In this system, PCL, as a coating component, was able to improve the brittleness and low strength of the HA scaffolds, while the particles in the coating were to improve the osteoconductivity and bioactivity of the coating layer. More complex formulations, such as PDLLA/PCL/HA [333], PLLA/PCL/HA [334], FA-HA/PCL [335], chitosan/PCL/PVA/HA [336], magnetic PCL/Fe-doped HA [337], supramolecular PCL/functionalized HA [338, 339] and biocomposites with other compounds [340] have been prepared as well. Further details on both the PCL/HA biocomposites and the processing methodologies thereof might be found elsewhere [153, 320, 324].

To finalize this section, one should mention on HA biocomposites with other polymers [341-348]. Among them, PTMC/HA biocomposites were proposed to be suitable for stereolithographic manufacturing of flexible patient-specific implants [346]. In addition, some of them may exhibit interesting properties. Namely, it is worth mentioning on polymers with shape-memory characteristics [349]. Therefore, $CaPO_4$-based composites with such polymers appear to have the shape-memory properties as well [342-344]. More complicated formulations are known as well. Namely, biopolymers/CHDA biocomposites with inclusions of magnetic nanoparticles were prepared recently [350]. Further details on HA/polymer biocomposites are available in other reviews [321-354].

5.1.2. TCP-based Formulations

Both α-TCP and β-TCP have a higher solubility than HA (Table 3). Besides, they are faster resorbed *in vivo* (however, there are some reports about a lack of TCP biodegradation after implantation in calvarial defects [355]). Therefore, α-TCP and β-TCP were widely used instead of apatites to prepare completely biodegradable biocomposites [356-377]. For example, a biodegradable and osteoconductive biocomposite made of β-TCP particles and gelatin was proposed [361]. This material was tested *in vivo* with good results. It was found to be biocompatible, osteoconductive and biodegradable with no need for a second surgical operation to remove the device after healing occurred. Both herbal extracts [362] and K_2HPO_4 [363] might be added to this formulation. Another research group prepared biocomposites of crosslinked gelatin with β-TCP and both a good biocompatibility and bone formation upon subcutaneous implantation in rats were found [364]. Yang, et al., [368] extended this to porous (porosity ~ 75%) β-TCP/gelatin biocomposites those also contained BMP-4. Furthermore, cell-compatible and possessive some osteoinductive properties porous β-TCP/alginate-gelatin composite scaffolds were prepared and successfully tested *in vitro* [365]. In addition, the $CaPO_4$ fillers were found to have a reinforcing effect [378]. More to the point, biocomposites of β-TCP with PLLA [294, 356-359] and PLGC [360] were prepared. Although β-TCP was able to counter the acidic degradation of the polyester to some extent, it did not prevent a pH drop down to ~ 6. Nevertheless, implantation of this biocomposite in beagles' mandibular bones was successful [360]. α-TCP/gelatin formulations are known as well [371].

Based on a self-reinforcement concept, biocomposites of TCP with polylactides were prepared and studied using conventional mechanical testing [379]. Resorbable scaffolds were fabricated from such biocomposites [380]. Chitosan was also used as the matrix for the incorporation of β-TCP by a solid/liquid phase separation of the polymer solution and subsequent sublimation of the solvent. Due to complexation of the functional groups of chitosan with calcium ions of β-TCP, these biocomposites had high compressive modulus and strength [381]. PCL/β-TCP biocomposites were developed in other studies [324, 325, 382-386] and their *in vitro* degradation behavior was systematically monitored by immersion in simulated body fluid at 37°C [384]. To extend this topic further, PCL/β-TCP biocomposites might be loaded by drugs [324, 385].

An *in vitro* study with primary rat calvarial osteoblasts showed an increased cellular activity in the BMP-loaded samples [368]. Other researchers investigated BMP-2-loaded porous β-TCP/gelatin biocomposites (porosity ~ 95%, average pore size 180 – 200 μm) [387] and confirmed the precious study. A long-term implantation study of PDLLA/α-TCP composites in a loaded sheep implant model showed good results after 12 months but a strong osteolytic reaction after 24 months. This was ascribed to the almost complete dissolution of α-TCP to this time and an adverse reaction of the remaining PDLLA [388].

More complex $CaPO_4$-based formulations are known as well. For example, there is a biocomposite consisting of three interpenetrating networks: TCP, CDHA and PLGA [389]. Firstly, a porous TCP network was produced by coating a PU foam by hydrolysable α-TCP slurry. Then, a CDHA network was derived from self-setting $CaPO_4$ formulations filled in the porous TCP network. Finally, the remaining open pore network in the CDHA/α-TCP structures was infiltrated with PLGA. This biocomposite consists of three phases with different degradation behavior. It was postulated that bone would grow on the fastest degrading network of PLGA, while the remaining $CaPO_4$ phases would remain intact thus maintaining their geometry and load bearing capability [389]. In addition, PCL/TCP/boron nitride biocomposites were prepared as well [390].

5.1.3. Formulations Based on Other Types of $CaPO_4$

The number of research publications devoted to formulations based on other types of $CaPO_4$ is substantially lesser than those devoted to apatites and TCP. Biphasic calcium phosphate (BCP), which is a solid composite of HA and β-TCP (however, similar formulations of HA and α-TCP, as well as of α-TCP and β-TCP are known, as well [391]) appears to be most popular among the remaining types of $CaPO_4$. For example, collagen coated BCP ceramics was studied and the biocompatibility towards osteoblasts was found to increase upon coating with collagen [392]. Another research group created porous PDLLA/BCP scaffolds and coated them with a hydrophilic PEG/vancomycin composite for both drug delivery purposes and surface modification [393]. More to the point, both PLGA/BCP [394, 395] and PLLA/BCP [396] biocomposites were fabricated and their cytotoxicity and fibroblast properties were found to be acceptable for natural bone tissue reparation, filling and augmentation [397, 398]. Besides, PCL/BCP [399, 400], PTMC/BCP [401] and gelatin/BCP [402, 403] biocomposites are known as well.

A choice of DCPD-based biocomposites of DCPD, albumin and duplex DNA was prepared by water/oil/water interfacial reaction method [143]. Core-shell type DCPD/chitosan biocomposite fibers were prepared by a wet spinning method in another study [404]. The energy-dispersive X-ray spectroscopy analysis indicated that Ca and P atoms were mainly distributed on the outer layer of the composite fibers; however, a little amount of P atoms remained inside the fibers. This indicated that the composite fibers formed a unique core-shell structure with shell of $CaPO_4$ and core of chitosan [404]. A similar formulation was prepared for further applications in self-setting biocomposites [405]. DCPA/BSA biocomposites were synthesized through the co-precipitation of BSA on the nanodimensional particles of DCPA performed in ethanol [406], while DCPA/ethyl cellulose ones were prepared by a sol-gel technique [407]. Nanodimensional DCPA was synthesized and incorporated into dental resins to form dental biocomposites [408-411], while porous DCPA/dextran/CMC nanocomposite scaffolds were prepared by lyophilization in a freeze-dryer [412]. Although, this is beyond the biomedical subject, it is interesting to mention that some DCPD/polymer composites could be used as proton conductors in battery devices [413, 414]. Nothing has been reported on their biocompatibility but, perhaps, sometime the improved formulations will be used to fabricate biocompatible batteries for implantable electronic devices.

To treat the consequences of dental caries, various ACP-based biocomposites for dental applications were developed [415, 416]. Among them there are rechargeable ones with sustained ion release and re-release, which were introduced in 2016 [417-420]. Besides, several ACP-based formulations were investigated as potential biocomposites for bone grafting [315, 421-423] and drug delivery [424, 425]. Namely, ACP/PPF biocomposites were prepared by *in situ* precipitation [422], while

PHB/carbonated ACP and PHBHV/carbonated ACP biocomposites appeared to be well suited as slowly biodegradable bone substitution material [315].

OCP-based biocomposites with polymers are known as well. For example, granules of OCP were combined with gelatin and the prepared composites were onlaid on rat calvaria. These composites were found to enhance bone augmentation through balanced osteoblastic and osteoclastic cellular activities [426]. Besides, there is an excellent review on OCP/polymer biocomposites with macro- to nano-level controlled hierarchical structures [427]. Additional information on the biomedically relevant OCP-based formulations might be found in a topical review [428].

5.2. Self-setting Formulations

Inorganic self-setting $CaPO_4$ formulations (cements) hardening in the body were introduced in the early 1980-s. Since then, they have been broadly studied and many formulations have been proposed. These formulations set and harden due to various chemical interactions among $CaPO_4$ that finally lead to formation of a monolithic body consisting of either CDHA or DCPD with possible admixtures of other phases. Unfortunately, having a ceramic nature, the self-setting $CaPO_4$ formulations are brittle after hardening and the setting time is sometimes unsuitable for clinical procedures [429]. Therefore, to improve their properties, various attempts have been performed to transform them into biocomposites e.g., by adding PEG [430], PEG-PLLA both with and without 2-hydroxyethyl methacrylate [431], PCL [432], hydroxycarboxylic acids [433, 434], gelatin [367, 435-438], osteocalcin/collagen [439], alginate [440], chitin [441], chitosan [442], silk fibroin [443, 444], silanized HPMC [445], CMC [446], PVA fibers both with CMC [447] and without it [448], bioactive glasses [449-452], calcium silicates [453-455], magnetic nanoparticles [456], *etc.* Light-cured formulations have been prepared as well [457]. More to the point, various reinforcement additives of different shapes and nature are widely used to improve the mechanical properties of the self-setting formulations [458, 459]. Although the biomaterials community does not use this term, a substantial amount of the reinforced formulations might be defined as $CaPO_4$-based concretes [429]. The idea behind the concretes is simple: if a strong filler is present in the matrix, it might stop crack propagation.

Various apatite-containing formulations with PMMA [460-468] and polyethylmethacrylate [469, 470] have been already developed. Such biocomposites might be prepared by dispersion of apatite powder into a viscous fluid of the polymer [471] and used for drug delivery purposes [472]. When the mechanical properties of the concretes composed of PMMA matrix and HA particles of various sizes were tested, the tensile results showed that strength was independent on particle sizes. After immersion into Ringer´s solution, the tensile strength was not altered whereas the fatigue properties were significantly reduced. The biocompatibility of PMMA/HA formulations was tested *in vivo* and enhanced osteogenic properties of the implants compared to single-phase PMMA were observed [461-464]. It was shown that not only the mechanical properties of PMMA were improved but the osteoblast response of PMMA was also enhanced with addition of HA [462]. Thereby, by adding of $CaPO_4$, a non-biodegradable PMMA was made more bioactive and osteoconductive, yielding a well-processible biocomposite concrete. As a drawback, the PMMA/HA formulations possess a low flexural, compressive and tensile strength.

Biocomposites made from $CaPO_4$ and various types of resins appeared to possess comparable mechanical and biological properties to typical PMMA cement, leading to potential uses for implant fixation [165, 167, 473-475]. To improve the mechanical properties of the self-setting $CaPO_4$-based formulations and stabilize them at the implant site, various researchers have resorted to formulations that set *in situ*, primarily through crosslinking reactions of the polymeric matrix. For example, TTCP was reacted with PAA, forming a crosslinked CDHA/calcium polyacrylate biocomposite [476]. In aqueous solutions, TTCP hydrolyzes to CDHA [16] and the liberated calcium cations react with PAA, forming the crosslinked network [476]. Reed et al., synthesized a dicarboxy polyphosphazene that can be crosslinked by calcium cations and CDHA/polyphosphazene biocomposites with a compressive strength ~ 10 MPa and of ~ 65% porosity were prepared as a result [477]. To mimic PMMA cements, PPF/β-TCP biocomposites were prepared with addition of vinyl monomer to crosslink PPF. As a result, quick setting and degradable formulations with a low heat output and compressive strengths in the range of 1 – 12 MPa were prepared by varying the molecular weight of PPF, as well as the contents of the monomer, β-TCP, initiator and NaCl, as a porogen [478, 479]. An acrylic formulation with Sr-containing HA as a filler [100], an injectable polydimethylsiloxane/HA formulation [480], biocomposites consisting of PLGA microspheres and self-setting $CaPO_4$-based formulations [481, 482], as well as ones of chitosan oligosaccharide/gelatin/$CaPO_4$ [483] were prepared as well.

To improve the mechanical properties of self-setting formulations, numerous researchers blended various polymers to them. For example, gelatin might be added to self-setting formulations, primarily to stabilize the paste in aqueous solution before it develops adequate rigidity and, secondly, to improve the compressive strength [367, 435, 484]. Adding rod-like fillers to the self-setting formulations also caused an improvement in the mechanical properties [484]. For example, PAA and PVA were successfully used to improve the mechanical properties of a TTCP + DCPD formulation but, unfortunately, with an inevitable and unacceptable reduction of both workability and setting time [485, 486]. Similar findings were reported in the presence of sodium alginate and sodium polyacrylate [487]. Other polymers, such as polyphosphazene might be used as well [488-490]. Other examples of polymer/$CaPO_4$ self-setting formulations might be found elsewhere [491, 492]. Metallic wires might be used as reinforcements as well [493].

Porous $CaPO_4$ scaffolds with interconnected macropores (~ 1 mm), micropores (~ 5 μm) and of high porosity (~ 80%) were prepared by coating PU foams with a TTCP + DCPA formulation, followed by firing at 1200 °C. To improve the mechanical properties of the scaffolds, the open micropores of the struts were then infiltrated by a PLGA solution to achieve an interpenetrating bioactive ceramic/biodegradable polymer composite structure. The PLGA filled struts were further coated with a 58S bioactive glass/PLGA composite coating. The obtained complex porous biocomposites could be used as tissue engineering scaffolds for low-load bearing applications [494]. A more complicated construction, in which the PLGA macroporous phase has been reinforced with a bioresorbable TTCP + DCPA formulation, followed by surface coating of the entire construct by a non-stoichiomentic CDHA layer, has been designed as well [495]. Furthermore, a moldable $CaPO_4$-based self-setting composite formulation was successfully constructed by infiltrating wollastonite/$CaPO_4$ paste into a 3D plotted PLGA network. After being heated at 50 °C, the PLGA/wollastonite/$CaPO_4$ composite appeared to be sufficiently plastic for successful adaptation to the complex-shaped bone defects [496]. In one more study, α-TCP and poly(ethylenephosphate) sodium salt-coated PLGA microparticles were mixed with castor oil and water to form an oil-in-water emulsion. The resulting emulsion spontaneously set in a humidified atmosphere at ambient temperature. The PLGA particles incorporated within the set formulations were hydrolytically degraded, which resulted in the formation of an interconnected microporosity [497].

A porosity level of 42–80% was introduced into self-setting $CaPO_4$/chitosan biocomposites by addition of the water-soluble mannitol [498]. Chitosan significantly improved the mechanical strength of the entire biocomposite [499]. A similar approach was used by other researchers who studied the effect of the addition of PLGA microparticles [500-503] (which can also be loaded with drugs or growth factors [504-506]) to self-setting $CaPO_4$ formulations. These biocomposites were implanted into cranial defects of rats and a content of ~ 30 wt. % of the microparticles was found to give the best results [500], while the addition of a growth factor to the biocomposites significantly increased bone contact at 2 weeks and enhanced new bone formation at 8 weeks [505]. The *in vivo* rabbit femur implant tests showed that self-setting PLGA/$CaPO_4$ formulations exhibited outstanding biocompatibility and bioactivity, as well as a better osteoconduction and degradability than self-setting formulations consisted solely from $CaPO_4$ [501]. In 2018, researches succeeded to produce bionic biocomposites with the Harvarsin microstructure from a self-setting TTCP + DCPD formulation combined with small intestinal submucosa [507]. Further details on this topic are available in literature [508].

5.3. Biocomposites with Collagen

The unique characteristics of bones are the spatial orientation between the nanodimentional crystals of biological apatite and collagen macromolecules at the nano-scale, where the crystals (about 50 nm length) are aligned parallel to the collagen fibrils, which is believed to be the source of the mechanical strength of bones (Figure 3a) [509]. The collagen molecules and the crystals of biological apatite assembled into mineralized fibrils are approximately 6 nm in diameter and 300 nm long. Although the complete mechanisms involved in the bone building strategy are still unclear, the strengthening effect of nanodimentional crystals of biological apatite in calcified tissues might be explained by the fact that the collagen matrix is a load transfer medium and thus transfers the load to the intrinsically rigid inorganic crystals [14, 15, 19-21]. Furthermore, the crystals of biological apatite located in between tangled fibrils cross-link the fibers either through a mechanical interlocking or by forming calcium ion bridges, thus increasing deformation resistance of the collagenous fiber network [510].

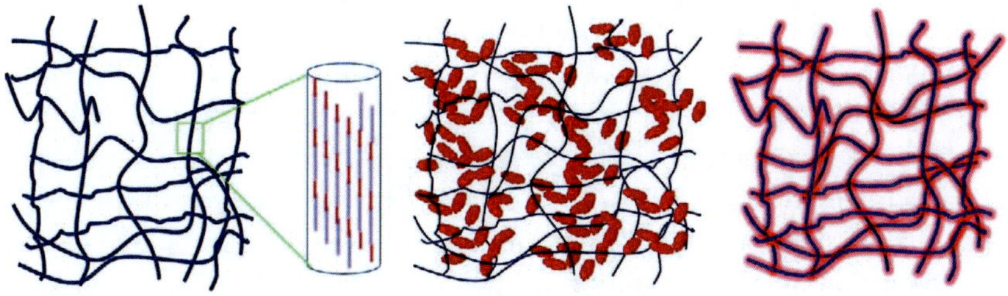

Figure 3. Schematic illustrations of collagen–$CaPO_4$ nanocomposites: (a) structure of bone: nanodimensional crystals of biological apatite aligned along with the collagen fibrils; linking occurs between the apatite and polar groups on collagen chains in bone, (b) nanodimensional crystals of $CaPO_4$ physically "trapped" within the collagen matrix, (c) nanodimensional crystals of $CaPO_4$ deposited on and covered the collagen fibrils surface. Reprinted from Ref. [509] with permission.

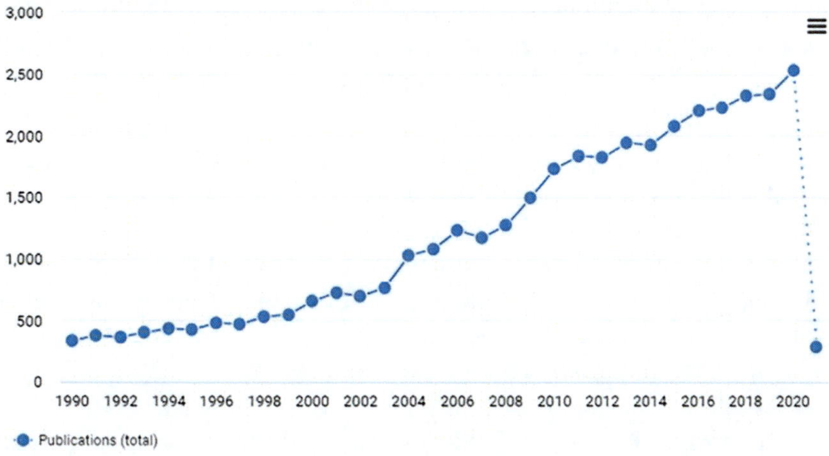

Figure 4. The amounts of publications issued each year with the following keywords: hydroxyapatite and collagen. Reprinted from Ref. [511] with permission.

Since biological apatite and collagen are the major constituents of calcified tissues of mammals, many attempts have been performed to combine these 2 substances into biocomposites in attempts to mimic both structure and composition of bones. The authors of a recent publication created a plot, showing that each passing year since 1990, the number of published works has increased exponentially (Figure 4) [511]. As seen from this figure, HA/collagen biocomposits are widely explored in the literature. Their physical forms include fibrous matrix, powders, dense materials, films and gels. The details and references to preparation techniques of all these forms are well described in a recent review [511].

When $CaPO_4$ are combined with collagen in a laboratory, one should aim to produce a material that closely resembles the chemical composition, micro and macro-structure and porosity of bones; alas, the prepared biocomposites appear to be substantially different from bone tissues (Figure 3) due to a lack of real interaction between the two components, i.e., the interactions those can modify the intrinsic characteristics of the singular components themselves. The main characteristics of the route, by which the mineralized hard tissues are formed *in vivo*, is that the organic matrix is laid down first and the inorganic reinforcing phase grows within this organic matrix [14, 15, 20, 21]. Although to date, neither the elegance of the biomineral assembly mechanisms nor the intricate composite nano-sized architectures have been duplicated by non-biological methods, the best way to mimic bone is to copy the way it is formed, namely by nucleation and growth of CDHA nano-sized crystals from a supersaturated solution both onto and within the collagen fibrils. Such syntheses were denoted as "biologically inspired" which means they reproduce an ordered pattern and an environment very similar to natural ones [512-514]. The biologically inspired biocomposites of collagen and $CaPO_4$ (mainly, apatites) for bone substitute have a long history [515-528] and started from the pioneering study by Banks, et al., who grew CDHA on reconstituted calfskin collagen tapes in 1977 [531 673]. Such combinations were found to be bioactive, osteoconductive, osteoinductive [532-534] and, in general, artificial grafts manufactured from this type of the biocomposites are likely to behave similarly to bones and be of more use in surgery than those prepared from any other materials. Indeed, data are available on the superiority of $CaPO_4$/collagen biocomposite scaffolds over the artificial polymeric and $CaPO_4$ bioceramic scaffolds individually [535]. In addition, multi-scale simulations of apatite/collagen composites were performed [536].

It has been found that $CaPO_4$ may be successfully precipitated onto a collagen substrate of whatever form or source [529, 537, 538]. However, adherence of $CaPO_4$ crystals to collagen did depend on how much the collagen had been denatured: the more fibrillar the collagen, the greater attachment. The first report on a biocomposite produced by DCPD precipitation onto a collagen matrix with the aid of phosphorylated amino acids was published in 1993 [515]. Further, self-setting CDHA forming formulations (DCPD + TTCP) were mixed with a collagen suspension, hydrated and allowed to set. CDHA crystals were found to nucleate on the collagen fibril network, giving a material with the mechanical properties weaker than those reported for bone.

Alas, the prepared biocomposites did not have a bone-like nanostructure [516, 539]. The oriented growth of OCP crystals on collagen was achieved by an experimental device in which Ca^{2+} and PO_4^{3-} ions diffused into a collagen disc from the opposite directions [540, 541]. Unfortunately, these experiments were designed to simulate the mechanism of *in vivo* precipitation of biological apatite only; due to this reason, the mechanical properties of the biocomposites were not tested [542].

Conventionally, collagen/$CaPO_4$ biocomposites can be prepared by blending or mixing of collagen and $CaPO_4$, as well as by biomimetic methods [512-516, 526, 537, 542-554]. For example, bone like apatite/collagen biocomposites were prepared by two different methodologies: (1) dispersion of apatite in a collagen aqueous suspension and then freeze dried and (2) direct nucleation of an apatitic phase on assembling collagen fibrils [514]. Biocomposites obtained using the first way were similar to uncalcified natural collagen. The crystallite sizes were not uniform and were often aggregated and randomly distributed into the matrix (Figure 3b), proving that there was no real interaction between apatite and collagen fibers. However, the second method allowed the direct nucleation of nano-sized crystals of apatite on self-assembled collagen fibers. In this case, the two components (CDHA and collagen) exhibited strong interactions (Figure 3c), highlighted by several analysis techniques, which showed a complete analogy of the composite with calcified natural tissue [514]. Other production techniques are also possible. For example, using a polymer-induced liquid-precursor process, collagen/apatite biocomposites mimicking the nanostructure of bones, wherein nano-sized crystals of apatite were embedded within the collagen fibrils, were prepared [551]. Much better results were obtained by other researchers, who produced multilevel hierarchical HA/collagen biocomposites scaffolds with a biomimetic bone Haversian microstructure by a combination of electrospinning, biomimetic mineralization and rolling techniques for a bottom-up synthesis from nano-level to micro-level and then to macro-level. MicroCT scanning and scanning electron microscopy confirmed the successful preparation of the HA/collagen scaffolds composed of fully assembled microscopic fibers, while the mechanical properties of the scaffolds were double reinforced by many newborn HA nano particles and the chemical bonds [554]. More complicated compositional formulations, such as a magnetite enriched HA/collagen [555], HA/collagen/PVA [556], HA/collagen/chitosan [557], HA/collagen/acidic gelatin/basic fibroblast growth factor (b-FGF) [558] and calcium silicate/β-TCP/collagen [559], were developed as well. Recent investigations revealed that $CaPO_4$/collagen formulations might be printed by 3D printers [560].

In addition, collagen might be incorporated into various self-setting $CaPO_4$-based formulations [516, 539, 561-565]. Typically, a type I collagen sponge is presoaked in PO_4^{3-}-containing a highly basic aqueous solution and then is immersed into a Ca^{2+}-containing solution to allow mineral deposition. Also, collagen I fibers might be dissolved in acetic acid and then this solution is added to phosphoric acid, followed by a neutralization synthesis (performed at 25°C and solution pH within 9–10) between an aqueous suspension of $Ca(OH)_2$ and the H_3PO_4/collagen solution [512, 513]. To ensure the quality of the final product, it is necessary to control the Ca/P ionic ratio in the reaction solution. One way to do this is to dissolve a commercial $CaPO_4$ in an acid; another one is to add Ca^{2+} and PO_4^{3-} ions in a certain ratio to the solution and after that induce the reaction [566]. Biomimetically, one can achieve an oriented growth of CDHA crystals onto dissolved collagen fibrils in aqueous solutions via a self-organization mechanism [561]. More to the point, collagen might be first dispersed in an acidic solution, followed by addition of calcium and orthophosphate ions and then co-precipitation of collagen and CDHA might be induced by either increasing the solution pH or adding mixing agents. Although it resulted in biocomposites with poor mechanical properties, pressing of the apatite/collagen mixtures at 40°C under 200 MPa for several days is also known [567]. Attempts have been performed for a computer simulation of apatite/collagen composite formation process [568]. It is interesting to note, that such biocomposites were found to possess some piezoelectric properties [569].

As most of the collagen/HA biocomposites are conventionally processed by anchoring micron-sized HA particles into collagen matrix, it makes quite difficult to obtain a uniform and homogeneous composite graft. Besides, such biocomposites have inadequate mechanical properties; over and above, the proper pore sizes have not been achieved either. Further, microcrystalline HA, which contrasts with nanocrystalline bone apatite, might take a longer time to be remodeled into a new bone tissue upon the implantation. In addition, some of the biocomposites exhibited very poor mechanical properties, probably due to a lack of strong interfacial bonding between the constituents. Furthermore, in all blended composites, the crystallite sizes of $CaPO_4$ were not uniform and the crystals were often aggregated and randomly distributed within a fibrous matrix of collagen. Therefore, no structural similarity to natural bone was obtained and only a compositional similarity to that of natural bone was achieved. The data clearly demonstrate that the chemical composition like that of bones is insufficient for manufacturing the proper grafts; both the mechanical properties and mimetic of the bone nanostructure are necessary to function as bone in recipient sites. There is a chance for improving osteointegration by reducing the grain size of HA crystals by activating of ultrafine apatite growth into the matrix. This may lead to enhance the mechanical properties and osteointegration with improved biological and biochemical affinity to the host bone. Besides, the porosity was found to have a positive influence on the ingrowth of the surrounding tissues into the pores of collagen/HA biocomposites [570, 571].

Bovine collagen might be mixed with $CaPO_4$ and such biocomposites are marketed commercially as bone-graft substitutes those further can be combined with bone marrow aspirated from the iliac crest of the site of the fracture. Bioimplant®, Bio-Oss Collagen®, Boneject®, Collagraft®, CollapAn®, Healos®, Integra Mozaik® and LitAr® are several examples of the commercially available $CaPO_4$/collagen grafts for the clinical use (Table 5). Application of these materials was compared with autografts for the management of acute fractures of long bones with defects, which had been stabilized by internal or external fixation [572, 573]. These biocomposites are osteogenic, osteoinductive and osteoconductive; however, they lack the structural strength and require a harvest of the patient's bone marrow. OCP/collagen biocomposites have been investigated [574] and clinically tested [575].

Table 5. A list of known commercially produced $CaPO_4$ biocomposites with natural and synthetic polymers

Calcium orthophosphate	Trade name and producer (when available)
HA embedded or suspended in a gel	Bio-Gel HT hydroxyapatite (Bio-Rad, CA, USA)
	Coaptite (Boston Scientific, MA, USA)
	Facetem (Daewoong, Korea)
	NanoBone (Artoss, Germany)
	Nanogel (Teknimed, France)
	Radiesse (Merz Aesthetics, Germany)
	Renú Calcium Hydroxylapatite Implant (Cytophil, WI, USA)
HA/collagen, CDHA/collagen and/or carbonate apatite/collagen	AUGMATRIX (Wright Medical Technology, TN, USA)
	Bioimplant (Connectbiopharm, Russia)
	Bio-Oss Collagen (Geitslich, Switzerland)
	Boneject (Koken, Japan)
	COL.HAP-91 (JHS Biomateriais, Brazil)
	Collagraft (Zimmer and Collagen Corporation, USA)
	CollaOss (SK Bioland, Korea)
	CollapAn (Intermedapatite, Russia)

Table 5. (Continued)

Calcium orthophosphate	Trade name and producer (when available)
	COLLAPAT (Symatese, France)
	DualPor collagen (OssGen, Korea)
	G-Graft (Surgiwear, India)
	HAPCOL (Polystom, Russia)
	Healos (DePuy Spine, USA)
	LitAr (LitAr, Russia)
	Ossbone Collagen (SK Bioland, Korea)
	OssFill (Sewon Cellontech, Korea)
	OssiMend (Collagen Matrix, NJ, USA)
	OSSIX Bone (Dentsply Sirona, NC, USA)
	Osteomatrix (Connectbiopharm, Russia)
	OsteoTape (Impladent, NY, USA)
	ReFit (HOYA Technosurgical, Japan
	RegenOss (JRI Orthopaedics, UK)
	RegenerOss Synthetic (Zimmer Dental, CA, USA)
	SmartBone (IBI, Switzerland)
	Straumann XenoFlex (Straumann, Switzerland)
	SwissBone (IBI, Switzerland)
HA/sodium alginate	Bialgin (Biomed, Russia)
rhBMP-2 + HA carrier	NOVOSIS (CGBIO, Korea)
HA/poly-L-lactic acid	Biosteon (Biocomposites, UK)
	ReOss (ReOss, Germany)
	OSTEOTRANS MX (Teijin Medical Technologies, Japan)
	SuperFIXSORB30 (Takiron, Japan)
HA/polyethylene	HAPEX (Gyrus, TN, USA)
HA/$CaSO_4$	BioWrist Bone Void Filler (Skeletal Kinetics, CA, USA)
	Bond Apatite (Augma Biomaterials, NJ, USA)
	Hapset (LifeCore, MN, USA)
	PerOssal (aap Implantate, Germany)
Calcium orthophosphate	Trade name and producer (when available)
HA/$CaSO_4$ powders suspended in a liquid	CERAMENT (BONESUPPORT, Sweden)
HA/glass	Bonelike (unknwn producer0
bovine bone (unsintered)	Unilab Surgibone (Unilab, NJ, USA)
bovine bone (unsintered) + polymer	Alpha-Bio's Graft (Alpha-Bio Tec, Israel)
	C-Graft Putty (unknwn producer0
human bone allograft	ALLOPURE (Wright Medical Technology, TN, USA)
	AlloOss (ACE Surgical Supply, MA, USA)
	Allosorb (Curasan, Germany)
	CancellOss (Impladent, NY, USA)
	CurOss (Impladent, NY, USA)
	J Bone Block (Impladent, NY, USA)
	Maxgraft (botiss, Germany)
	Mega-Oss (Megagen Implant, Korea)
	NonDemin (Impladent, NY, USA)

Calcium orthophosphate	Trade name and producer (when available)
	Osnatal (aap Implantate, Germany)
	OsteoDemin (Impladent, NY, USA)
	Osteopure (O.S.T Laboratories, France)
	OsteoWrap (Curasan, Germany)
	OVIS ALLO (DENTIS, Korea)
	PentOS OI (Citagenix, QC, Canada)
	Puros (Zimmer Biomet, USA)
	RAPTOS (Citagenix, QC, Canada)
	Straumann AlloGraft (Straumann, Switzerland)
	TenFUSE (Wright Medical Technology, TN, USA)
β-TCP/$CaSO_4$	Fortoss vital (Biocomposites, UK)
	Genex (Biocomposites, UK)
β-TCP/poly-lactic acid	Bilok (Biocomposites, UK)
	Duosorb (SBM, France)
	Matryx Interference Screws (Conmed, USA)
β-TCP/poly-lactic-*co*-glycolic acid	Evolvemer TCP30PLGA (Arctic Biomaterials, Finland)
β-TCP/polymer	AttraX putty (NuVasive, CA, USA)
	ExcelOs-inject (CGBIO, Korea)
	Therigraft (Therics, OH, USA)
β-TCP/bone marrow aspirate	Induce (Skeletal Kinetics, CA, USA)
β-TCP/collagen	Integra Mozaik (Integra Orthobiologics, CA, USA)
β-TCP/growth-factor	GEM 21S (Lynch Biologics, USA)
β-TCP/rhPDGF-BB solution	AUGMENT Bone Graft (Wright Medical Group, TN, USA)
BCP (HA + β-TCP)/collagen	Allograft (Zimmer, IN, USA)
	collacone max (botiss, Germany)
	Collagraft (Zimmer, IN, USA)
	Cross.Bone Matrix (Biotech Dental, France)
	Indost (Polystom, Russia)
	MasterGraft (Medtronic Sofamor Danek, TN, USA)
	MATRI BONE (Biom'Up, France)
	Osteon III collagen (GENOSS, Korea)
	SynergOss (Nobil Bio Ricerche, Italy)
	without trade name (MedicalGroup, France)
BCP (HA + β-TCP)/hydrogel	4MATRIX+ (MIS Implants, Israel)
	Eclipse (Citagenix, QC, Canada)
BCP (HA + β-TCP)/polymer	In'Oss (Biomatlante, France)
	Hydros (Biomatlante, France)
	Octane HyperPEEK (Octane, ON, Canada)
	Osteocaf (Texas Innovative Medical Devices, TX, USA)
	Osteotwin (Biomatlante, France)
BCP (HA + β-TCP)/chitosan	k-IBS (Bioceramed, Portugal)
BCP (HA + β-TCP)/fibrin	TricOS (Baxter BioScience, France)
BCP (HA + β-TCP)/silicon	FlexHA (Xomed, FL, USA)
bioglass + α-TCP + β-TCP + HA + polymers	OsteoFlo NanoPutty (SurGenTec, FL, USA)

Table 5. (Continued)

Calcium orthophosphate	Trade name and producer (when available)
DCPA + $MgHPO_4 \cdot 3H_2O$ + SiO_2 + carboxymethyl cellulose	Novogro (OsteoNovus, OH, USA)
DCPD/collagen	CopiOs Bone Void Filler (Zimmer, IN, USA)
DCPD + β-TCP/$CaSO_4$	PRO-DENSE (Wright Medical Group, TN, USA)
DCPD + β-TCP/$CaSO_4$ + collagen	PRO-STIM (Wright Medical Group, TN, USA)
OCP/fibrin	FibroFor (BioNova, Russia)
OCP/collagen	Bonarc (Toyobo, Japan)
undisclosed $CaPO_4$ + biologics	i-FACTOR (Cerapedics, CO, USA)
3D-printed Ti mesh embedded in MCPM+β-TCP+Ca pyrophosphate	OssDsign (OssDsign AB, Sweden)

Collagen sponges with an open porosity (30 – 100 μm) were prepared by a freeze-drying technique and then their surface was coated by a 10-μm layer of biomimetic apatite precipitated from simulated body fluid [576]. The researchers found a good *in vitro* performance with fibroblast cell culture. Other preparation techniques are also possible [577]. Collagen/HA microspheres or gel beads have been prepared in the intention of making injectable bone fillers [578-580]. Liao, et al. succeeded in mimicking the bone structure by blending carbonateapatite with collagen [581]. A similar material (mineralized collagen) was implanted into femur of rats and excellent clinical results were observed after 12 weeks [582]. Collagen/HA biocomposites were prepared and their mechanical performance was increased by cross-linking the collagen fibers with glutaraldehyde. Those biocomposites were tested in rabbits and showed a good biological performance, osteoconductivity and biodegradation. A similar approach was selected to prepare HA/collagen microspheres (diameter ~ 5 μm) by a water-oil emulsion technique in which the surface was also cross-linked by glutaraldehyde [579]. That material showed a good *in vitro* performance with osteoblast cell culture. A porous bone graft substitute was formed from a nano-sized HA/collagen biocomposite combined with PLA by a freeze-drying method; the resulting material was found to mimic natural bones at several hierarchical levels. Porous (porosity level ~ 95% with interconnected pores of 50 – 100 μm) biocomposites of collagen (crosslinked with glutaraldehyde) and β-TCP have been prepared by a freeze-drying technique, followed by sublimation of the solvent; the biocomposites showed a good biocompatibility upon implantation in the rabbit jaw [583].

Since biocomposites of $CaPO_4$ with collagen were found to be able to form bioconjugates (bioconjugation is a chemical strategy to form a stable covalent link between two molecules, at least one of which is a biomolecule [584]), they appeared to be useful for delivery of drugs, growth factors and other important biomolecules [518, 565, 585, 586]. Besides, a two-layered collagen/β-TCP implant augmented with chondral inductive growth factors was developed for repair of osteochondral defects in the trochlear groove of minipigs [585]. Further details on this topic are available in recent review [586].

Although biocomposites of apatites with collagen appear to be a very hot topic of the research (Figures 4 and Table 5), not many papers are devoted to biocomposites of other $CaPO_4$ with collagen [544-547, 585, 587-591]. Among them there are ones with a very complicated composition. For example, collagen hydrogel incorporating platelet-derived growth factor-loaded photopolymerizable zinc-based zeolitic-imidazolate-frameworks-polydopamine nanoparticles (PDGF@ZIF-8-PDA@COL hydrogel) was prepared and then perfused into PLGA-TCP scaffolds. The resulting PDGF@ZIF-8-PDA@COL/PLGA-TCP biocomposite scaffolds sustainably released PDGF and showed an excellent antibacterial activity against both the Gram-negative bacterium *E. coli* and the Gram-positive bacterium *S. aureus*. Further, the scaffolds exhibited excellent osteoinductive and osteoconductive capacity *in vitro* and accelerated bone formation *in vivo* in a rat model of a cranial defect [591].

To conclude, currently known biocomposites of $CaPO_4$ with collagen mimic natural bones to some extent, while their subsequent biological evaluation suggests that they are readily incorporated into the bone metabolism in a way like bone remodeling, instead of acting as permanent implant [531]. However, the performance of such biocomposites depends on the source of collagen from which it was processed. Several attempts have been made to simulate the collagen-HA interfacial behavior in real bone by means of crosslinking agents such as glutaraldehyde [537, 579, 583] with the purpose to improve the mechanical properties of these biocomposites. Unfortunately, a further progress in this direction is restricted by a high cost, difficulty to control cross-infection, a poor definition of commercial sources of collagens, as well as by a lack of an appropriate technology to fabricate bone-resembling microstructures. Nevertheless, since collagen is the protein that constitutes the majority of organic bone matrix, the synthesis of $CaPO_4$/collagen biocomposites has been studied extensively for decades. Further details on this topic are available elsewhere [509, 511, 521, 592].

5.4. Formulations with Other Bioorganic Compounds and/or Biological Macromolecules

The biggest practical problems with collagen type I are its cost and the poor definition of commercial sources of this material, which makes it difficult to follow up on well controlled processing. Therefore, collagen type I can be replaced by other compounds. One should notice, that, besides collagen, both human and mammalian bodies contain many types of various bioorganic compounds, proteins and biological macromolecules. Therefore, the substantial amount of them potentially might be used to prepare biocomposites with $CaPO_4$. For example, a biologically strong adhesion (to prevent invasion of bacteria) between teeth and the surrounding epithelial tissues is attributed to a cell-adhesive protein, laminin [593]. To mimic the nature, a laminin/apatite biocomposite layer was successfully created on the surface of both titanium [594] and EVOH [595, 596] using the biomimetic approach. A more complicated laminin/DNA/apatite biocomposite layer was found to be an efficient gene transfer system [597]. Further details on this subject are available in a topical review [598].

$CaPO_4$/gelatin biocomposites are widely investigated as potential bone replacement biomaterials [152, 361-368, 387, 435-437, 599-604]. For example, gelatin foams were successfully mechanically reinforced by HA and then crosslinked by a carbodiimide derivative [152]. Such foams were shown to be a good carrier for antibiotic tetracycline [600]. Several biocomposites of $CaPO_4$ with alginates have been prepared [365, 513, 605-609]. For example, porous HA/alginate composites based on hydrogels were prepared both biomimetically [513] and by using a freeze-drying technique [605]. More complicated formulations have been developed as well [610, 611].

Various biocomposites of $CaPO_4$ with chitosan [381, 404, 421, 442, 460, 498, 612-621] and chitin [370, 441, 622-626] are also very popular. For example, a solution-based method was developed to combine HA powders with chitin, in which the ceramic particles were uniformly dispersed [622, 623]. Unfortunately, it was difficult to obtain the uniform dispersions. The mechanical properties of the final biocomposites were not very good; due to a poor adhesion between the filler and the matrix both the tensile strength and modulus were found to decrease with increase of the HA amount. Microscopic examination revealed that HA particles were intervened between the polymer chains, weakening their interactions and decreasing the entire strength [622, 623]. Other manufacturing techniques might be found in the references; I just would like to mention on an interesting approach, in which a HA/chitosan biocomposite was produced by a hydrothermal process from natural $CaCO_3$/chitosan biocomposite of crab shells [615]. Similarly, HA/chitosan biocomposite was produced by a hydrothermal process from previously prepared DCPD/chitosan biocomposite [616]. Biocomposites of natural HA with chitosan were found to possess both a good hard tissue biocompatibility and an excellent osteoconductivity, which is suitable for artificial bone implants and frame materials of tissue engineering [613]. Data are available that addition of $CaPO_4$ into chitosan improved cell attachment, provided a higher cell proliferation and well-spread morphology when compared to chitosan alone [617]. More complex biocomposites, such as silk fibers reinforced HA/chitosan [627], HA/collagen/chitosan [628], HA/silk fibroin/sodium alginate [629], CDHA/chitosan loaded with ciprofloxacin [620] and chitosan/silk fibroin/nitrogen-doped carbon quantum dot/α-TCP [630], have been studied as well. Besides biomedical applications, biocomposites of nano-sized HA with chitin/chitosan might be used for removal of Fe(III) [631] and fluorides [632, 633] from aqueous solutions. Further details on this topic are available in a recent review [634].

Biocomposites of CDHA with water-soluble proteins, such as BSA, might be prepared by a precipitation method [491, 635-638]. In such biocomposites, BSA is not strongly fixed to solid CDHA, which is useful for a sustained release. However, this is not the case if a water/oil/water interfacial reaction route has been used [143]. To extend this subject, inclusion of DNA into CDHA/BSA biocomposites was claimed [143, 639-641]. Furthermore, biocomposites of an unspecified $CaPO_4$ with DNA [642, 643], as well as those of nano-sized crystals of biomimetic apatite with C_{60} fullerene and Au-DNA nano-sized particles [644] were prepared as well.

Akashi and co-workers developed a procedure to prepare $CaPO_4$-based biocomposites by soaking hydrogels in supersaturated by Ca^{2+} and PO_4^{3-} ions solutions to precipitate CDHA in the hydrogels (up to 70% by weight of CDHA could be added to these biocomposites) [645]. This procedure was applied to chitosan; the 3D shape of the resulting biocomposite was controlled by the shape of the starting chitosan hydrogel [646]. Another research group developed biocomposites based on *in situ* $CaPO_4$ mineralization of self-assembled supramolecular hydrogels [647]. Other experimental approaches are also possible [648, 649].

Various biocomposites of CDHA with glutamic and aspartic amino acids, as well as poly-glutamic and poly-aspartic amino acids have been prepared and investigated [245, 246, 650-653]. These (poly)amino acids were quantitatively incorporated into CDHA crystals, provoking a reduction of the coherent length of the crystalline domains and decreasing the crystal sizes. The relative amounts of the (poly)amino acid content in the solid phase, determined through high-performance liquid chromatography analysis, increased with their concentration in solution up to a maximum of about 7.8 wt. % for CDHA/aspartic acid and 4.3 wt. % for CDHA/glutamic acid biocomposites. The small crystal dimensions, which implied a great surface area, and the presence of (poly)amino acids were suggested to be relevant for possible application of these biocomposites for hard tissues replacement [245, 246, 650-653]. A schematic description of a biocomposite formation from amino acids and ACP is shown in Figure 5 [654].

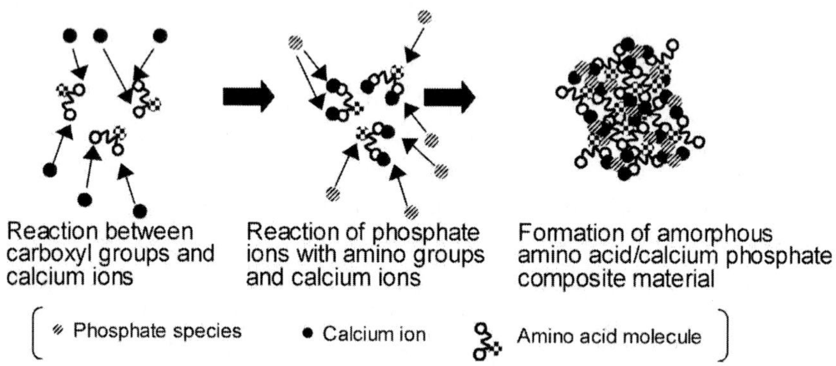

Figure 5. A proposed mechanism for the formation of ACP/amino acid biocomposites in aqueous solutions. Reprinted from Ref. [654] with permission.

Furthermore, BCP (HA + β-TCP)/agarose macroporous scaffolds with controlled and complete interconnection, high porosity, thoroughly open pores and tailored pore size were prepared for tissue engineering application [655, 656]. Agarose, a biodegradable polymer, was selected as the organic matrix because it was a biocompatible hydrogel, which acted as gelling agent leading to strong gels and fast room temperature polymerization. Porous scaffolds with the designed architecture were manufactured by combining a low temperature shaping method with stereo-lithography and two drying techniques. The biocompatibility of this BCP/agarose system was tested with mouse L929 fibroblasts and human SAOS-2 osteoblasts during different colonization times [657].

Fibrin sealants are non-cytotoxic, fully resorbable, biological matrices that simulate the last stages of a natural coagulation cascade, forming a structured fibrin clot like a physiological clot [658]. Biocomposites of $CaPO_4$ with fibrin sealants might develop the clinical applications of bone substitutes.

The 3D mesh of fibrin sealant interpenetrates the macro- and micro-porous structure of $CaPO_4$ ceramics. The physical, chemical and biological properties of $CaPO_4$ bioceramics and the fibrin glue might be cumulated in biocomposites, suitable for preparation of advanced bone grafts [659-665].

In addition, there are biocomposites of $CaPO_4$ with bisphosphonates [666], silk fibroin [141, 667-674], cellulose [675], chitosan + silk fibroin [676], chitosan + zein [677], chitosan derivatives [678], fibronectin [679], triazine-trione [680], chondroitin sulfate [533, 681], casein phosphopeptides [682], okra hydrocolloids [683], keratin [684], gellan gum [685] and graphene oxide/chitosan [686] hydrogels, amyloid fibrils [687], agarose [688], glycine [689] and vitamins [690]. Photopolymerisable formulations have been developed as well [691]. Besides, the reader's attention is pointed out to an interesting approach to crystallize CDHA inside poly(allylamine)/poly(styrene sulfonate) polyelectrolyte capsules resulting in empty biocomposite spheres of micron size. Depending on the amount of precipitated CDHA, the thickness of the shell of biocomposite spheres can be varied between 25 and 150 nm. These biocomposite capsules might find application as medical agents for bone repairing and catalytic microreactors [692].

An interesting phenomenon of fractal growth of FA/gelatin composite crystals (Figure 6) was achieved by diffusion of calcium- and orthophosphate + fluoride-containing solutions from the opposite sides into a tube filled with a gelatin gel [693-699]. The reasons of this phenomenon are not quite clear yet. Other types of $CaPO_4$-based composites, based on DCPD and OCP, were grown by the similar technique in an iota-carrageenan gel [700]. Up to now, nothing has been reported yet on a possible biomedical application of such unusual structural composites.

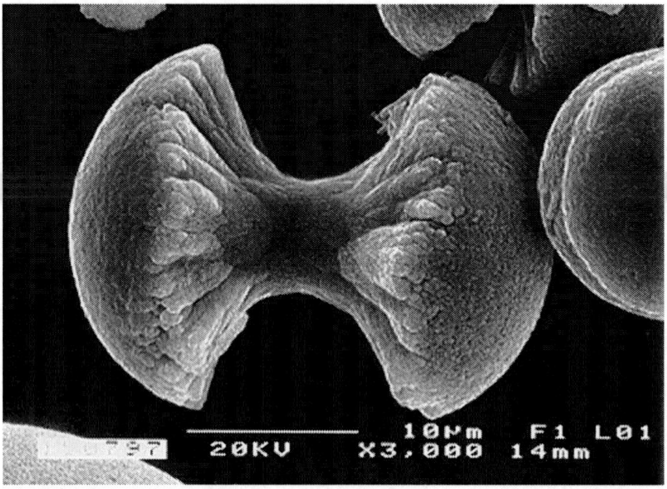

Figure 6. A biomimetically grown aggregate of FA that was crystallized in a gelatin matrix. Its shape can be explained and simulated by a fractal growth mechanism. Scale bar: 10 µm. Reprinted from Ref. [693] with permission.

5.5. Injectable Bone Substitutes (IBS)

Injectable bone substitutes (IBS) represent compelling options for bone regenerative medicine as they can be used to optimally fill complex bone defects through minimally invasive intervention. Therefore, IBS appear to be a convenient alternative to solid bone-filling grafts. They represent ready-to-use suspensions of $CaPO_4$ microspheres [701, 702], nano-sized rods [703] or powder(s) blended with viscous solutions of macromolecules acting as a liquid carrier phase. However, addition of other phases, such as calcium sulfate [704], is also possible. Generally, IBS look like opaque viscous pastes with the rheological properties, sufficient to inject them into bone defects by means of surgical syringes and needles. Occasionally, their constituents may require to be mixed in the operating room and, in the latter case, it is sometimes possible to produce or combine the IBS with blood, bone marrow or platelet-rich plasma. Besides, IBS could be easily produced in a sterile stage. Their stable composition and mechanical properties are suitable for reproducibility of the biological response [705, 706]. All types of IBS are divided into 2 major groups: self-setting formulations (see section 5.2. above) and those, which do not set. The latter ones are described here.

IBS requires suitable rheological properties to ensure bonding and cohesion of the mineral phase *in situ* with good cell permeability. Usually, the necessary level of viscosity is created by addition of suitable water-soluble polymers [92, 147, 707-716]. However, IBS based on hydrogels are known as well [717-720]. Some of them possess self-gelling properties [720]. Therefore, most $CaPO_4$-containing IBS formulations might be considered as a subgroup of $CaPO_4$/polymer biocomposites. For example, an IBS was described that involved a silanized hydroxyethylcellulose carrier with BCP (HA + β-TCP). The suspension is liquid at pH within 10 – 12, but gels quickly at pH < 9 [709]. Injectable composites can be formed with β-TCP to improve mechanical integrity [478]. Similarly, a polydioxanone-*co*-glycolide-based biocomposite reinforced with HA or β-TCP can be used as an injectable or moldable putty [710]. During the crosslinking reaction following injection, carbon dioxide is released allowing the formation of interconnected pores. Furthermore, HA/poly(L-lactide-*co*-ε-caprolactone) biocomposite microparticles were fabricated as an injectable scaffold via the Pickering emulsion route in the absence of any molecular surfactants. A stable injectable oil-in-water emulsion was obtained using water dispersed HA nano-sized crystals as the particulate emulsifier and a dichloromethane solution of poly(L-lactide-*co*-ε-caprolactone) as an oil phase [147]. $CaPO_4$-containing IBS based on other (bio)organic compounds, such as CMC [713], oligo(poly(ethylene glycol) fumarate) [711, 712], chitosan + collagen [714] have been developed as well. In addition, photocrosslinkable formulations are known [715].

Daculsi, et al. developed viscous IBS biocomposites based on BCP (60% HA + 40% β-TCP) and 2% aqueous solution of HPMC that was said to be perfectly biocompatible, resorbable and easily fitted bone defects (due to an initial plasticity) [708, 721-728]. The best ratio BCP/HPMC aqueous solution was found to be at ~ 65/35 w/w. To extend this subject further, IBS might be loaded by cells [729, 730], radiopaque elements [731] or microparticles [732], as well as functionalized by nucleic acids [743]. Self-hardening formulations, based on Si-HPMC hydrogel, are known as well [729]. The list of the commercially available $CaPO_4$-based IBS formulations is presented in Table 6 [733].

Table 6. A list of several commercial non-setting $CaPO_4$-based IBS and pastes with indication of producer, product name, composition (when available) and form [733]

Producer	Product name	Composition	Form
ApaTech (UK)	Actifuse™	HA, polymer and aqueous solution	pre-mixed
	Actifuse™ Shape Actifuse™ ABX	Si-substituted $CaPO_4$ and a polymer	pre-mixed
Baxter (US)	TricOs T TricOs	BCP (60% HA, 40% β-TCP) granules and Tissucol (fibrin glue)	to be mixed
Berkeley Advanced Biomaterials	Bi-Ostetic Putty	not disclosed	not disclosed
BioForm (US)	Calcium hydroxylapatite implant	HA powder embedded in a mixture of glycerine, water and CMC	pre-mixed
Biomatlante (FR)	In'Oss™	BCP granules (60% HA, 40% β-TCP; 0.08 – 0.2 mm) and 2% HPMC	pre-mixed
	MBCP Gel®	BCP granules (60% HA, 40% β-TCP; 0.08 – 0.2 mm) and 2% HPMC	pre-mixed
	Hydr'Os	BCP granules (60% HA, 40% β-TCP; micro- and nano-sized particles) and saline solution	pre-mixed
Biocerамed (PT)	k-IBS®	BCP granules (75% HA, 25% β-TCP; 0.125 – 0.355 mm) and chitosan dissolved in PEG	pre-mixed
	n-IBS®	Aqueous suspension of HA powders	pre-mixed
Degradable solutions (CH)	Easy graft™	β-TCP or BCP granules (0.45 – 1.0 mm) coated with 10 μm PLGA, N-methyl-2-pyrrolydone	to be mixed
Dentsply (US)	Pepgen P-15® flow	HA (0.25 – 0.42 mm), P-15 peptide and aqueous Na hyaluronate solution	to be mixed
DePuy Spine (US)	Healos® Fx	HA (20 – 30%) and collagen	to be mixed
Fluidinova (P)	nanoXIM TCP	β-TCP (5 or 15%) and water	pre-mixed
	nanoXIM HA	HA (5, 15, 30 or 40%) and water	pre-mixed
Integra LifeSciences (US)	Mozaik Osteoconductive Scaffold	β-TCP (80%) and type 1 collagen (20%)	to be mixed
Mathys Ltd (CH)	Ceros® Putty/cyclOS® Putty	β-TCP granules (0.125 – 0.71 mm; 94%) and recombinant Na hyaluronate powder (6%)	to be mixed
Medtronic (US)	Mastergraft®	BCP (85% HA, 15% β-TCP) and bovine collagen	to be mixed
Merz Aesthetics (GER)	RADIESSE®	HA particles suspended in a gel	pre-mixed
NuVasive (USA)	AttraX® putty	β-TCP and alkylene oxide copolymer	pre-mixed

Producer	Product name	Composition	Form
Osartis/AAP (GER)	Ostim®	Nanocrystalline HA (35%) and water (65%)	pre-mixed
Smith & Nephew (US)	JAXTCP	β-TCP granules and an aqueous solution of 1.75% CMC and 10% glycerol	to be mixed
Stryker (US)	Calstrux™	β-TCP granules and CMC	to be mixed
Teknimed (FR)	Nanogel	HA (100 – 200 nm) (30%) and water (70%)	pre-mixed
Therics (US)	Therigraft™ Putty	β-TCP granules and polymer	pre-mixed
Zimmer (US)	Collagraft	BCP granules (65% HA, 35% β-TCP; 0.5 – 1.0 mm), bovine collagen and bone marrow aspirate	to be mixed

The advanced characteristics of IBS come from their good rheological properties and biocompatibility and the ease of tissue regeneration. Although the fabrication of IBS biocomposites in most cases improved the mechanical properties of the system and provided the material with resistance to fluids penetration, these achievements were limited by the amount of polymer that can be added to the paste. For instance, Mickiewicz, et al. reported that after a critical concentration (that depended on the type and molecular weight of the polymer, but was always around 10%), the polymer started forming a thick coating on the crystal clusters, preventing them from interlocking, originating plastic flow and, consequently, decreasing mechanical properties [491]. More to the point, Fujishiro, et al. reported a decrease in mechanical properties with higher amounts of gel, which was attributed to formation of pores due to leaching of gelatin in solution [484]. Therefore, it seems that mechanical properties, although improved by the addition of polymers, are still a limitation for the application of $CaPO_4$-based IBS formulations in load-bearing sites [125]. Further details on this topic are available elsewhere [705, 706].

Conclusion

All types of calcified tissues of mammals appear to possess complex hierarchical structures and, from the material point of view, they are multi-phase organic/inorganic biocomposites. Their mechanical properties are outstanding (considering weak constituents from which they are assembled) and far beyond those, that can be achieved using the same synthetic materials with present technologies. This is because biological organisms produce biocomposites that are organized in terms of both composition and structure, containing both brittle $CaPO_4$ and ductile bioorganic components in very complex structures, hierarchically organized at the nano-, micro- and meso-levels. Additionally, the calcified tissues are always multifunctional: for example, bones provide structural support for the body and are responsible for blood cell formation. The third defining characteristic of biological systems, in contrast with current synthetic systems, is their self-healing ability, which is nearly universal in nature. These complex structures, which have risen from millions of years of evolution, inspire materials scientists in the design of novel biomaterials [734].

Obviously, no single-phase biomaterial can provide all the essential features of bones and/or other calcified tissues and therefore, there is a great need to engineer multi-phase biocomposites with the structure and composition mimicking those of the calcified tissues. This resulted in an idea of adding $CaPO_4$ to polymers, which further led from the bioinert monolithic materials used in the 1970s and 1980s to the bioactive biodegradable biocomposites used now as modulus-matched biomaterials and since recently as the scaffolds for tissue engineering [735]. This approach resulted in applications of conventional composite manufacturing techniques to the field of biomaterials and the studies summarized in this review have shown that the proper combination of a ductile matrix with a brittle, hard and bioactive $CaPO_4$ filler offers many advantages for biomedical applications. Namely, the desirable properties of some components can compensate for a poor mechanical behavior of $CaPO_4$ bioceramics, while in turn the desirable bioactive properties of $CaPO_4$ improve those of other phases, thus expanding the possible application of each material within the body [66, 67]. Furthermore, following the modern tendency of tissue engineering, a novel generation of $CaPO_4$ biocomposites with polymers should also contain a biological living part, including various types of biologically active substances, such as hormones, bioactive factors, drugs, *etc.* and living cells [736-738]. Thus, due to their documented biocompatibility, osteoconductivity and osteointegration potential, they represent a promising class of biomaterials for biomedical applications.

To finalize, the future of the $CaPO_4$ biocomposites with polymers is now directly dependent on the formation of multidisciplinary teams composed of experts ready to collaborate with others and thus be able to deal efficiently with the complexity of the human organism. The physical chemistries of solids, solid surfaces, polymer dispersion and solutions, as well as material-cell interactions are among the phenomena to be tackled. Furthermore, much work remains to be done on a long way from a laboratory to clinics and the success depends on the effective co-operation of clinicians, chemists, biologists, bioengineers and materials scientists.

References

[1] Chau, A. M. T., Mobbs, R. J. Bone graft substitutes in anterior cervical discectomy and fusion. *Eur. Spine J.* 2009, 18, 449-464.

[2] Kaveh, K., Ibrahim, R., Bakar, M. Z. A., Ibrahim, T. A. Bone grafting and bone graft substitutes. *J. Anim. Vet. Adv.* 2010, 9, 1055-1067.

[3] Shibuya, N., Jupiter, D. C. Bone graft substitute: allograft and xenograft. *Clin. Podiatr. Med. Surg.* 2015, 32, 21-34.

[4] Conway, J. D. Autograft and nonunions: morbidity with intramedullary bone graft versus iliac crest bone graft. *Orthop. Clin. North Am.* 2010, 41, 75-84.

[5] Li, S., Chen, Y., Lin, Z., Cui, W., Zhao, J., Su, W. A systematic review of randomized controlled clinical trials comparing hamstring autografts versus bone-patellar tendon-bone autografts for the reconstruction of the anterior cruciate ligament. *Arch. Orthop. Trauma Surg.* 2012, 132, 1287-1297.

[6] Keller, E. E., Triplett, W. W. Iliac crest bone grafting: review of 160 consecutive cases. *J. Oral Maxillofac. Surg.* 1987, 45, 11-14.

[7] Schaaf, H., Lendeckel, S., Howaldt, H. P., Streckbein, P. Donor site morbidity after bone harvesting from the anterior iliac crest. *Oral Surg. Oral Med. Oral Pathol. Oral Radiol. Endod.* 2010, 109, 52-58.

[8] Carlsen, A., Gorst-Rasmussen, A., Jensen, T. Donor site morbidity associated with autogenous bone harvesting from the ascending mandibular ramus. *Implant Dent.* 2013, 22, 503-506.

[9] Qvick, L. M., Ritter, C. A., Mutty, C. E., Rohrbacher, B. J., Buyea, C. M., Anders, M. J. Donor site morbidity with reamer-irrigator-aspirator (RIA) use for autogenous bone graft harvesting in a single centre 204 case series. *Injury* 2013, 44, 1263-1269.

[10] Li, Z., Kawashita, M. Current progress in inorganic artificial biomaterials. *J. Artif. Organ.* 2011, 14, 163-170.

[11] Bojar, W., Kucharska, M., Ciach, T., Koperski, Ł., Jastrzębski, Z., Szałwiński, M. Bone regeneration potential of the new chitosan-based alloplastic biomaterial. *J. Biomater. Appl.* 2014, 28, 1060-1068.

[12] Panchbhavi, V. K. Synthetic bone grafting in foot and ankle surgery. *Foot Ankle Clin.* 2010, 15, 559-576.

[13] Dinopoulos, H., Dimitriou, R., Giannoudis, P. V. Bone graft substitutes: what are the options? *Surgeon* 2012, 10, 230-239.

[14] Weiner, S., Wagner, H. D. The material bone: structure-mechanical function relations. *Ann. Rev. Mater. Sci.* 1998, 28, 271-298.

[15] Rey, C., Combes, C., Drouet, C., Glimcher, M. J. Bone mineral: update on chemical composition and structure. *Osteoporos. Int.* 2009, 20, 1013-1021.

[16] Dorozhkin, S. V. *Calcium orthophosphates: applications in nature, biology, and medicine.* Pan Stanford, Singapore, 2012; 854 pp.

[17] Dorozhkin, S. V. *Calcium orthophosphate-based bioceramics and biocomposites.* Wiley-VCH: Weinheim, Germany, 2016; 405 pp.

[18] Burr, D. B. The contribution of the organic matrix to bone's material properties. *Bone* 2002, 31, 8-11.

[19] Fratzl, P., Gupta, H. S., Paschalis, E. P., Roschger, P. Structure and mechanical quality of the collagen-mineral nano-composite in bone. J. Mater. Chem. 2004, 14, 2115-2123.

[20] Olszta, M. J., Cheng, X. G., Jee, S. S., Kumar, B. R., Kim, Y. Y., Kaufman, M. J., Douglas, E. P., Gower, L. B. Bone structure and formation: a new perspective. Mater. Sci. Eng. R 2007, 58, 77-116.

[21] Fonseca, H., Moreira-Gonçalves, D., Coriolano, H. J. A., Duarte, J. A. Bone quality: the determinants of bone strength and fragility. Sports Med. 2014, 44, 37-53.

[22] Murugan, R., Ramakrishna, S. Development of nanocomposites for bone grafting. Compos. Sci. Technol. 2005, 65, 2385-2406.

[23] Suchanek, W., Yoshimura, M. Processing and properties of hydroxyapatite-based biomaterials for use as hard tissue replacement implants. J. Mater. Res. 1998, 13, 94-117.

[24] Vallet-Regi, M., Arcos, D. Nanostructured hybrid materials for bone tissue regeneration. Curr. Nanosci. 2006, 2, 179-189.

[25] Doblaré, M., Garcia, J. M., Gómez, M. J. Modelling bone tissue fracture and healing: a review. Eng. Fract. Mech. 2004, 71, 1809-1840.

[26] Vallet-Regi, M. Revisiting ceramics for medical applications. Dalton Trans. 2006, 5211-5220.

[27] Pioletti, D. P. Biomechanics in bone tissue engineering. Comput. Methods Biomech. Biomed. Engin. 2010, 13, 837-846.

[28] Huiskes, R., Ruimerman, R., van Lenthe, H. G., Janssen, J. D. Effects of mechanical forces on maintenance and adaptation of form in trabecular bone. Nature 2000, 405, 704-706.

[29] Boccaccini, A. R., Blaker, J. J. Bioactive composite materials for tissue engineering scaffolds. Expert Rev. Med. Dev. 2005, 2, 303-317.

[30] Hutmacher, D. W., Schantz, J. T., Lam, C. X. F., Tan, K. C., Lim, T. C. State of the art and future directions of scaffold-based bone engineering from a biomaterials perspective. J. Tissue Eng. Regen. Med. 2007, 1, 245-260.

[31] Guarino, V., Causa, F., Ambrosio, L. Bioactive scaffolds for bone and ligament tissue. Expert Rev. Med. Dev. 2007, 4, 405-418.

[32] Yunos, D. M., Bretcanu, O., Boccaccini, A. R. Polymer-bioceramic composites for tissue engineering scaffolds. J. Mater. Sci. 2008, 43, 4433-4442.

[33] Zhao, H. X. Progress of study on drug-loaded chitosan/hydroxyapatite composite in bone tissue engineering. J. Funct. Mater. 2014, 45, 13006-13012+13020.

[34] Hench, L. L., Polak, J. M. Third-generation biomedical materials. Science 2002, 295, 1014-1017.
[35] Mathijsen, A. Nieuwe Wijze van Aanwending van het Gips-Verband bij Beenbreuken. J. B. van Loghem, Haarlem, Netherlands, 1852, 21 pp.
[36] Dreesman, H. Über Knochenplombierung. Beitr. Klin. Chir. 1892, 9, 804-810.
[37] Wang, M. Developing bioactive composite materials for tissue replacement. Biomaterials 2003, 24, 2133-2151.
[38] https://en.wikipedia.org/wiki/Composite_material (assessed in August 2023).
[39] Gibson, R. F. A review of recent research on mechanics of multifunctional composite materials and structures. Compos. Struct. 2010, 92, 2793-2810.
[40] Evans, S. L., Gregson, P. J. Composite technology in load-bearing orthopaedic implants. Biomaterials 1998, 19, 1329-1342.
[41] Wan, Y. Z., Hong, L., Jia, S. R., Huang, Y., Zhu, Y., Wang, Y. L., Jiang, H. J. Synthesis and characterization of hydroxyapatite-bacterial cellulose nanocomposites. Compos. Sci. Technol. 2006, 66, 1825-1832.
[42] Wan, Y. Z., Huang, Y., Yuan, C. D., Raman, S., Zhu, Y., Jiang, H. J., He, F., Gao, C. Biomimetic synthesis of hydroxyapatite/bacterial cellulose nanocomposites for biomedical applications. Mater. Sci. Eng. C 2007, 27, 855-864.
[43] Ohtsuki, C., Kamitakahara, M., Miyazaki, T. Coating bone-like apatite onto organic substrates using solutions mimicking body fluid. J. Tissue Eng. Regen. Med. 2007, 1, 33-38.
[44] Oyane, A. Development of apatite-based composites by a biomimetic process for biomedical applications. J. Ceram. Soc. Jpn. 2010, 118, 77-81.
[45] Dorozhkin, S. V. Calcium orthophosphate deposits: preparation, properties and biomedical applications. Mater. Sci. Eng. C 2015, 55, 272-326.
[46] Surmenev, R. A., Surmeneva, M. A., Ivanova, A. A. Significance of calcium phosphate coatings for the enhancement of new bone osteogenesis - a review. Acta Biomater. 2014, 10, 557-579.
[47] Dorozhkin, S. V. Calcium orthophosphate coatings on magnesium and its biodegradable alloys. Acta Biomater. 2014, 10, 2919-2934.
[48] Zhao, J., Guo, L. Y., Yang, X. B., Weng, J. Preparation of bioactive porous HA/PCL composite scaffolds. Appl. Surf. Sci. 2008, 255, 2942-2946.
[49] Dorozhkin, S., Ajaal, T. Toughening of porous bioceramic scaffolds by bioresorbable polymeric coatings. Proc. Inst. Mech. Eng. H 2009, 223, 459-470.
[50] Woo, A. S., Jang, J. L., Liberman, R. F., Weinzweig, J. Creation of a vascularized composite graft with acellular dermal matrix and hydroxyapatite. Plast. Reconstr. Surg. 2010, 125, 1661-1669.
[51] Furko, M., Detsch, R., Tolnai, I., Tolnai, I., Balázsi, K., Boccaccini, A. R., Balázsi, C. Biomimetic mineralized amorphous carbonated calcium phosphate-polycaprolactone bioadhesive composites as potential coatings on implant materials. Ceram. Int. 2023, 49, Part B, 18565-18576.
[52] Dong, J., Uemura, T., Kojima, H., Kikuchi, M., Tanaka, J., Tateishi, T. Application of low-pressure system to sustain in vivo bone formation in osteoblast/porous hydroxyapatite composite. Mater. Sci. Eng. C 2001, 17, 37-43.

[53] Mikán, J., Villamil, M., Montes, T., Carretero, C., Bernal, C., Torres, M. L., Zakaria, F. A. Porcine model for hybrid material of carbonated apatite and osteoprogenitor cells. Mater. Res. Innov. 2009, 13, 323-326.
[54] Oe, K., Miwa, M., Nagamune, K., Sakai, Y., Lee, S. Y., Niikura, T., Iwakura, T., Hasegawa, T., Shibanuma, N., Hata, Y., Kuroda, R, Kurosaka, M. Nondestructive evaluation of cell numbers in bone marrow stromal cell/β-tricalcium phosphate composites using ultrasound. Tissue Eng. C 2010, 16, 347-353.
[55] Krout, A., Wen, H. B., Hippensteel, E., Li, P. A hybrid coating of biomimetic apatite and osteocalcin. J. Biomed. Mater. Res. A 2005, 73A, 377-387.
[56] Kundu, B., Soundrapandian, C., Nandi, S. K., Mukherjee, P., Dandapat, N., Roy, S., Datta, B. K., Mandal, T. K., Basu, D., Bhattacharya, R. N. Development of new localized drug delivery system based on ceftriaxone-sulbactam composite drug impregnated porous hydroxyapatite: a systematic approach for in vitro and in vivo animal trial. Pharm. Res. 2010, 27, 1659-1676.
[57] Holzmeister, I., Schamel, M., Groll, J., Gbureck, U., Vorndran, E. Artificial inorganic biohybrids: the functional combination of microorganisms and cells with inorganic materials. Acta Biomater. 2018, 74, 17-35.
[58] Kickelbick, G. (Ed.) Hybrid materials. Synthesis, characterization, and applications, Wiley-VCH Verlag: Weinheim, Germany, 2007; 498 pp.
[59] Matthews, F. L., Rawlings, R. D. Composite materials: engineering and science. CRC Press: Boca Raton, FL, USA, 2000; 480 pp.
[60] Xia, Z., Riester, L., Curtin, W. A., Li, H., Sheldon, B. W., Liang, J., Chang, B., Xu, J. M. Direct observation of toughening mechanisms in carbon nanotube ceramic matrix composites. Acta Mater. 2004, 52, 931-944.
[61] Tavares, M. I. B., Ferreira, O., Preto, M., Miguez, E., Soares, I. L., da Silva, E. P. Evaluation of composites miscibility by low field NMR. Int. J. Polym. Mater. 2007, 56, 1113-1118.
[62] Kiran, E. Polymer miscibility, phase separation, morphological modifications and polymorphic transformations in dense fluids. J. Supercrit. Fluids 2009, 47, 466-483.
[63] Šupová, M. Problem of hydroxyapatite dispersion in polymer matrices: a review. J. Mater. Sci. Mater. Med. 2009, 20, 1201-1213.
[64] Böstman, O., Pihlajamäki, H. Clinical biocompatibility of biodegradable orthopaedic implants for internal fixation: a review. Biomaterials 2000, 21, 2615-2621.
[65] Zhang, X. D., Williams, D. F. (Eds.) Definitions of biomaterials for the twenty-first century. Proceedings of a consensus conference held in Chengdu, China, June 11th and 12th 2018. Materials Today; Elsevier, 2019, 290 pp.
[66] John, M. J., Thomas, S. Biofibres and biocomposites. Carbohydr. Polym. 2008, 71, 343-364.
[67] Tanner, K. E. Bioactive ceramic-reinforced composites for bone augmentation. J. R. Soc. Interface 2010, 7, S541-S557.
[68] Gravitis, Y. A., Tééyaér, R. E., Kallavus, U. L., Andersons, B. A., Ozol'-Kalnin, V. G., Kokorevich, A. G., Érin'sh, P. P., Veveris, G. P. Biocomposite structure of

wood cell membranes and their destruction by explosive autohydrolysis. Mech. Compos. Mater. 1987, 22, 721-725.
[69] Bernard, S. L., Picha, G. J. The use of coralline hydroxyapatite in a 'biocomposite' free flap. Plast. Reconstr. Surg. 1991, 87, 96-107.
[70] Rao, H., Thompson, W. A., Katz, J. L., Harper, R. A. Elastic constants of the composite system hydroxyapatite-dicalcium phosphate dihydrate. J. Dent. Res. 1976, 55, 708.
[71] Dorozhkin, S. V. Calcium orthophosphates and human beings. A historical perspective from the 1770s until 1940. Biomatter 2012, 2, 53-70.
[72] Dorozhkin, S. V. A detailed history of calcium orthophosphates from 1770s till 1950. Mater. Sci. Eng. C 2013, 33, 3085-3110.
[73] Hing, K. A. Bioceramic bone graft substitutes: influence of porosity and chemistry. Int. J. Appl. Ceram. Technol. 2005, 2, 184-199.
[74] Naqshbandi, A. R., Sopyan, I., Gunawan, Development of porous calcium phosphate bioceramics for bone implant applications: a review. Rec. Pat. Mater. Sci. 2013, 6, 238-252.
[75] LeGeros, R. Z. Calcium phosphates in oral biology and medicine. Monographs in oral science. Myers, H. M. (Ed.); Karger: Basel, Switzerland, 1991; Vol. 15, 201 pp.
[76] Elliott, J. C. Structure and chemistry of the apatites and other calcium orthophosphates, Studies in inorganic chemistry. Elsevier: Amsterdam, Netherlands, 1994; Vol. 18, 389 pp.
[77] Heimann, R. B. (Ed.) Calcium phosphate: structure, synthesis, properties, and applications. Nova Science, NY, USA, 2012; 498 pp.
[78] Gshalaev, V. S., Demirchan, A. C. (Eds.) Hydroxyapatite: synthesis, properties and applications. Nova Science, NY, USA, 2012; 477 pp.
[79] Carraher, C. E., Jr. Introduction to polymer chemistry. 2nd Ed., CRC Press: Boca Raton, FL, USA, 2010; 534 pp.
[80] Young, R. J., Lovell, P. A. Introduction to polymers. 3rd Ed., CRC Press: Boca Raton, FL, USA, 2011; 688 pp.
[81] Thomson, R. C., Ak, S., Yaszemski, M. J., Mikos, A. G. Polymer scaffold processing. In: Principles of tissue engineering, Academic Press: NY, USA, 2000; pp. 251-262.
[82] Ramakrishna, S., Mayer, J., Wintermantel, E., Leong, K. W. Biomedical applications of polymer-composite materials: a review. Compos. Sci. Technol. 2001, 61, 1189-1224.
[83] Shastri, V. P. Non-degradable biocompatible polymers in medicine: past, present and future. Curr. Pharm. Biotechnol. 2003, 4, 331-337.
[84] Chen, H., Yuan, L., Song, W., Wu, Z., Li, D. Biocompatible polymer materials: role of protein-surface interactions. Prog. Polym. Sci. 2008, 33, 1059-1087.
[85] Tanaka, M., Sato, K., Kitakami, E., Kobayashi, S., Hoshiba, T., Fukushima, K. Design of biocompatible and biodegradable polymers based on intermediate water concept. Polymer J. 2015, 47, 114-121.
[86] Lanza, R. P., Hayes, J. L., Chick, W. L. Encapsulated cell technology. Nature Biotechnol. 1996, 14, 1107-1111.

[87] Shukla, S. C., Singh, A., Pandey, A. K., Mishra, A. Review on production and medical applications of ε-polylysine. Biochem. Eng. J. 2012, 65, 70-81.

[88] Agrawal, C. M., Ray, R. B. Biodegradable polymeric scaffolds for musculoskeletal tissue engineering. J. Biomed. Mater. Res. 2001, 55, 141-150.

[89] Kweon, H., Yoo, M., Park, I., Kim, T., Lee, H., Lee, S., Oh, J., Akaike, T., Cho, C. A novel degradable polycaprolactone network for tissue engineering. Biomaterials 2003, 24, 801-808.

[90] Wang, Y. C., Zhang, P. H. Electrospun absorbable polycaprolactone (PCL) scaffolds for medical applications. Adv. Mater. Res. 2014, 906, 221-225.

[91] Sartori, S., Chiono, V., Tonda-Turo, C., Mattu, C., Gianluca, C. Biomimetic polyurethanes in nano and regenerative medicine. J. Mater. Chem. B 2014, 2, 5128-5144.

[92] Temenoff, J. S., Mikos, A. G. Injectable biodegradable materials for orthopedic tissue engineering. Biomaterials 2000, 21, 2405-2412.

[93] Behravesh, E., Yasko, A. W., Engel, P. S., Mikos, A. G. Synthetic biodegradable polymers for orthopaedic applications. Clin. Orthop. Rel. Res. 1999, 367S, S118-S125.

[94] Lewandrowski, K. U., Gresser, J. D., Wise, D. L., White, R. L., Trantolo, D. J. Osteoconductivity of an injectable and bioresorbable poly(propyleneglycol-co-fumaric acid) bone cement. Biomaterials 2000, 21, 293-298.

[95] Lee, K. W., Wang, S., Fox, B. C., Ritman, E. L., Yaszemski, M. J., Lu, L. Poly(propylene fumarate) bone tissue engineering scaffold fabrication using stereolithography: effects of resin formulations and laser parameters. Biomacromolecules 2007, 8, 1077-1084.

[96] Xu, J., Feng, E., Song, J. Renaissance of aliphatic polycarbonates: new techniques and biomedical applications. J. Appl. Polym. Sci. 2014, 131, 39822.

[97] Boland, E. D., Coleman, B. D., Barnes, C. P., Simpson, D. G., Wnek, G. E., Bowlin, G. L. Electrospinning polydioxanone for biomedical applications. Acta Biomater. 2005, 1, 115-123.

[98] Gilbert, J. L. Acrylics in biomedical engineering. In: Encyclopedia of materials: science and technology. Elsevier: Amsterdam, Netherlands, 2001; pp. 11-18.

[99] Frazer, R. Q., Byron, R. T., Osborne, P. B., West, K. P. PMMA: an essential material in medicine and dentistry. J. Long-Term Eff. Med. Implants 2005, 15, 629-639.

[100] Li, Y. W., Leong, J. C. Y., Lu, W. W., Luk, K. D. K., Cheung, K. M. C., Chiu, K. Y., Chow, S. P. A novel injectable bioactive bone cement for spinal surgery: a developmental and preclinical study. J. Biomed. Mater. Res. 2000, 52, 164-170.

[101] Mckellop, H., Shen, F., Lu, B., Campbell, P., Salovey, R. Development of an extremely wear resistant UHMW polyethylene for total hip replacements. J. Orthop. Res. 1999, 17, 157-167.

[102] Kurtz, S. M., Muratoglu, O. K., Evans, M., Edidin, A. A. Advances in the processing, sterilization and crosslinking of ultra-high molecular weight polyethylene for total joint arthroplasty. Biomaterials 1999, 20, 1659-1688.

[103] Laurencin, C. T., Ambrosio, M. A., Borden, M. D., Cooper, J. A., Jr. Tissue engineering: orthopedic applications. Ann. Rev. Biomed. Eng. 1999, 1, 19-46.

[104] Meijer, G. J., Cune, M. S., van Dooren, M., de Putter, C., van Blitterswijk, C. A. A comparative study of flexible (Polyactive™) versus rigid (hydroxylapatite) permucosal dental implants. I. Clinical aspects. J. Oral Rehabil. 1997, 24, 85-92.
[105] Meijer, G. J., Dalmeijer, R. A., de Putter, C., van Blitterswijk, C. A. A comparative study of flexible (Polyactive™) versus rigid (hydroxylapatite) permucosal dental implants. II. Histological aspects. J. Oral Rehabil. 1997, 24, 93-101.
[106] Waris, E., Ashammakhi, N., Lehtimäki, M., Tulamo, R. M., Törmälä, P., Kellomäki, M., Konttinen, Y. T. Long-term bone tissue reaction to polyethylene oxide/polybutylene terephthalate copolymer (Polyactive®) in metacarpophalangeal joint reconstruction. Biomaterials 2008, 29, 2509-2515.
[107] Svensson, A., Nicklasson, E., Harrah, T., Panilaitis, B., Kaplan, D. L., Brittberg, M., Gatenholm, P. Bacterial cellulose as a potential scaffold for tissue engineering of cartilage. Biomaterials 2005, 26, 419-431.
[108] Rampinelli, G., di Landro, L., Fujii, T. Characterization of biomaterials based on microfibrillated cellulose with different modifications. J. Reinf. Plast. Compos. 2010, 29, 1793-1803.
[109] Granja, P. L., Barbosa, M. A., Pouysége, L., de Jéso, B., Rouais, F., Baquuey, C. Cellulose phosphates as biomaterials. Mineralization of chemically modified regenerated cellulose hydrogels. J. Mater. Sci. 2001, 36, 2163-2172.
[110] Granja, P. L., Jéso, B. D., Bareille, R., Rouais, F., Baquey, C., Barbosa, M. A. Cellulose phosphates as biomaterials. In vitro biocompatibility studies. React. Funct. Polym. 2006, 66, 728-739.
[111] Thomas, V., Dean, D. R., Vohra, Y. K. Nanostructured biomaterials for regenerative medicine. Curr. Nanosci. 2006, 2, 155-177.
[112] Dee, K. C., Bizios, R. Mini-review: proactive biomaterials and bone tissue engineering. Biotechnol. Bioeng. 1996, 50, 438-442.
[113] Ashammakhi, N., Rokkanen, P. Absorbable polyglycolide devices in trauma and bone surgery. Biomaterials 1997, 18, 3-9.
[114] Boyan, B., Lohmann, C., Somers, A., Neiderauer, G., Wozney, J., Dean, D., Carnes, D., Schwartz, Z. Potential of porous poly-D,L-lactide-co-glycolide particles as a carrier for recombinant human bone morphogenetic protein-2 during osteoinduction in vivo. J. Biomed. Mater. Res. 1999, 46, 51-59.
[115] Hollinger, J. O., Leong, K. Poly(α-hydroxyacids): carriers for bone morphogenetic proteins. Biomaterials 1996, 17, 187-194.
[116] Griffith, L. G. Polymeric biomaterials. Acta Mater. 2000, 48, 263-277.
[117] Peter, S. J., Miller, M. J., Yasko, A. W., Yaszemski, M. J., Mikos, A. G. Polymer concepts in tissue engineering. J. Biomed. Mater. Res. 1998, 43, 422-427.
[118] Ishuang, S. L., Payne, R. G., Yaszemski, M. J., Aufdemorte, T. B., Bizios, R., Mikos, A. G. Osteoblast migration on poly(α-hydroxy esters). Biotechnol. Bioeng. 1996, 50, 443-451.
[119] Shikinami, Y., Okuno, M. Bioresorbable devices made of forged composites of hydroxyapatite (HA) particles and poly-L-lactide (PLLA): Part I. Basic characteristics. Biomaterials 1999, 20, 859-877.
[120] Khor, E., Lim, L. Y. Implantable applications of chitin and chitosan. Biomaterials 2003, 24, 2339-2349.

[121] di Martino, A., Sittinger, M., Risbud, M. V. Chitosan: a versatile biopolymer for orthopaedic tissue-engineering. Biomaterials 2005, 26, 5983-5990.

[122] Piskin, E., Bölgen, N., Egri, S., Isoglu, I. A. Electrospun matrices made of poly(α-hydroxy acids) for medical use. Nanomedicine 2007, 2, 441-457.

[123] Rezwana, K., Chena, Q. Z., Blakera, J. J., Boccaccini, A. R. Biodegradable and bioactive porous polymer/inorganic composite scaffolds for bone tissue engineering. Biomaterials 2006, 27, 3413-3431.

[124] Seal, B. L., Otero, T. C., Panitch, A. Polymeric biomaterials for tissue and organ regeneration. Mater. Sci. Eng. R 2001, 34, 147-230.

[125] Mano, J. F., Sousa, R. A., Boesel, L. F., Neves, N. M., Reis, R. L. Bioinert, biodegradable and injectable polymeric matrix composites for hard tissue replacement: state of the art and recent developments. Compos. Sci. Technol. 2004, 64, 789-817.

[126] Middleton, J., Tipton, A. Synthetic biodegradable polymers as orthopedic devices. Biomaterials 2000, 21, 2335-2346.

[127] Coombes, A. G., Meikle, M. C. Resorbable synthetic polymers as replacements for bone graft. Clin. Mater. 2004, 17, 35-67.

[128] de las Heras Alarcón, C., Pennadam, S., Alexander, C. Stimuli responsive polymers for biomedical applications. Chem. Soc. Rev. 2005, 34, 276-285.

[129] Kohane, D. S., Langer, R. Polymeric biomaterials in tissue engineering. Pediatric Res. 2008, 63, 487-491.

[130] Okada, M. Chemical syntheses of biodegradable polymers. Prog. Polym. Sci. 2002, 27, 87-133.

[131] Jordan, J., Jacob, K. I., Tannenbaum, R., Sharaf, M. A., Jasiuk, I. Experimental trends in polymer nanocomposites – a review. Mater. Sci. Eng. A 2005, 393, 1-11.

[132] Mathieu, L. M., Bourban, P. E., Manson, J. A. E. Processing of homogeneous ceramic/polymer blends for bioresorbable composites. Compos. Sci. Technol. 2006, 66, 1606-1614.

[133] Schappo, H., Giry, K., Salmoria, G., Damia, C., Hotza, D. Polymer/calcium phosphate biocomposites manufactured by selective laser sintering: an overview. Prog. Additive Manuf. 2023, 8, 285-301.

[134] Gao, X., Wang, H., Luan, S., Zhou, G. Biofabrication of poly(aryletherketone)-hydroxyapatite composite scaffolds via low-temperature printing. Adv. Mater. Technol. 2023, 8, 2201676.

[135] Pezzotti, G., Asmus, S. M. F. Fracture behavior of hydroxyapatite/polymer interpenetrating network composites prepared by in situ polymerization process. Mater. Sci. Eng. A 2001, 316, 231-237.

[136] Weickmann, H., Gurr, M., Meincke, O., Thomann, R., Mülhaupt, R. A versatile solvent-free "one-pot" route to polymer nanocomposites and the in situ formation of calcium phosphate/layered silicate hybrid nanoparticles. Adv. Funct. Mater. 2010, 20, 1778-1786.

[137] Wei, G., Gong, C., Hu, K., Wang, Y., Zhang, Y. Biomimetic hydroxyapatite on graphene supports for biomedical applications: a review. Nanomaterials 2019, 9, 1435.

[138] Boyapati, P. C. S., Srinivas, K., Akhil, S., Bollikolla, H. B., Chandu, B. A comprehensive review on novel graphene-hydroxyapatite nanocomposites for potential bioimplant applications. ChemistrySelect 2023, 8, e202204585.
[139] Wang, X. J., Li, Y., Wei, J., de Groot, K. Development of biomimetic nano-hydroxyapatite/poly(hexamethylene adipamide) composites. Biomaterials 2002, 23, 4787-4791.
[140] Wei, J., Li, Y. Tissue engineering scaffold material of nano-apatite crystals and polyamide composite. Eur. Polym. J. 2004, 40, 509-515.
[141] Memoto, R., Nakamura, S., Isobe, T., Senna, M. Direct synthesis of hydroxyapatite-silk fibroin nano-composite sol via a mechano-chemical route. J. Sol Gel Sci. Technol. 2001, 21, 7-12.
[142] Yoshida, A., Miyazaki, T., Ashizuka, M., Ishida, E. Bioactivity and mechanical properties of cellulose/carbonate hydroxyapatite composites prepared in situ through mechanochemical reaction. J. Biomater. Appl. 2006, 21, 179-194.
[143] Fujiwara, M., Shiokawa, K., Morigaki, K., Tatsu, Y., Nakahara, Y. Calcium phosphate composite materials including inorganic powders, BSA or duplex DNA prepared by W/O/W interfacial reaction method. Mater. Sci. Eng. C 2008, 28, 280-288.
[144] Nagata, F., Miyajima, T., Yokogawa, Y. A method to fabricate hydroxyapatite/poly(lactic acid) microspheres intended for biomedical application. J. Eur. Ceram. Soc. 2006, 26, 533-535.
[145] Russias, J., Saiz, E., Nalla, R. K., Tomsia, A. P. Microspheres as building blocks for hydroxyapatite/polylactide biodegradable composites. J. Mater. Sci. 2006, 41, 5127-5133.
[146] Khan, Y. M., Cushnie, E. K., Kelleher, J. K., Laurencin, C. T. In situ synthesized ceramic-polymer composites for bone tissue engineering: bioactivity and degradation studies. J. Mater. Sci. 2007, 42, 4183-4190.
[147] Liu, X., Okada, M., Maeda, H., Fujii, S., Furuzono, T. Hydroxyapatite/biodegradable poly(L-lactide-co-ε-caprolactone) composite microparticles as injectable scaffolds by a Pickering emulsion route. Acta Biomater. 2011, 7, 821-828.
[148] Hu, Y., Zou, S., Chen, W., Tong, Z., Wang, C. Mineralization and drug release of hydroxyapatite/poly(L-lactic acid) nanocomposite scaffolds prepared by Pickering emulsion templating. Colloid Surface B 2014, 122, 559-565.
[149] Kim, H. W., Knowles, J. C., Kim, H. E. Hydroxyapatite and gelatin composite foams processed via novel freeze-drying and crosslinking for use as temporary hard tissue scaffolds. J. Biomed. Mater. Res. A 2005, 72A, 136-145.
[150] Mohandes, F., Salavati-Niasari, M. Freeze-drying synthesis, characterization and in vitro bioactivity of chitosan/graphene oxide/hydroxyapatite nanocomposite. RSC Adv. 2014, 4, 25993-26001.
[151] Sinha, A., Das, G., Sharma, B. K., Roy, R. P., Pramanick, A. K., Nayar, S. Poly(vinyl alcohol)-hydroxyapatite biomimetic scaffold for tissue regeneration. Mater. Sci. Eng. C 2007, 27, 70-74.

[152] Sugawara, A., Yamane, S., Akiyoshi, K. Nanogel-templated mineralization: polymer-calcium phosphate hybrid nanomaterials. Macromol. Rapid Commun. 2006, 27, 441-446.

[153] Baji, A., Wong, S. C., Srivatsan, T. S., Njus, G. O., Mathur, G. Processing methodologies for polycaprolactone-hydroxyapatite composites: a review. Mater. Manuf. Process. 2006, 21, 211-218.

[154] Guan, L., Davies, J. E. Preparation and characterization of a highly macroporous biodegradable composite tissue engineering scaffold. J. Biomed. Mater. Res. A 2004, 71A, 480-487.

[155] Sun, F., Zhou, H., Lee, J. Various preparation methods of highly porous hydroxyapatite/polymer nanoscale biocomposites for bone regeneration. Acta Biomater. 2011, 7, 3813-3828.

[156] Kumar, A., Negi, Y. S., Choudhary, V., Bhardwaj, N. K. Microstructural and mechanical properties of porous biocomposite scaffolds based on polyvinyl alcohol, nano-hydroxyapatite and cellulose nanocrystals. Cellulose 2015, 21, 3409-3426.

[157] Teng, X. R., Ren, J., Gu, S. Y. Preparation and characterization of porous PDLLA/HA composite foams by supercritical carbon dioxide technology. J. Biomed. Mater. Res. B Appl. Biomater. 2007, 81B, 185-193.

[158] Ren, J., Zhao, P., Ren, T., Gu, S., Pan, K. Poly (D,L-lactide)/nano-hydroxyapatite composite scaffolds for bone tissue engineering and biocompatibility evaluation. J. Mater. Sci. Mater. Med. 2008, 19, 1075-1082.

[159] Yang, X., Qi, D., Xu, K., Alimu, A., Cao, L. Bio-compatible n-HAPs/polymer monolithic composites templated from CO_2-in-water high internal phase emulsions. React. Funct. Polym. 2022, 181, 105417.

[160] Dorozhkin, S. V. Calcium orthophosphate ($CaPO_4$)-based bioceramics: preparation, properties, and applications. Coatings 2022, 12, 1380.

[161] Babilotte, J., Guduric, V., le Nihouannen, D., Naveau, A., Fricain, J. C., Catros, S. 3D printed polymer-mineral composite biomaterials for bone tissue engineering: fabrication and characterization. J. Biomed. Mater. Res. B Appl. Biomater. 2019, 107B, 2579-2595.

[162] Converse, G. L., Yue, W., Roeder, R. K. Processing and tensile properties of hydroxyapatite-whisker-reinforced polyetheretherketone. Biomaterials 2007, 28, 927-935.

[163] Converse, G. L., Roeder, R. K. Tensile properties of hydroxyapatite whisker reinforced polyetheretherketone. Mater. Res. Soc. Symp. Proc. 2005, 898, 44-49.

[164] Choi, W. Y., Kim, H. E., Kim, M. J., Kim, U. C., Kim, J. H., Koh, Y. H. Production and characterization of calcium phosphate (CaP) whisker-reinforced poly(ε-caprolactone) composites as bone regenerative. Mater. Sci. Eng. C 2010, 30, 1280-1284.

[165] Zhang, H., Darvell, B. W. Failure and behavior in water of hydroxyapatite whisker-reinforced bis-GMA-based resin composites. J. Mech. Behav. Biomed. Mater. 2012, 10, 39-47.

[166] Liu, F., Wang, R., Cheng, Y., Jiang, X., Zhang, Q., Zhu, M. Polymer grafted hydroxyapatite whisker as a filler for dental composite resin with enhanced physical and mechanical properties. Mater. Sci. Eng. C 2013, 33, 4994-5000.

[167] Liu, F. W., Bao, S., Jin, Y., Jiang, X. Z., Zhu, M. F. Novel bionic dental resin composite reinforced by hydroxyapatite whisker. Mater. Res. Innov. 2014, 18, S4854-S4858.

[168] Nouri-Felekori, M., Mesgar, A. S. M., Mohammadi, Z. Development of composite scaffolds in the system of gelatin – calcium phosphate whiskers/fibrous spherulites for bone tissue engineering. Ceram. Int. 2015, 41, 6013-6019.

[169] Watanabe, T., Ban, S., Ito, T., Tsuruta, S., Kawai, T., Nakamura, H. Biocompatibility of composite membrane consisting of oriented needle-like apatite and biodegradable copolymer with soft and hard tissues in rats. Dental Mater. J. 2004, 23, 609-612.

[170] Li, H., Chen, Y., Xie, Y. Photo-crosslinking polymerization to prepare polyanhydride/needle-like hydroxyapatite biodegradable nanocomposite for orthopedic application. Mater. Lett. 2003, 57, 2848-2854.

[171] Nejati, E., Firouzdor, V., Eslaminejad, M. B., Bagheri, F. Needle-like nano hydroxyapatite/poly(L-lactide acid) composite scaffold for bone tissue engineering application. Mater. Sci. Eng. C 2009, 29, 942-949.

[172] Sun, S. P., Wei, M., Olson, J. R., Shaw, M. T. A modified pultrusion process for preparing composites reinforced with continuous fibers and aligned hydroxyapatite nano needles. Polym. Composite 2015, 36, 931-938.

[173] Kasuga, T., Ota, Y., Nogami, M., Abe, Y. Preparation and mechanical properties of polylactic acid composites containing hydroxyapatite fibers. Biomaterials 2000, 22, 19-23.

[174] Smith, L. Ceramic-plastic material as a bone substitute. Arch. Surg. 1963, 87, 653-661.

[175] Bonfield, W., Grynpas, M. D., Tully, A. E., Bowman, J., Abram, J. Hydroxyapatite reinforced polyethylene – a mechanically compatible implant material for bone replacement. Biomaterials 1981, 2, 185-189.

[176] Bonfield, W., Bowman, J., Grynpas, M. D. Composite material for use in orthopaedics. UK Patent 8032647, 1981.

[177] Bonfield, W. Composites for bone replacement. J. Biomed. Eng. 1988, 10, 522-526.

[178] Guild, F. J., Bonfield, W. Predictive character of hydroxyapatite-polyethelene HAPEX™ composite. Biomaterials 1993, 14, 985-993.

[179] Huang, J., di Silvio, L., Wang, M., Tanner, K. E., Bonfield, W. In vitro mechanical and biological assessment of hydroxyapatite-reinforced polyethylene composite. J. Mater. Sci. Mater. Med. 1997, 8, 775-779.

[180] Wang, M., Joseph, R., Bonfield, W. Hydroxyapatite-polyethylene composites for bone substitution: effect of ceramic particle size and morphology. Biomaterials 1998, 19, 2357-2366.

[181] Ladizesky, N. H., Ward, I. M., Bonfield, W. Hydroxyapatite/high-performance polyethylene fiber composites for high load bearing bone replacement materials. J. Appl. Polym. Sci. 1997, 65, 1865-1882.

[182] Nazhat, S. N., Joseph, R., Wang, M., Smith, R., Tanner, K. E., Bonfield, W. Dynamic mechanical characterisation of hydroxyapatite reinforced polyethylene: effect of particle size. J. Mater. Sci. Mater. Med. 2000, 11, 621-628.

[183] Guild, F. J., Bonfield, W. Predictive modelling of the mechanical properties and failure processes of hydroxyapatite-polyethylene (HAPEX™) composite. J. Mater. Sci. Mater. Med. 1998, 9, 497-502.

[184] Wang M., Ladizesky N. H., Tanner, K. E., Ward IM., Bonfield, W. Hydrostatically extruded HAPEX™. J. Mater. Sci. 2000, 35, 1023-1030.

[185] That P. T., Tanner, K. E., Bonfield, W. Fatigue characterization of a hydroxyapatite-reinforced polyethylene composite. I. Uniaxial fatigue. J. Biomed. Mater. Res. 2000, 51, 453-460.

[186] That P. T., Tanner, K. E., Bonfield, W. Fatigue characterization of a hydroxyapatite-reinforced polyethylene composite. II. Biaxial fatigue. J. Biomed. Mater. Res. 2000, 51, 461-468.

[187] Bonner M., Saunders L. S., Ward I. M., Davies G. W., Wang M., Tanner, K. E., Bonfield, W. Anisotropic mechanical properties of oriented HAPEX™. J. Mater. Sci. 2002, 37, 325-334.

[188] di Silvio, L., Dalby, M. J., Bonfield, W. Osteoblast behaviour on HA/PE composite surfaces with different HA volumes. Biomaterials 2002, 23, 101-107.

[189] Dalby, M. J., Kayser, M. V., Bonfield, W., di Silvio, L. Initial attachment of osteoblasts to an optimised HAPEX™ topography. Biomaterials 2002, 23, 681-690.

[190] Zhang, Y., Tanner, K. E., Gurav, N., di Silvio, L. In vitro osteoblastic response to 30 vol% hydroxyapatite-polyethylene composite. J. Biomed. Mater. Res. A 2007, 81A, 409-417.

[191] Rea, S. M., Brooks, R. A., Schneider, A., Best, S. M., Bonfield, W. Osteoblast-like cell response to bioactive composites-surface-topography and composition effects. J. Biomed. Mater. Res. B Appl. Biomater. 2004, 70B, 250-261.

[192] Salernitano, E., Migliaresi, C. Composite materials for biomedical applications: a review. J. Appl. Biomater. Biomech. 2003, 1, 3-18.

[193] Pandey, A., Jan, E., Aswath, P. B. Physical and mechanical behavior of hot rolled HDPE/HA composites. J. Mater. Sci. 2006, 41, 3369-3376.

[194] Bonner M., Ward I. M., McGregor W., Tanner, K. E., Bonfield, W. Hydroxyapatite/polypropylene composite: a novel bone substitute material. J. Mater. Sci. Lett. 2001, 20, 2049-2052.

[195] Suppakarn, N., Sanmaung, S., Ruksakulpiwa, Y., Sutapun, W. Effect of surface modification on properties of natural hydroxyapatite/polypropylene composites. Key Eng. Mater. 2008, 361-363, 511-514.

[196] Younesi, M., Bahrololoom, M. E. Formulating the effects of applied temperature and pressure of hot pressing process on the mechanical properties of polypropylene-hydroxyapatite bio-composites by response surface methodology. Mater. Des. 2010, 31, 4621-4630.

[197] Younesi, M., Bahrololoom, M. E. Effect of polypropylene molecular weight, hydroxyapatite particle size, and Ringer's solution on creep and impact behavior of

polypropylene-surface treated hydroxyapatite biocomposites. J. Compos. Mater. 2011, 45, 513-523.
[198] Sousa, R. A., Reis, R. L., Cunha, A. M., Bevis, M. J. Processing and properties of bone-analogue biodegradable and bioinert polymeric composites. Compos. Sci. Technol. 2003, 63, 389-402.
[199] Wang, M., Deb, S., Bonfield, W. Chemically coupled hydroxyapatite-polyethylene composites: processing and characterisation. Mater. Lett. 2000, 44, 119-124.
[200] Wang, M., Bonfield, W. Chemically coupled hydroxyapatite-polyethylene composites: structure and properties. Biomaterials 2001, 22, 1311-1320.
[201] Homaeigohar, S. S., Shokrgozar, M. A., Khavandi, A., Sadi, A. Y. In vitro biological evaluation of β-TCP/HDPE – a novel orthopedic composite: a survey using human osteoblast and fibroblast bone cells. J. Biomed. Mater. Res. A 2008, 84A, 491-499.
[202] Sadi, A. Y., Homaeigohar, S. Sh., Khavandi, A. R., Javadpour, J. The effect of partially stabilized zirconia on the mechanical properties of the hydroxyapatite-polyethylene composites. J. Mater. Sci. Mater. Med. 2004, 15, 853-858.
[203] Nath, S., Bodhak, S., Basu, B. HDPE-Al_2O_3-HAp composites for biomedical applications: processing and characterizations. J. Biomed. Mater. Res. B Appl. Biomater. 2009, 88B, 1-11.
[204] Downes, R. N., Vardy, S., Tanner, K. E., Bonfield, W. Hydroxyapatite-polyethylene composite in orbital surgery. Bioceramics 4. 1991; pp. 239-246.
[205] Dornhoffer, H. L. Hearing results with the dornhoffer ossicular replacement prostheses. Laryngoscope 1998, 108, 531-536.
[206] Swain, R. E., Wang, M., Beale, B., Bonfield, W. HAPEX™ for otologic applications. Biomed. Eng. Appl. Basis Commun. 1999, 11, 315-320.
[207] Yi, Z., Li, Y., Jidong, L., Xiang, Z., Hongbing, L., Yuanyuan, W., Weihu, Y. Novel bio-composite of hydroxyapatite reinforced polyamide and polyethylene: composition and properties. Mater. Sci. Eng. A 2007, 452-453, 512-517.
[208] Unwin, A. P., Ward, I., M., Ukleja, P., Weng, J. The role of pressure annealing in improving the stiffness of polyethylene/hydroxyapatite composites. J. Mater. Sci. 2001, 36, 3165-3177.
[209] Fang, L. M., Leng, Y., Gao, P. Processing and mechanical properties of HA/UHMWPE nanocomposites. Biomaterials 2006, 27, 3701-3707.
[210] Fang, L. M., Gao, P., Leng, Y. High strength and bioactive hydroxyapatite nano-particles reinforced ultrahigh molecular weight polyethylene. Composites B 2007, 38, 345-351.
[211] Fang, L. M., Leng, Y., Gao, P. Processing of hydroxyapatite reinforced ultrahigh molecular weight polyethylene for biomedical applications. Biomaterials 2005, 26, 3471-3478.
[212] Selvin, T. P., Seno, J., Murukan, B., Santhosh, A. A., Sabu, T., Weimin, Y., Sri, B. Poly(ethylene-co-vinyl acetate)/calcium phosphate nanocomposites: thermo mechanical and gas permeability measurements. Polym. Composite 2010, 31, 1011-1019.

[213] Reis, R. L., Cunha, A. M., Oliveira, M. J., Campos, A. R., Bevis, M. J. Relationship between processing and mechanical properties of injection molded high molecular mass polyethylene + hydroxyapatite composites. Mater. Res. Inn. 2001, 4, 263-272.

[214] Sousa, R. A., Reis, R. L., Cunha, A. M., Bevis, M. J. Structure development and interfacial interactions in high-density polyethylene/hydroxyapatite (HDPE/HA) composites molded with preferred orientation. J. Appl. Polym. Sci. 2002, 86, 2873-2886.

[215] Mirsalehi, S. A., Khavandi, A., Mirdamadi, S., Naimi-Jamal, M. R., Kalantari, S. M. Nanomechanical and tribological behavior of hydroxyapatite reinforced ultrahigh molecular weight polyethylene nanocomposites for biomedical applications. J. Appl. Polym. Sci. 2015, 132, 42052.

[216] Donners, J. J. J. M., Nolte, R. J. M., Sommerdijk, N. A. J. M. Dendrimer-based hydroxyapatite composites with remarkable materials properties. Adv. Mater. 2003, 15, 313-316.

[217] Schneider, O. D., Stepuk, A., Mohn, D., Luechinger, N. A., Feldman, K., Stark, W. J. Light-curable polymer/calcium phosphate nanocomposite glue for bone defect treatment. Acta Biomater. 2010, 6, 2704-2710.

[218] Ignjatovic, N. L., Plavsic, M., Miljkovic, M. S., Zivkovic, L. M., Uskokovic, D. P. Microstructural characteristics of calcium hydroxyapatite/poly-L-lactide based composites. J. Microsc. 1999, 196, 243-248.

[219] Skrtic, D., Antonucci, J. M., Eanes, E. D. Amorphous calcium phosphate-based bioactive polymeric composites for mineralized tissue regeneration. J. Res. Natl. Inst. Stand. Technol. 2003, 108, 167-182.

[220] Rizzi, S. C., Heath, D. J., Coombes, A. G., A., Bock, N., Textor, M., Downes, S. Biodegradable polymer/hydroxyapatite composites: surface analysis and initial attachment of human osteoblasts. J. Biomed. Mater. Res. 2001, 55, 475-486.

[221] Navarro, M., Planell, J. A. Bioactive composites based on calcium phosphates for bone regeneration. Key Eng. Mater. 2010, 441, 203-233.

[222] Zhang, R. Y., Ma, P. X. Porous poly(L-lactic acid)/apatite composites created by biomimetic process. J. Biomed. Mater. Res. 1999, 45, 285-293.

[223] Liu, Q., de Wijn, J. R., van Blitterswijk, C. A. Composite biomaterials with chemical bonding between hydroxyapatite filler particles and PEG/PBT copolymer matrix. J. Biomed. Mater. Res. 1998, 40, 490-497.

[224] Cerrai, P., Guerra, G. D., Tricoli, M., Krajewski, A., Ravaglioli, A., Martinetti, R., Dolcini, L. Fini, M., Scarano, A., Piattelli, A. Periodontal membranes from composites of hydroxyapatite and bioresorbable block copolymers. J. Mater. Sci. Mater. Med. 1999, 10, 677-682.

[225] Roeder, R. K., Sproul, M. M., Turner, C. H. Hydroxyapatite whiskers provide improved mechanical properties in reinforced polymer composites. J. Biomed. Mater. Res. A 2003, 67A, 801-812.

[226] Wagoner Johnson, A. J., Herschler, B. A. A review of the mechanical behavior of CaP and CaP/polymer composites for applications in bone replacement and repair. Acta Biomater. 2011, 7, 16-30.

[227] Hutmacher, D. W. Scaffolds in tissue engineering bone and cartilage. Biomaterials 2000, 21, 2529-2543.

[228] Owji, N., Mandakhbayar, N., Cha, J. R., Padalhin, A. R., Erdogan, Z. K., Aldaadaa, A., Shakouri, T., Sawadkar, P. Frost, O., Kim, H. W., García-Gareta, E., Knowles, J. C. Inclusion of calcium phosphate does not further improve in vitro and in vivo osteogenesis in a novel, highly biocompatible, mechanically stable and 3D printable polymer. Sci. Rep. 2022, 12, 16977.

[229] Kickelbick, G. Concepts for the incorporation of inorganic building blocks into organic polymers on a nanoscale. Prog. Polym. Sci. 2003, 28, 83-114.

[230] Liu, Q., de Wijn, J. R., van Blitterswijk, C. A. Nanoapatite/polymer composites: mechanical and physicochemical characteristics. Biomaterials 1997, 18, 1263-1270.

[231] Uskokovic, P. S., Tang, C. Y., Tsui, C. P., Ignjatovic, N., Uskokovic, D. P. Micromechanical properties of a hydroxyapatite/poly-L-lactide biocomposite using nanoindentation and modulus mapping. J. Eur. Ceram. Soc. 2007, 27, 1559-1564.

[232] Todo, M., Kagawa, T. Improvement of fracture energy of HA/PLLA biocomposite material due to press processing. J. Mater. Sci. 2008, 43, 799-801.

[233] Shuai, C., Yu, L., Feng, P., Peng, S., Pan, H., Bai, X. Construction of a stereocomplex between poly(D-lactide) grafted hydroxyapatite and poly(L-lactide): toward a bioactive composite scaffold with enhanced interfacial bonding. J. Mater. Chem. B 2022, 10, 214-223.

[234] Ielo, I., Calabrese, G., de Luca, G., Conoci, S. Recent advances in hydroxyapatite-based biocomposites for bone tissue regeneration in orthopedics. Int. J. Mol. Sci. 2022, 23, 9721.

[235] Wang, M., Yue, C. Y., Chua, B. Production and evaluation of hydroxyapatite reinforced polysulfone for tissue replacement. J. Mater. Sci. Mater. Med. 2001, 12, 821-826.

[236] Chlopek, J., Rosol, P., Morawska-Chochol, A. Durability of polymer-ceramics composite implants determined in creep tests. Compos. Sci. Technol. 2006, 66, 1615-1622.

[237] Robinson, P., Wilson, C., Mecholsky, J. Processing and mechanical properties of hydroxyapatite-polysulfone laminated composites. J. Eur. Ceram. Soc. 2014, 34, 1387-1396.

[238] Xu, F., Li, Y., Yao, X., Liao, H., Zhang, L. Preparation and in vivo investigation of artificial cornea made of nano-hydroxyapatite/poly (vinyl alcohol) hydrogel composite. J. Mater. Sci. Mater. Med. 2007, 18, 635-640.

[239] Xu, F., Li, Y., Deng, Y., Xiong, G. Porous nano-hydroxyapatite/poly(vinyl alcohol) composite hydrogel as artificial cornea fringe: characterization and evaluation in vitro. J. Biomater. Sci. Polym. Edn. 2008, 19, 431-439.

[240] Nayar, S., Pramanick, A. K., Sharma, B. K., Das, G., Kumar, B. R., Sinha, A. Biomimetically synthesized polymer-hydroxyapatite sheet like nano-composite. J. Mater. Sci. Mater. Med. 2008, 19, 301-304.

[241] Poursamar, S. A., Orang, F., Bonakdar, S., Savar, M. K. Preparation and characterisation of poly vinyl alcohol/hydroxyapatite nanocomposite via in situ synthesis: a potential material as bone tissue engineering scaffolds. Int. J. Nanomanuf. 2010, 5, 330-334.

[242] Guha, A., Nayar, S., Thatoi, H. N. Microwave irradiation enhances kinetics of the biomimetic process of hydroxyapatite nanocomposites. Bioinspir. Biomim. 2010, 5, 024001.

[243] Timofejeva, A., Loca, D. Hydroxyapatite/polyvinyl alcohol composite hydrogels for bone and cartilage tissue engineering. Key Eng. Mater. 2018, 762, 54-58.

[244] Pramanik, N., Biswas, S. K., Pramanik, P. Synthesis and characterization of hydroxyapatite/poly(vinyl alcohol phosphate) nanocomposite biomaterials. Int. J. Appl. Ceram. Technol. 2008, 5, 20-28.

[245] Bigi, A., Boanini, E., Gazzano, M., Rubini, K. Structural and morphological modifications of hydroxyapatite-polyaspartate composite crystals induced by heat treatment. Cryst. Res. Technol. 2005, 40, 1094-1098.

[246] Bertoni, E., Bigi, A., Falini, G., Panzavolta, S., Roveri, N. Hydroxyapatite polyacrylic acid nanocrystals. J. Mater. Chem. 1999, 9, 779-782.

[247] Qiu, H. J., Yang, J., Kodali, P., Koh, J., Ameer, G. A. A citric acid-based hydroxyapatite composite for orthopedic implants. Biomaterials 2006, 27, 5845-5854.

[248] Greish, Y. E., Brown, P. W. Chemically formed HAp-Ca poly(vinyl phosphonate) composites. Biomaterials 2001, 22, 807-816.

[249] Greish, Y. E., Brown, P. W. Preparation and characterization of calcium phosphate-poly(vinyl phosphonic acid) composites. J. Mater. Sci. Mater. Med. 2001, 12, 407-411.

[250] Greish, Y. E., Brown, P. W. Formation and properties of hydroxyapatite-calcium poly(vinyl phosphonate) composites. J. Am. Ceram. Soc. 2002, 85, 1738-1744.

[251] Sailaja, G. S., Velayudhan, S., Sunny, M. C., Sreenivasan, K., Varma, H. K., Ramesh, P. Hydroxyapatite filled chitosan-polyacrylic acid polyelectrolyte complexes. J. Mater. Sci. 2003, 38, 3653-3662.

[252] Piticescu, R. M., Chitanu, G. C., Albulescu, M., Giurginca, M., Popescu, M. L., Łojkowski, W. Hybrid HAp-maleic anhydride copolymer nanocomposites obtained by in-situ functionalisation. Solid State Phenom. 2005, 106, 47-56.

[253] Song, J., Saiz, E., Bertozzi, C. R. A new approach to mineralization of biocompatible hydrogel scaffolds: an efficient process toward 3-dimensional bonelike composites. J. Am. Chem. Soc. 2003, 125, 1236-1243.

[254] Kutikov, A. B., Song J. An amphiphilic degradable polymer/hydroxyapatite composite with enhanced handling characteristics promotes osteogenic gene expression in bone marrow stromal cells. Acta Biomater. 2013, 9, 8354-8364.

[255] Wu, S., Wang, J., Zou, L., Jin, L., Wang, Z., Li, Y. A three-dimensional hydroxyapatite/polyacrylonitrile composite scaffold designed for bone tissue engineering. RSC Adv. 2018, 8, 1730-1736.

[256] Borkowski, L., Lübek, T., Jojczuk, M., Nogalski, A., Belcarz, A., Hajnos, M., Ginalska, G. Behavior of new hydroxyapatite/glucan composite in human serum. J. Biomed. Mater. Res. B Appl. Biomater. 2018, 106B, 2653-2664.

[257] Abu Bakar, M. S., Cheng, M. H. W., Tang, S. M., Yu, S. C., Liao, K., Tan, C. T., Khor, K. A., Cheang, P. Tensile properties, tension-tension fatigue and biological response of polyetheretherketone-hydroxyapatite composites for load-bearing orthopedic implants. Biomaterials 2003, 24, 2245-2250.

[258] Abu Bakar, M. S., Cheang, P., Khor, K. A. Mechanical properties of injection molded hydroxyapatite-polyetheretherketone biocomposites. Compos. Sci. Technol. 2003, 63, 421-425.
[259] Abu Bakar, M. S., Cheang, P., Khor, K. A. Tensile properties and microstructural analysis of spheroidized hydroxyapatite-poly(etheretherketone) biocomposites. Mater. Sci. Eng. A 2003, 345, 55-63.
[260] Fan, J. P., Tsui, C. P., Tang, C. Y. Modeling of the mechanical behavior of HA/PEEK biocomposite under quasi-static tensile load. Mater. Sci. Eng. A 2004, 382, 341-350.
[261] Wang, L., He, S., Wu, X., Liang, S., Mu, Z., Wei, J., Deng, F., Deng, Y., Wei, S. Polyetheretherketone/nano-fluorohydroxyapatite composite with antimicrobial activity and osseointegration properties. Biomaterials 2014, 35, 6758-6775.
[262] Qi, D., Wang, N., Wang, S., Liu, L., Zhu, S., She, P., Yue, X. High-strength porous polyetheretherketone/hydroxyapatite composite for the treatment of bone defect. Compos. Comm. 2023, 38, 101473.
[263] Gong, X. H., Tang, C. Y., Hu, H. C., Zhou, X. P. Improved mechanical properties of HIPS/hydroxyapatite composites by surface modification of hydroxyapatite via in situ polymerization of styrene. J. Mater. Sci. Mater. Med. 2004, 15, 1141-1146.
[264] Fu, G., Xia, Z., Jiang, J., Jing, B., Zhang, X. Fabrication and characterization of nanocomposites with high-impact polystyrene and hydroxyapatite with well-defined polystyrene via ATRP. J. Reinf. Plast. Comp. 2011, 30, 1445-1453.
[265] Petricca, S. E., Marra, K. G., Kumta, P. N. Chemical synthesis of poly(lactic-co-glycolic acid)/hydroxyapatite composites for orthopaedic applications. Acta Biomater. 2006, 2, 277-286.
[266] Kim, S. S., Ahn, K. M., Park, M. S., Lee, J. H., Choi, C. Y., Kim, B. S. A poly(lactide-co-glycolide)/hydroxyapatite composite scaffold with enhanced osteoconductivity. J. Biomed. Mater. Res. A 2007, 80A, 206-215.
[267] Oliveira, J., Miyazaki, T., Lopes, M., Ohtsuki, C., Santos, J. Bonelike®/PLGA hybrid materials for bone regeneration: preparation route and physicochemical characterization. J. Mater. Sci. Mater. Med. 2005, 16, 253-259.
[268] Aboudzadeh, N., Imani, M., Shokrgozar, M. A., Khavandi, A., Javadpour, J., Shafieyan, Y., Farokhi, M. Fabrication and characterization of poly(D,L-lactide-co-glycolide)/ hydroxyapatite nanocomposite scaffolds for bone tissue regeneration. J. Biomed. Mater. Res. A 2010, 94A, 137-145.
[269] Zhou, H., Lawrence, J. G., Bhaduri, S. B. Fabrication aspects of PLA-CaP/PLGA-CaP composites for orthopedic applications: a review. Acta Biomater. 2012, 8, 1999-2016.
[270] Hoekstra, J. W. M., Ma, J., Plachokova, A. S., Bronkhorst, E. M., Bohner, M., Pan, J., Meijer, G. J., Jansen, J. A., van den Beucken, J. J. J. P. The in vivo performance of CaP/PLGA composites with varied PLGA microsphere sizes and inorganic compositions. Acta Biomater. 2013, 9, 7518-7526.
[271] Leung, L. H. Naguib, H. E. Characterizing the viscoelastic behaviour of poly(lactide-co-glycolide acid)–hydroxyapatite foams. J. Cell. Plast. 2013, 49, 497-505.

[272] Takeoka, Y., Hayashi, M., Sugiyama, N., Yoshizawa-Fujita, M., Aizawa, M., Rikukawa, M. In situ preparation of poly(l-lactic acid-co-glycolic acid)/hydroxyapatite composites as artificial bone materials. Polym. J. 2015, 47, 164-170.

[273] Fisher, P. D., Venugopal, G., Milbrandt, T. A., Hilt, J. Z., Puleo, D. A. Hydroxyapatite-reinforced in situ forming PLGA systems for intraosseous injection. J. Biomed. Mater. Res. A 2015, 103A, 2365-2373.

[274] Athanasiou, K. A., Schmitz, J. P., Agrawal, C. M. The effects of porosity on in vitro degradation of polylactic acid- polyglycolic acid implants used in repair of articular cartilage. Tissue Eng. 1998, 4, 53-63.

[275] Verheyen, C. C. P. M., Klein, C. P. A. T., de Blieck-Hogervorst, J. M. A., Wolke, J. G. C., de Wijin, J. R., van Blitterswijk, C. A., de Groot, K. Evaluation of hydroxylapatite poly(L-lactide) composites: physico-chemical properties. J. Mater. Sci. Mater. Med. 1993, 4, 58-65.

[276] Agrawal, C. M., Athanasiou, K. A. Technique to control pH in vicinity of biodegrading PLA-PGA implants. J. Biomed. Mater. Res. Appl. Biomater. 1997, 38, 105-114.

[277] Li, H., Chang, J. pH-compensation effect of bioactive inorganic fillers on the degradation of PLGA. Compos. Sci. Technol. 2005, 65, 2226-2232.

[278] Peter, S. J., Miller, S. T., Zhu, G., Yasko, A. W., Mikos, A. G. In vivo degradation of a poly(propylene fumarate)/β-tricalcium phosphate injectable composite scaffold. J. Biomed. Mater. Res. 1998, 41, 1-7.

[279] Ara, M., Watanabe, M., Imai, Y. Effect of blending calcium compounds on hydrolitic degradation of poly(D,L-lactic acid-co-glycolic acid). Biomaterials 2002, 23, 2479-2483.

[280] Linhart, W., Peters, F., Lehmann, W., Schwarz, K., Schilling, A., Amling, M., Rueger, J. M., Epple, M. Biologically and chemically optimized composites of carbonated apatite and polyglycolide as bone substitution materials. J. Biomed. Mater. Res. 2001, 54, 162-171.

[281] Schiller, C., Epple, M. Carbonated apatites can be used as pH-stabilizing filler for biodegradable polyesters. Biomaterials 2003, 24, 2037-2043.

[282] Schiller, C., Rasche, C., Wehmöller, M., Beckmann, F., Eufinger, H., Epple, M., Weihe, S. Geometrically structured implants for cranial reconstruction made of biodegradable polyesters and calcium phosphate/calcium carbonate. Biomaterials 2004, 25, 1239-1247.

[283] Shikinami, Y., Okuno, M. Bioresorbable devices made of forged composites of hydroxyapatite (HA) particles and poly L-lactide (PLLA). Part II: practical properties of miniscrews and miniplates. Biomaterials 2001, 22, 3197-3211.

[284] Russias, J., Saiz, E., Nalla, R. K., Gryn, K., Ritchie, R. O., Tomsia, A. P. Fabrication and mechanical properties of PLA/HA composites: a study of in vitro degradation. Mater. Sci. Eng. C 2006, 26, 1289-1295.

[285] Akagi, H., Iwata, M., Ichinohe, T., Amimoto, H., Hayashi, Y., Kannno, N., Ochi, H., Fujita, Y., Harada, Y., Tagawa, M., Hara, Y. Hydroxyapatite/poly-L-lactide acid screws have better biocompatibility and femoral burr hole closure than does poly-L-lactide acid alone. J. Biomater. Appl. 2014, 28, 954-962.

[286] Kim, H. W., Lee, H. H., Knowles, J. C. Electrospinning biomedical nanocomposite fibers of hydroxyapaite/poly(lactic acid) for bone regeneration. J. Biomed. Mater. Res. A 2006, 79A, 643-649.
[287] Gross, K. A., Rodríguez-Lorenzo, L. M. Biodegradable composite scaffolds with an interconnected spherical network for bone tissue engineering. Biomaterials 2004, 25, 4955-4962.
[288] Zhang, H., Chen, Z. Fabrication and characterization of electrospun PLGA/MWNTs/ hydroxyapatite biocomposite scaffolds for bone tissue engineering. J. Bioact. Compat. Polym. 2010, 25, 241-259.
[289] Durucan, C., Brown, P. W. Low temperature formation of calcium-deficient hydroxyapatite-PLA/PLGA composites. J. Biomed. Mater. Res. 2000, 51, 717-725.
[290] Durucan, C., Brown, P. W. Calcium-deficient hydroxyapatite-PLGA composites: mechanical and microstructural investigation. J. Biomed. Mater. Res. 2000, 51, 726-734.
[291] Durucan, C., Brown, P. W. Biodegradable hydroxyapatite-polymer composites. Adv. Eng. Mater. 2001, 3, 227-231.
[292] Nazhat, S. N., Kellomäki, M., Törmälä, P., Tanner, K. E., Bonfield, W. Dynamic mechanical characterization of biodegradable composites of hydroxyapatite and polylactides. J. Biomed. Mater. Res. 2001, 58, 335-343.
[293] Ignjatovic, N., Suljovrujic, E., Biudinski-Simendic, J., Krakovsky, I., Uskokovic, D. Evaluation of hot-presses hydroxyapatite/poly-L-lactide composite biomaterial characteristics. J. Biomed. Mater. Res. B Appl. Biomater. 2004, 71B, 284-294.
[294] Wang, X., Lou, T., Yang, J., Yang, Z., He, K. Preparation of PLLA/HAP/β-TCP composite scaffold for bone tissue engineering. Appl. Mech. Mater. 2014, 513-517, 143-146.
[295] Hasegawa, S., Tamura, J., Neo, M., Goto, K., Shikinami, Y., Saito, M., Kita, M., Nakamura, T. In vivo evaluation of a porous hydroxyapatite/poly-D,L-lactide composite for use as a bone substitute. J. Biomed. Mater. Res. A 2005, 75A, 567-579.
[296] Hasegawa, S., Neo, M., Tamura, J., Fujibayashi, S., Takemoto, M., Shikinami, Y., Okazaki, K., Nakamura, T. In vivo evaluation of a porous hydroxyapatite/poly-D,L-lactide composite for bone tissue engineering. J. Biomed. Mater. Res. A 2007, 81A, 930-938.
[297] Kim, S. S., Park, M. S., Jeon, Q., Choi, C. Y., Kim, B. S. Poly(lactide-co-glycolide)/hydroxyapatite composite scaffolds for bone tissue engineering. Biomaterials 2006, 27, 1399-1409.
[298] Mao, D., Li, Q., Bai, N., Dong, H., Li, D. Porous stable poly(lactic acid)/ethyl cellulose/hydroxyapatite composite scaffolds prepared by a combined method for bone regeneration. Carbohyd. Polym. 2018, 180, 104-111.
[299] Carvalho, T. S. S., Ribeiro, N., Torres, P. M. C., Almeida, J. C., Belo, J. H., Araújo J. P., Ramos, A., Oliveira, M., Olhero, S. M. Magnetic polylactic acid-calcium phosphate-based biocomposite as a potential biomaterial for tissue engineering applications. Mater. Chem. Phys. 2023, 296, 127175.
[300] Reis, R. L., Cunha, A. M. New degradable load-bearing biomaterials composed of reinforced starch based blends. J. Appl. Med. Polym. 2000, 4, 1-5.

[301] Sousa, R. A., Mano, J. F., Reis, R. L., Cunha, A. M., Bevis, M. J. Mechanical performance of starch based bioactive composites moulded with preferred orientation for potential medical applications. Polym. Eng. Sci. 2002, 42, 1032-1045.

[302] Marques, A. P., Reis, R. L. Hydroxyapatite reinforcement of different starch-based polymers affects osteoblast-like cells adhesion/spreading and proliferation. Mater. Sci. Eng. C 2005, 25, 215-229.

[303] Reis, R. L., Cunha, A. M., Allan, P. S., Bevis, M. J. Structure development and control of injection-molded hydroxylapatite-reinforced starch/EVOH composites. Adv. Polym. Tech. 1997, 16, 263-277.

[304] Vaz, C. M., Reis, R. L., Cunha, A. M. Use of coupling agents to enhance the interfacial interactions in starch-EVOH/hydroxylapatite composites. Biomaterials 2002, 23, 629-635.

[305] Leonor, I. B., Ito, A., Onuma, K., Kanzaki, N., Reis, R. L. In vitro bioactivity of starch thermoplastic/hydroxyapatite composite biomaterials: an in situ study using atomic force microscopy. Biomaterials 2003, 24, 579-585.

[306] Vaz, C. M., Reis, R. L., Cunha, A. M. Degradation model of starch-EVOH+HA composites. Mater. Res. Innov. 2001, 4, 375-380.

[307] Chen, L. J., Wang, M. Production and evaluation of biodegradable composites based on PHB-PHV copolymer. Biomaterials 2002, 23, 2631-2639.

[308] Ni, J., Wang, M. In vitro evaluation of hydroxyapatite reinforced polyhydroxybutyrate composite. Mater. Sci. Eng. C 2002, 20, 101-109.

[309] Carlo, E. C., Borges, A. P. B., Del Carlo, R. J., Martinez, M. M. M., Oliveira, P. M., Morato, G. O, Eleotério, R. B., Reis, M. S. Comparison of in vivo properties of hydroxyapatite-polyhydroxybutyrate composites assessed for bone substitution. J. Craniofac. Surg. 2009, 20, 853-859.

[310] Reis, E. C. C., Borges, A. P. B., Fonseca, C. C., Martinez, M. M. M., Eleotério, R. B., Morato, G. O., Oliveira, P. M. Biocompatibility, osteointegration, osteoconduction, and biodegradation of a hydroxyapatite-polyhydroxybutyrate composite. Braz. Arch. Biol. Technol. 2010, 53, 817-826.

[311] Sadat-Shojai, M., Khorasani, M. T., Jamshidi, A., Irani, S. Nano-hydroxyapatite reinforced polyhydroxybutyrate composites: a comprehensive study on the structural and in vitro biological properties. Mater. Sci. Eng. C 2013, 33, 2776-2787.

[312] Chen, D. Z., Tang, C. Y., Chan, K. C., Tsui, C. P., Yu, P. H. F., Leung, M. C. P., Uskokovic, P. S. Dynamic mechanical properties and in vitro bioactivity of PHBHV/HA nanocomposite. Compos. Sci. Technol. 2007, 67, 1617-1626.

[313] Rai, B., Noohom, W., Kithva, P. H., Grøndahl, L., Trau, M. Bionanohydroxyapatite/poly(3-hydroxybutyrate-co-3-hydroxyvalerate) composites with improved particle dispersion and superior mechanical properties. Chem. Mater. 2008, 20, 2802-2808.

[314] Wang, Y. W., Wu, Q., Chen, J., Chen, G. Q. Evaluation of three-dimensional scaffolds made of blends of hydroxyapatite and poly(3-hydroxybutyrate-co-3-hydroxyhexanoate) for bone reconstruction. Biomaterials 2005, 26, 899-904.

[315] Linhart, W., Lehmann, W., Siedler, M., Peters, F., Schilling, A. F., Schwarz, K., Amling, M., Rueger, J. M., Epple, M. Composites of amorphous calcium phosphate and poly(hydroxybutyrate) and poly(hydroxybutyrate-co-hydroxyvalerate) for bone substitution: assessment of the biocompatibility. J. Mater. Sci. 2006, 41, 4806-4813.

[316] Azevedo, M., Reis, R. L., Claase, M., Grijpma, D., Feijen, J. Development and properties of polycaprolactone/hydroxyapatite composite biomaterials. J. Mater. Sci. Mater. Med. 2003, 14, 103-107.

[317] Walsh, D., Furuzono, T., Tanaka, J. Preparation of porous composite implant materials by in situ polymerization of porous apatite containing ε-caprolactone or methyl methacrylate. Biomaterials 2001, 22, 1205-1212.

[318] Kim, H. W. Biomedical nanocomposites of hydroxyapatite/polycaprolactone obtained by surfactant mediation. J. Biomed. Mater. Res. A 2007, 83A, 169-177.

[319] Heo, S. J., Kim, S. E., Wei, J., Hyun, Y. T., Yun, H. S., Kim, D. H., Shin, J. W., Shin, J. W. Fabrication and characterization of novel nano- and micro-HA/PCL composite scaffolds using a modified rapid prototyping process. J. Biomed. Mater. Res. A 2009, 89A, 108-116.

[320] Chuenjitkuntaworn, B., Inrung, W., Damrongsri, D., Mekaapiruk, K., Supaphol, P., Pavasant, P. Polycaprolactone/hydroxyapatite composite scaffolds: preparation, characterization, and in vitro and in vivo biological responses of human primary bone cells. J. Biomed. Mater. Res. A 2010, 94A, 241-251.

[321] Bang, L. T., Kawachi, G., Nakagawa, M., Munar, M., Ishikawa, K., Othman, R. The use of poly (ε-caprolactone) to enhance the mechanical strength of porous Si-substituted carbonate apatite. J. Appl. Polym. Sci. 2013, 130, 426-433.

[322] Mohd Yusoff, M. F., Abdul Kadir, M. R., Iqbal, N., Hassan, M. A., Hussain, R. Dipcoating of poly (ε-caprolactone)/hydroxyapatite composite coating on Ti6Al4V for enhanced corrosion protection. Surf. Coat. Technol. 2014, 245, 102-107.

[323] Kim, B. S., Yang, S. S., Lee, J. A polycaprolactone/cuttlefish bone-derived hydroxyapatite composite porous scaffold for bone tissue engineering. J. Biomed. Mater. Res. B Appl. Biomater. 2014, 102B, 943-951.

[324] Hajiali, F., Tajbakhsh, S., Shojaei, A. Fabrication and properties of polycaprolactone composites containing calcium phosphate-based ceramics and bioactive glasses in bone tissue engineering: a review. Polym. Rev. 2017, 58, 164-207.

[325] Huang, B., Caetano, G., Vyas, C., Blaker, J. J., Diver, C., Bártolo, P. Polymer-ceramic composite scaffolds: the effect of hydroxyapatite and β-tri-calcium phosphate. Materials 2018, 11, 129.

[326] Causa, F., Netti, P. A., Ambrosio, L., Ciapetti, G., Baldini, N., Pagani, S., Martini, D., Giunti, A. Poly-ε-caprolactone/hydroxyapatite composites for bone regeneration: in vitro characterization and human osteoblast response. J. Biomed. Mater. Res. A 2006, 76A, 151-162.

[327] Thomas, V., Jagani, S., Johnson, K., Jose, M. V., Dean, D. R., Vohra, Y. K., Nyairo, E. Electrospun bioactive nanocomposite scaffolds of polycaprolactone and nanohydroxyapatite for bone tissue engineering. J. Nanosci. Nanotechol. 2006, 6, 487-493.

[328] Marra, K. G., Szem, J. W., Kumta, P. N., DiMilla, P. A., Weiss, L. E. In vitro analysis of biodegradable polymer blend/hydroxyapatite composites for bone tissue engineering. J. Biomed. Mater. Res. 1999, 47, 324-335.

[329] Dunn, A., Campbell, P., Marra, K. G. The influence of polymer blend composition on the degradation of polymer/hydroxyapatite biomaterials. J. Mater. Sci. Mater. Med. 2001, 12, 673-677.

[330] Calandrelli, L., Immirzi, B., Malinconico, M., Volpe, M., Oliva, A., Ragione, F. Preparation and characterization of composites based on biodegradable polymers for in vivo application. Polymer 2000, 41, 8027-8033.

[331] Chen, B., Sun, K. Poly(ε-caprolactone)/hydroxyapatite composites: effects of particle size, molecular weight distribution and irradiation on interfacial interaction and properties. Polym. Test. 2005, 24, 64-70.

[332] Kim, H. W., Knowles, J. C., Kim, H. E. Hydroxyapatite/poly(ε-caprolactone) composite coatings on hydroxyapatite porous bone scaffold for drug delivery. Biomaterials 2004, 25, 1279-1287.

[333] Ural, E., Kesenci, K., Fambri, L., Migliaresi, C., Piskin, E. Poly(D,L-lactide/ε-caprolactone)/hydroxyapatite composites. Biomaterials 2000, 21, 2147-2154.

[334] Rodenas-Rochina, J., Vidaurre, A., Cortázar, I. C., Lebourg, M. Effects of hydroxyapatite filler on long-term hydrolytic degradation of PLLA/PCL porous scaffolds. Polym. Degrad. Stabil. 2015, 119, 121-131.

[335] Kim, H. W., Lee, E. J., Kim, H. E., Salih, V., Knowles, J. C. Effect of fluoridation of hydroxyapatite in hydroxyapatite/polycaprolactone composites on osteoblast activity. Biomaterials 2005, 26, 4395-4404.

[336] Reyna-Urrutia, V. A., Estevez, M., González-González, A. M., Rosales-Ibáñez, R. 3D scaffolds of caprolactone/chitosan/polyvinyl alcohol/hydroxyapatite stabilized by physical bonds seeded with swine dental pulp stem cell for bone tissue engineering. J. Mater. Sci. Mater. Med. 2022, 33, 81.

[337] Gloria, A., Russo, T., D'Amora, U., Zeppetelli, S., D'Alessandro, T., Sandri, M., Bañobre-López, M., Piñeiro-Redondo, Y., Uhlarz, M., Tampieri, A., Rivas, J., Herrmannsdörfer, T., Dediu, V. A., Ambrosio, L., de Santis, R. Magnetic poly(ε-caprolactone)/iron-doped hydroxyapatite nanocomposite substrates for advanced bone tissue engineering. J. R. Soc. Interface 2013, 10, 20120833.

[338] Shokrollahi, P., Mirzadeh, H., Scherman, O. A., Huck, W. T. S. Biological and mechanical properties of novel composites based on supramolecular polycaprolactone and functionalized hydroxyapatite. J. Biomed. Mater. Res. A 2010, 95A, 209-221.

[339] Mehmanchi, M., Shokrollahi, P., Atai, M., Omidian, H., Bagheri, R. Supramolecular polycaprolactone nanocomposite based on functionalized hydroxyapatite. J. Bioact. Compat. Polym. 2012, 27, 467-480.

[340] Cojocaru, F. D., Gardikiotis, I., Dodi, G., Rotaru, A., Balan, V., Rezus, E., Verestiuc, L. Polysaccharides-calcium phosphates composite beads as bone substitutes for fractures repair and regeneration. Polymers 2023, 15, 1509.

[341] Yu, J., Xu, Y., Li, S., Seifert, G. V., Becker, M. L. Three-dimensional printing of nano hydroxyapatite/poly(ester urea) composite scaffolds with enhanced bioactivity. Biomacromolecules 2017, 18, 4171-4183.

[342] Deng, C., Wang, B., Dongqin, X., Zhou, S., Duan, K., Weng, J. Preparation and shape memory property of hydroxyapatite/poly (vinyl alcohol) composite. Polym. – Plast. Technol. Eng. 2012, 51, 1315-1318.

[343] Kutikov, A. B., Reyer, K. A., Song, J. Shape-memory performance of thermoplastic amphiphilic triblock copolymer poly(D,L-lactic acid-co-ethylene glycol-co-D,L-lactic acid) (PELA)/hydroxyapatite composites. Macromol. Chem. Phys. 2014, 215, 2482-2490.

[344] Wong, T. W., Wahit, M. U., Abdul Kadir, M. R., Soheilmoghaddam, M., Balakrishnan, H. A novel poly(xylitol-co-dodecanedioate)/hydroxyapatite composite with shape-memory behaviour. Mater. Lett. 2014, 126, 105-108.

[345] Degli Esposti, M., Chiellini, F., Bondioli, F., Morselli, D., Fabbri, P. Highly porous PHB-based bioactive scaffolds for bone tissue engineering by in situ synthesis of hydroxyapatite. Mater. Sci. Eng. C 2019, 100, 286-296.

[346] Guillaume, O., Geven, M. A., Varjas, V., Gehweiler, D., Stadelmann, V. A., Smidt, T., Zeiter, S., Sprecher, C., Bos, R. R. M., Grijpma, D. W., Alini, M., Yuan, H., Richards, G. R., Tang, T., Qin, L., Yuxiao, L, Jiang, P., Eglin, D. Orbital floor repair using patient specific osteoinductive implant made by stereolithography. Biomaterials 2020, 233, 119721.

[347] Sikkema, R., Keohan, B., Zhitomirsky, I. Alginic acid polymer-hydroxyapatite composites for bone tissue engineering. Polymers 2021, 13, 3070.

[348] Novesar, J., Dinda, A., Rahmayeni, Upita, S., Vivi, S. The effect of temperature on the synthesis and characterization of hydroxyapatite-polyethylene glycol composites by in-situ process. Hybrid Adv. 2023, 2, 100031.

[349] Behl, M., Razzaq, M. Y., Lendlein, A. Multifunctional shape-memory polymers. Adv. Mater. 2010, 22, 3388-3410.

[350] Cojocaru, F. D., Balan, V., Popa, M. I., Lobiuc, A., Antoniac, A., Antoniac, I. V., Verestiuc, L. Biopolymers – calcium phosphates composites with inclusions of magnetic nanoparticles for bone tissue engineering. Int. J. Biol. Macromol. 2019, 125, 612-620.

[351] Wang, R., Sun, K., Wang, J., He, Y., Song, P., Xiong, Y. Preparation and application of natural polymer/hydroxyapatite composite. Prog. Chem. 2016, 28, 885-895.

[352] Ramesh, N., Moratti, S. C., Dias, G. J. Hydroxyapatite-polymer biocomposites for bone regeneration: a review of current trends. J. Biomed. Mater. Res. B Appl. Biomater. 2018, 106B, 2046-2057.

[353] Ielo, I., Calabrese, G., de Luca, G., Conoci, S. Recent advances in hydroxyapatite-based biocomposites for bone tissue regeneration in orthopedics. Int. J. Mol. Sci. 2022, 23, 9721.

[354] Furko, M., Balázsi, K., Balázsi, C. Calcium phosphate loaded biopolymer composites –a comprehensive review on the most recent progress and promising trends. Coatings 2023, 13, 360.

[355] Handschel, J., Wiesmann, H. P., Stratmann, U., Kleinheinz, J., Meyer, U., Joos, U. TCP is hardly resorbed and not osteoconductive in a non-loading calvarial model. Biomaterials 2002, 23, 1689-1695.

[356] Kikuchi, M., Tanaka, J. Chemical interaction in β-tricalcium phosphate/copolymerized poly-L-lactide composites. J. Ceram. Soc. Jpn. 2000, 108, 642-645.
[357] Aunoble, S., Clement, D., Frayssinet, P., Harmand, M. F., le Huec, J. C. Biological performance of a new β-TCP/PLLA composite material for applications in spine surgery: in vitro and in vivo studies. J. Biomed. Mater. Res. A 2006, 78A, 416-422.
[358] Haaparanta, A. M., Haimi, S., Ellä, V., Hopper, N., Miettinen, S., Suuronen, R., Kellomäki, M. Porous polylactide/β-tricalcium phosphate composite scaffolds for tissue engineering applications. J. Tissue Eng. Regen. Med. 2010, 4, 366-373.
[359] Shin, D. Y., Kang, M. H., Kang, I. G., Kim, H. E., Jeong, S. H. In vitro and in vivo evaluation of polylactic acid-based composite with tricalcium phosphate microsphere for enhanced biodegradability and osseointegration. J. Biomater. Appl. 2018, 32, 1360-1370.
[360] Kikuchi, M., Koyama, Y., Yamada, T., Imamura, Y., Okada, T., Shirahama, N., Akita, K., Takakuda, K., Tanaka, J. Development of guided bone regeneration membrane composed of β-tricalcium phosphate and poly(L-lactide-co-glycolide-co-ε-caprolactone) composites. Biomaterials 2004, 25, 5979-5986.
[361] Chen, T. M., Yao, C. H., Wang, H. J., Chou, G. H., Lee, T. W., Lin, F. H. Evaluation of a novel malleable, biodegradable osteoconductive composite in a rabbit cranial defect model. Mater. Chem. Phys. 1998, 55, 44-50.
[362] Dong, G. C., Chen, H. M., Yao, C. H. A novel bone substitute composite composed of tricalcium phosphate, gelatin and drynaria fortunei herbal extract. J. Biomed. Mater. Res. A 2008, 84A, 167-177.
[363] Ji, J., Yuan, X., Xia, Z., Liu, P., Chen, J. Porous β-tricalcium phosphate composite scaffold reinforced by K_2HPO_4 and gelatin. Key Eng. Mater. 2010, 434-435, 620-623.
[364] Yao, C. H., Liu, B. S., Hsu, S. H., Chen, Y. S., Tsai, C. C. Biocompatibility and biodegradation of a bone composite containing tricalcium phosphate and genipin crosslinked gelatin. J. Biomed. Mater. Res. A 2004, 69A, 709-717.
[365] Eslaminejad, M. B., Mirzadeh, H., Mohamadi, Y., Nickmahzar, A. Bone differentiation of marrow-derived mesenchymal stem cells using β-tricalcium phosphate-alginate-gelatin hybrid scaffolds. J. Tissue Eng. Regen. Med. 2007, 1, 417-424.
[366] Takahashi, Y., Yamamoto, M., Tabata, Y. Osteogenic differentiation of mesenchymal stem cells in biodegradable sponges composed of gelatin and β-tricalcium phosphate. Biomaterials 2005, 26, 3587-3596.
[367] Bigi, A., Cantelli, I., Panzavolta, S., Rubini, K. α-tricalcium phosphate-gelatin composite cements. J. Appl. Biomater. Biomech. 2004, 2, 81-87.
[368] Yang, S. H., Hsu, C. K., Wang, K. C., Hou, S. M., Lin, F. H. Tricalcium phosphate and glutaraldehyde crosslinked gelatin incorporating bone morphogenetic protein – a viable scaffold for bone tissue engineering. J. Biomed. Mater. Res. B Appl. Biomater. 2005, 74B, 468-475.
[369] Kato, M., Namikawa, T., Terai, H., Hoshino, M., Miyamoto, S., Takaoka, K. Ectopic bone formation in mice associated with a lactic acid/dioxanone/ethylene

glycol copolymer-tricalcium phosphate composite with added recombinant human bone morphogenetic protein-2. Biomaterials 2006, 27, 3927-3933.

[370] Muramatsu, K., Oba, K., Mukai, D., Hasegawa, K., Masuda, S., Yoshihara, Y. Subacute systemic toxicity assessment of β-tricalcium phosphate/carboxymethyl-chitin composite implanted in rat femur. J. Mater. Sci. Mater. Med. 2007, 18, 513-522.

[371] Panzavolta, S., Fini, M., Nicoletti, A., Bracci, B., Rubini, K., Giardino, R., Bigi, A. Porous composite scaffolds based on gelatin and partially hydrolyzed α-tricalcium phosphate. Acta Biomater. 2009, 5, 636-643.

[372] Uchino, T., Kamitakahara, M., Otsuka, M., Ohtsuki, C. Hydroxyapatite-forming capability and mechanical properties of organic-inorganic hybrids and α-tricalcium phosphate porous bodies. J. Ceram. Soc. Jpn. 2010, 118, 57-61.

[373] Boguń, M., Rabiej, S. The influence of fiber formation conditions on the structure and properties of nanocomposite alginate fibers containing tricalcium phosphate or montmorillonite. Polym. Composite 2010, 31, 1321-1331.

[374] Park, C. H., Kim, E. K., Tijing, L. D., Amarjargal, A., Pant, H. R., Kim, C. S., Shon, H. K. Preparation and characterization of LA/PCL composite fibers containing beta tricalcium phosphate (β-TCP) particles. Ceram. Int. 2014, 40, 5049-5054.

[375] Ngamviriyavong, P., Patntirapong, S., Janvikul, W., Arphavasin, S., Meesap, P., Singhatanadgit, W. Development of poly(butylene succinate)/calcium phosphate composites for bone engineering. Compos. Interface 2014, 21, 431-441.

[376] Flauder, S., Sajzew, R., Müller, F. A. Mechanical properties of porous β-tricalcium phosphate composites prepared by ice-templating and poly(σ-caprolactone) impregnation. ACS Appl. Mater. Interfaces 2015, 7, 845-851.

[377] Agyemang, F. O., Sheikh, F. A., Appiah-Ntiamoah, R., Chandradass, J., Kim, H. Synthesis and characterization of poly (vinylidene fluoride)–calcium phosphate composite for potential tissue engineering applications. Ceram. Int. 2015, 41, 7066-7072.

[378] Cohen, B., Panker, M., Zuckerman, E., Foox, M., Zilberman, M. Effect of calcium phosphate-based fillers on the structure and bonding strength of novel gelatin-alginate bioadhesives. J. Biomater. Appl. 2014, 28, 1366-1375.

[379] Bleach, N. C., Tanner, K. E., Kellomäki, M., Törmälä, P. Effect of filler type on the mechanical properties of self-reinforced polylactide-calcium phosphate composites. J. Mater. Sci. Mater. Med. 2001, 12, 911-915.

[380] Liu, L., Xiong, Z., Yan, Y. N., Hu, Y. Y., Zhang, R. J., Wang, S. G. Porous morphology, porosity, mechanical properties of poly(α-hydroxy acid)-tricalcium phosphate composite scaffolds fabricated by low-temperature deposition. J. Biomed. Mater. Res. A 2007, 82A, 618-629.

[381] Zhang, Y., Zhang, M. Q. Synthesis and characterization of macroporous chitosan/calcium phosphate composite scaffolds for tissue engineering. J. Biomed. Mater. Res. 2001, 55, 304-312.

[382] Rai, B., Teoh, S. H., Hutmacher, D. W., Cao, T., Ho, K. H. Novel PCL-based honeycomb scaffolds as drug delivery systems for rhBMP-2. Biomaterials 2005, 26, 3739-3748.

[383] Rai, B., Teoh, S. H., Ho, K. H., Hutmacher, D. W., Cao, T., Chen, F., Yacob, K. The effect of rhBMP-2 on canine osteoblasts seeded onto 3D bioactive polycaprolactone scaffolds. Biomaterials 2004, 25, 5499-5506.

[384] Lei, Y., Rai, B., Ho, K. H., Teoh, S. H. In vitro degradation of novel bioactive polycaprolactone – 20% tricalcium phosphate composite scaffolds for bone engineering. Mater. Sci. Eng. C 2007, 27, 293-298.

[385] Miyai, T., Ito, A., Tamazawa, G., Matsuno, T., Sogo, Y., Nakamura, C., Yamazaki, A., Satoh, T. Antibiotic-loaded poly-ε-caprolactone and porous β-tricalcium phosphate composite for treating osteomyelitis. Biomaterials 2008, 29, 350-358.

[386] Li, Y., Wu, Z. G., Li, X. K., Guo, Z., Wu, S. H., Zhang, Y. Q., Shi, L., Teoh, S. H., Liu, Y. C., Zhang, Z. Y. A polycaprolactone-tricalcium phosphate composite scaffold as an autograft-free spinal fusion cage in a sheep model. Biomaterials 2014, 35, 5647-5659.

[387] Takahashi, Y., Yamamoto, M., Tabata, Y. Enhanced osteoinduction by controlled release of bone morphogenetic protein-2 from biodegradable sponge composed of gelatin and β-tricalcium phosphate. Biomaterials 2005, 26, 4856-4865.

[388] Ignatius, A. A., Betz, O., Augat, P., Claes, L. E. In vivo investigations on composites made of resorbable ceramics and poly(lactide) used as bone graft substitutes. J. Biomed. Mater. Res. Appl. Biomater. 2001, 58, 701-709.

[389] Miao, X., Lim, W. K., Huang, X., Chen, Y. Preparation and characterization of interpenetrating phased TCP/HA/PLGA composites. Mater. Lett. 2005, 59, 4000-4005.

[390] Ozbek, B., Erdogan, B., Ekren, N., Oktar, F. N., Akyol, S., Ben-Nissan, B., Sasmazel, H. T., Kalkandelen, C., Mergen, A., Kuruca, S. E., Ozen, G., Gunduz, O. Production of the novel fibrous structure of poly(ε-caprolactone)/tri-calcium phosphate/hexagonal boron nitride composites for bone tissue engineering. J. Austral. Ceram. Soc. 2018, 54, 251-260.

[391] Dorozhkin, S. V. Multiphasic calcium orthophosphate ($CaPO_4$) bioceramics and their biomedical applications. Ceram. Int. 2016, 42, 6529-6554.

[392] Brodie, J. C., Goldie, E., Connel, G., Merry, J., Grant, M. H. Osteoblast interactions with calcium phosphate ceramics modified by coating with type I collagen. J. Biomed. Mater. Res. A 2005, 73A, 409-421.

[393] Zhang, L. F., Sun, R., Xu, L., Du, J., Xiong, Z. C., Chen, H. C., Xiong, C. D. Hydrophilic poly (ethylene glycol) coating on PDLLA/BCP bone scaffold for drug delivery and cell culture. Mater. Sci. Eng. C 2008, 28, 141-149.

[394] Ignjatovic, N., Ninkov, P., Ajdukovic, Z., Konstantinovic, V., Uskokovic, D. Biphasic calcium phosphate/poly-(D,L-lactide-co-glycolide) biocomposite as filler and blocks for reparation of bone tissue. Mater. Sci. Forum 2005, 494, 519-524.

[395] Ignjatovic, N., Ninkov, P., Ajdukovic, Z., Vasiljevic-Radovic, D., Uskokovic, D. Biphasic calcium phosphate coated with poly-D,L-lactide-co-glycolide biomaterial as a bone substitute. J. Eur. Ceram. Soc. 2007, 27, 1589-1594.

[396] Yang, W., Yin, G., Zhou, D., Gu, J., Li, Y. In vitro characteristics of surface-modified biphasic calcium phosphate/poly(L-Lactide) biocomposite. Adv. Eng. Mater. 2010, 12, B128-B132.

[397] Ignjatovic, N., Ninkov, P., Kojic, V., Bokurov, M., Srdic, V., Krnojelac, D., Selakovic, S., Uskokovic, D. Cytotoxicity and fibroblast properties during in vitro test of biphasic calcium phosphate/poly-D,L-lactide-co-glycolide biocomposites and different phosphate materials. Microsc. Res. Techniq. 2006, 69, 976-982.

[398] Ajdukovic, Z., Ignjatovic, N., Petrovic, D., Uskokovic, D. Substitution of osteoporotic alveolar bone by biphasic calcium phosphate/poly-D,L-lactide-co-glycolide biomaterials. J. Biomater. Appl. 2007, 21, 317-328.

[399] Kim, H. W., Knowles, J. C., Kim, H. E. Effect of biphasic calcium phosphates on drug release and biological and mechanical properties of poly(ε-caprolactone) composite membranes. J. Biomed. Mater. Res. A 2004, 70A, 467-479.

[400] Kwak, K. A., Jyoti, M. A., Song, H. Y. In vitro and in vivo studies of three dimensional porous composites of biphasic calcium phosphate/poly ε-caprolactone: effect of bio-functionalization for bone tissue engineering. Appl. Surf. Sci. 2014, 301, 307-314.

[401] van Leeuwen, A. C., Yuan, H., Passanisi, G., van der Meer, J. W., de Bruijn, J. D., van Kooten, T. G., Grijpma, D. W., Bos, R. R. Poly(trimethylene carbonate) and biphasic calcium phosphate composites for orbital floor reconstruction: a feasibility study in sheep. Eur. Cell. Mater. 2014, 27, 81-96; discussion 96-97.

[402] Bakhtiari, L., Rezai, H. R., Hosseinalipour, S. M., Shokrgozar, M. A. Investigation of biphasic calcium phosphate/gelatin nanocomposite scaffolds as a bone tissue engineering. Ceram. Int. 2010, 36, 2421-2426.

[403] Bakhtiari, L., Rezai, H. R., Hosseinalipour, S. M., Shokrgozar, M. A. Preparation of porous Biphasic calcium phosphate-gelatin nanocomposite for bone tissue engineering. J. Nano Res. 2010, 11, 67-72.

[404] Matsuda, A., Ikoma, T., Kobayashi, H., Tanaka, J. Preparation and mechanical property of core-shell type chitosan/calcium phosphate composite fiber. Mater. Sci. Eng. C 2004, 24, 723-728.

[405] Rattanachan, S., Lorprayoon, C., Boonphayak, P. Synthesis of chitosan/brushite powders for bone cement composites. J. Ceram. Soc. Jpn. 2008, 116, 36-41.

[406] Ohsawa, H., Ito, A., Sogo, Y., Yamazaki, A., Ohno, T. Synthesis of albumin/DCP nano-composite particles. Key Eng. Mater. 2007, 330-332, 239-242.

[407] Mohammad, F., Arfin, T., Al-Lohedan, H. A. Synthesis, characterization and applications of ethyl cellulose-based polymeric calcium (II) hydrogen phosphate composite. J. Electron. Mater. 2018, 47, 2954-2963.

[408] Xu, H. H. K., Sun, L., Weir, M. D., Antonucci, J. M., Takagi, S., Chow, L. C., Peltz, M. Nano DCPA-whisker composites with high strength and Ca and PO_4 release. J. Dent. Res. 2006, 85, 722-727.

[409] Xu, H. H. K., Weir, M. D., Sun, L., Takagi, S., Chow, L. C. Effects of calcium phosphate nanoparticles on Ca-PO_4 composite. J. Dent. Res. 2007, 86, 378-383.

[410] Xu, H. H. K., Weir, M. D., Sun, L. Nanocomposites with Ca and PO_4 release: effects of reinforcement, dicalcium phosphate particle size and silanization. Dent. Mater. 2007, 23, 1482-1491.

[411] Chen, W. C., Chang, K. C., Wu, H. Y., Ko, C. L., Huang, C. L. Thermal cycling effect of dicalcium phosphate-reinforced composites on auto-mineralized dental resin. Mater. Sci. Eng. C 2014, 45, 359-368.

[412] El-Meliegy, E., Mabrouk, M., Kamal, G. M., Awad, S. M., El-Tohamy, A. M., El Gohary, M. I. Anticancer drug carriers using dicalcium phosphate/dextran/CMCnanocomposite scaffolds. Journal of Drug Delivery Science and Technology 2018, 45, 315-322.

[413] Tortet, L., Gavarri, J. R., Nihoul, G., Dianoux, A. J. Proton mobilities in brushite and brushite/polymer composites. Solid State Ionics 1997, 97, 253-256.

[414] Tortet, L., Gavarri, J. R., Musso, J., Nihoul, G., Sarychev, A. K. Percolation and modeling of proton conduction in polymer/brushite composites. J. Solid State Chem. 1998, 141, 392-403.

[415] Dorozhkin, S. V. Synthetic amorphous calcium phosphates (ACPs): preparation, structure, properties, and biomedical applications. Biomater. Sci. 2021, 9, 7748-7798.

[416] Dorozhkin, S. V. Dental applications of calcium orthophosphates ($CaPO_4$). J. Dentistry Res. 2019, 1, 0007, 024-054.

[417] Zhang, L., Weir, M. D., Chow, L. C., Antonucci, J. M., Chen, J., Xu, H. H. K. Novel rechargeable calcium phosphate dental nanocomposite. Dent. Mater. 2016, 32, 285-293.

[418] Zhang, L., Weir, M. D., Chow, L. C., Reynolds, M. A., Xu, H. H. K. Rechargeable calcium phosphate orthodontic cement with sustained ion release and re-release. Sci. Rep. 2016, 6, 36476.

[419] Xie, X. J., Xing, D., Wang, L., Zhou, H., Weir, M. D., Bai, Y. X., Xu, H. H. K. Novel rechargeable calcium phosphate nanoparticle-containing orthodontic cement. Int. J. Oral Sci. 2018, 9, 24-32.

[420] Liang, K., Xiao, S., Wu, J., Li, J., Weir, M. D., Cheng, L., Reynolds, M. A., Zhou, X., Xu, H. H. K. Long-term dentin remineralization by poly(amido amine) and rechargeable calcium phosphate nanocomposite after fluid challenges. Dent. Mater. 2018, 34, 607-618.

[421] Gutierrez, M. C., Jobbágy, M., Ferrer, M. L., del Monte, F. Enzymatic synthesis of amorphous calcium phosphate-chitosan nanocomposites and their processing into hierarchical structures. Chem. Mater. 2008, 20, 11-13.

[422] Hakimimehr, D., Liu, D. M., Troczynski, T. In-situ preparation of poly(propylene fumarate) – hydroxyapatite composite. Biomaterials 2005, 26, 7297-7303.

[423] Antonucci, J. M., Regnault, W. F., Skrtic, D. Polymerization shrinkage and stress development in amorphous calcium phosphate/urethane dimethacrylate polymeric composites. J. Compos. Mater. 2010, 44, 355-367.

[424] Wang, K. W., Zhu, Y. J., Chen, F., Cao, S. W. Calcium phosphate/block copolymer hybrid porous nanospheres: preparation and application in drug delivery. Mater. Lett. 2010, 64, 2299-2301.

[425] Cao, S. W., Zhu, Y. J., Wu, J., Wang, K. W., Tang, Q. L. Preparation and sustained-release property of triblock copolymer/calcium phosphate nanocomposite as nanocarrier for hydrophobic drug. Nanoscale Res. Lett. 2010, 5, 781-785.

[426] Iwama, R., Anada, T., Shiwaku, Y., Tsuchiya, K., Takahashi, T., Suzuki, O. Osteogenic cellular activity around onlaid octacalcium phosphate-gelatin composite onto rat calvaria. J. Biomed. Mater. Res. A 2018, 106A, 1322-1333.

[427] Yokoi, T. The development of novel calcium phosphate-polymer composite biomaterials with macro- to nano-level controlled hierarchical structures. J. Ceram. Soc. Jpn. 2019, 127, 715-721.

[428] Suzuki, O. Octacalcium phosphate (OCP)-based bone substitute materials. Jpn. Dent. Sci. Rev. 2013, 49, 58-71.

[429] Dorozhkin, S. V. Self-setting calcium orthophosphate ($CaPO_4$) formulations and their biomedical applications. Adv. Nano-Bio. Mater. Dev. 2019, 3, 321-421.

[430] Brückner, T., Schamel, M., Kübler, A. C., Groll, J., Gbureck, U. Novel bone wax based on poly(ethylene glycol)–calcium phosphate cement mixtures. Acta Biomater. 2016, 33, 252-263.

[431] Rödel, M., Teßmar, J., Groll, J., Gbureck, U. Tough and elastic α-tricalcium phosphate cement composites with degradable PEG-based cross-linker. Materials 2019, 12, 53.

[432] Fuchs, A., Bartolf-Kopp, M., Böhm, H., Straub, A., Kübler, A. C., Linz, C., Gbureck, U. Composite grafts made of polycaprolactone fiber mats and oil-based calcium phosphate cement pastes for the reconstruction of cranial and maxillofacial defects. Clin. Oral Invest. 2023, 27, 3199-3209.

[433] Lou, C. W., Huang, C. C., Chen, W. C., Hu, J. J., Lu, C. T., Lin, J. H. Preliminary studies in composite scaffolds of calcium phosphate bone cement with polylactide. Appl. Mech. Mater. 2012, 184-185, 1098-1101.

[434] Zhong, M. L., Chen, X. Q., Fan, H. S., Zhang, X. D. Incorporation of salmon calcitonin-loaded poly(lactide-co-glycolide) (PLGA) microspheres into calcium phosphate bone cement and the biocompatibility evaluation in vitro. J. Bioact. Compat. Polym. 2012, 27, 133-147.

[435] Bigi, A., Bracci, B., Panzavolta, S. Effect of added gelatin on the properties of calcium phosphate cement. Biomaterials 2004, 25, 2893-2899.

[436] Bigi, A., Panzavolta, S., Sturba, L., Torricelli, P., Fini, M., Giardino, R. Normal and osteopenic bone-derived osteoblast response to a biomimetic gelatin-calcium phosphate bone cement. J. Biomed. Mater. Res. A 2006, 78A, 739-745.

[437] Panzavolta, S., Torricelli, P., Sturba, L., Bracci, B., Giardino, R., Bigi, A. Setting properties and in vitro bioactivity of strontium-enriched gelatin-calcium phosphate bone cements. J. Biomed. Mater. Res. A 2008, 84A, 965-972.

[438] Maazouz, Y., Montufar, E. B., Guillem-Marti, J., Fleps, I., Öhman, C., Persson, C., Ginebra, M. P. Robocasting of biomimetic hydroxyapatite scaffolds using self-setting inks. J. Mater. Chem. B 2014, 2, 5378-5386.

[439] Rammelt, S., Neumann, M., Hanisch, U., Reinstorf, A., Pompe, W., Zwipp, H., Biewener, A. Osteocalcin enhances bone remodeling around hydroxyapatite/collagen composites. J. Biomed. Mater. Res. A 2005, 73A, 284-294.

[440] Park, J. H., Lee, E. J., Knowles, J. C., Kim, H. W. Preparation of in situ hardening composite microcarriers: calcium phosphate cement combined with alginate for bone regeneration. J. Biomater. Appl. 2014, 28, 1079-1084.

[441] Wang, X., Ma, J., Feng, Q., Cui, F. Z. Skeletal repair in rabbits with calcium phosphate cements incorporated phosphorylated chitin. Biomaterials 2002, 23, 4591-4600.

[442] Xue, B., Zhang, C., Wang, Y., Wang, J., Zhang, J., Lu, M., Li, G., Cao, Z., Huang, Q. A novel controlled-release system for antibacterial enzyme lysostaphin delivery using hydroxyapatite/chitosan composite bone cement. PLoS ONE 2014, 9, e113797.

[443] Cao, C., Li, H., Li, J., Liu, C., Yang, H., Li, B. Mechanical reinforcement of injectable calcium phosphate cement/silk fibroin (SF) composite by mineralized SF. Ceram. Int. 2014, 40, 13987-13993.

[444] Han F., Shi C., Yang H., Li B. Calcium phosphate-silk fibroin composites: bone cement and beyond. In: Developments and applications of calcium phosphate bone cements. Liu C., He H. (Eds.); Springer series in biomaterials science and engineering, vol 9. Springer, Singapore 2018, pp. 449-472.

[445] Liu, W., Zhang, J., Rethore, G., Khairoun, K., Pilet, P., Tancret, F., Bouler, J. M., Weiss, P. A novel injectable, cohesive and toughened Si-HPMC (silanized-hydroxypropyl methylcellulose) composite calcium phosphate cement for bone substitution. Acta Biomater. 2014, 10, 3335-3345.

[446] Mirkiani, S., Mesgar, A. S., Mohammadi, Z., Matinfar, M. Synergetic reinforcement of brushite cements by monetite/apatite whisker-like fibers and carboxymethylcellulose. Materialia 2022, 21, 101329.

[447] de Lacerda Schickert, S., Pinto, J. C., Jansen, J., Leeuwenburgh, S. C. G., van den Beucken, J. J. J. P. Tough and injectable fiber reinforced calcium phosphate cement as an alternative to polymethylmethacrylate cement for vertebral augmentation: a biomechanical study. Biomater. Sci. 2020, 8, 4239.

[448] Luo, J., Faivre, J., Engqvist, H., Persson, C. The addition of poly(vinyl alcohol) fibers to apatitic calcium phosphate cement can improve its toughness. Materials 2019, 12, 1531.

[449] Sadiasa, A., Sarkar, S. K., Franco, R. A., Min, Y. K., Lee, B. T. Bioactive glass incorporation in calcium phosphate cement-based injectable bone substitute for improved in vitro biocompatibility and in vivo bone regeneration. J. Biomater. Appl. 2014, 28, 739-756.

[450] El-Fiqi, A., Kim, J. H., Kim, H. W. Highly bioactive bone cement microspheres based on α-tricalcium phosphate microparticles/mesoporous bioactive glass nanoparticles: formulation, physico-chemical characterization and in vivo bone regeneration. Colloids Surface B 2022, 217, 112650.

[451] Demir-Oğuz, Ö., Boccaccini, A. R., Loca, D. Injectable bone cements: what benefits the combination of calcium phosphates and bioactive glasses could bring? Bioact. Mater. 2023, 19, 217-236.

[452] Richter, R. F., Ahlfeld, T., Gelinsky, M., Lode, A. Composites consisting of calcium phosphate cements and mesoporous bioactive glasses as a 3D plottable drug delivery system. Acta Biomater. 2023, 156, 146-157.

[453] Zhou, S., Ma, J., Shen, Y., Haapasalo, M., Ruse, N. D., Yang, Q., Troczynski, T. In vitro studies of calcium phosphate silicate bone cements. J. Mater. Sci. Mater. Med. 2013, 24, 355-364.

[454] Motisuke, M., Santos, V. R., Bazanini, N. C., Bertran, C. A. Apatite bone cement reinforced with calcium silicate fibers. J. Mater. Sci. Mater. Med. 2014, 25, 2357-2363.

[455] Correa, D., Almirall, A., Carrodeguas, R. G., dos Santos, L. A., de Aza, A. H., Parra, J., Morejon, L., Delgado, J. A. α-Tricalcium phosphate cements modified with β-dicalcium silicate and tricalcium aluminate: physicochemical characterization, in vitro bioactivity and cytotoxicity. J. Biomed. Mater. Res. B Appl. Biomater. 2015, 103B, 72-83.
[456] Perez, R. A., Patel, K. D., Kim, H. W. Novel magnetic nanocomposite injectables: calcium phosphate cements impregnated with ultrafine magnetic nanoparticles for bone regeneration. RSC Adv. 2015, 5, 13411-13419.
[457] Elahpour, N., Rabiee, S. M., Ebrahimzadeh, M. H., Moradi, A. In-vitro formation and growth kinetics of apatite on a new light-cured composite calcium phosphate cement. Ceram. Int. 2018, 44, 15317-15322.
[458] Canal, C., Ginebra, M. P. Fibre-reinforced calcium phosphate cements: a review. J. Mech. Behav. Biomed. Mater. 2011, 4, 1658-1671.
[459] Hasan, M. S., Carpenter, N., Wei, T. L., McNally, D., Ahmed, I., Boszczyk, B. M. Effects of adding resorbable phosphate glass fibres and PLA to calcium phosphate bone cements. J. Appl. Biomater. Funct. Mater. 2014, 12, 203-209.
[460] Kim, S. B., Kim, Y. J., Yoon, T. L., Park, S. A., Cho, I. H., Kim, E. J., Kim, I. A., Shin, J. W. The characteristics of a hydroxyapatite-chitosan-PMMA bone cement. Biomaterials 2004, 25, 5715-5723.
[461] Vallo, C. I., Montemartini, P. E., Fanovich, M. A., Lópes, J. M. P., Cuadrado, T. R. Polymethylmethacrylate-based bone cement modified with hydroxyapatite. J. Biomed. Mater. Res. B Appl. Biomater. 1999, 48, 150-158.
[462] Moursi, A. M., Winnard, A. V., Winnard, P. L., Lannutti, J. J., Seghi, R. R. Enhanced osteoblast response to a PMMA – HA composite. Biomaterials 2002, 23, 133-144.
[463] Dalby, M. J., di Silvio, L., Harper, E. J., Bonfield, W. Increasing hydroxyapatite incorporation into poly(methylmethacrylate) cement increases osteoblast adhesion and response. Biomaterials 2002, 23, 569-576.
[464] Itokawa, H., Hiraide, T., Moriya, M., Fujimoto, M., Nagashima, G., Suzuki, R., Fujimoto, T. A 12 month in vivo study on the response of bone to a hydroxyapatite-polymethylmethacrylate cranioplasty composite. Biomaterials 2007, 28, 4922-4927.
[465] Tham, W. L., Chow, W. S., Ishak, Z. A. M. Flexural and morphological properties of poly(methyl methacrylate)/hydroxyapatite composites: effects of planetary ball mill grinding time. J. Reinf. Plastics Compos. 2010, 29, 2065-2075.
[466] Pattanayak, D. K., Rao, B. T., Mohan, T. R. R. Calcium phosphate bioceramics and bioceramic composites. J. Sol Gel Sci. Technol. 2011, 59, 432-447.
[467] Arabmotlagh, M., Bachmaier, S., Geiger, F., Rauschmann, M. PMMA-hydroxyapatite composite material retards fatigue failure of augmented bone compared to augmentation with plain PMMA: in vivo study using a sheep model. J. Biomed. Mater. Res. B Appl. Biomater. 2014, 102, 1613-1619.
[468] Bai, H., Walsh, F., Gludovatz, B., Delattre, B., Huang, C., Chen, Y., Tomsia, A. P., Ritchie, R. O. Bioinspired hydroxyapatite/poly(methyl methacrylate) composite with a nacre-mimetic architecture by a bidirectional freezing method. Adv. Mater. 2016, 28, 50-56.

[469] Harper, E. J., Behiri, J. C., Bonfield, W. Flexural and fatigue properties of a bone cement based upon polyethylmethacrylate and hydroxyapatite. J. Mater. Sci. Mater. Med. 1995, 6, 799-803.
[470] Arnold, J. C., Venditti, N. P. Prediction of the long-term creep behaviour of hydroxyapatite-filled polyethylmethacrylate bone cements. J. Mater. Sci. Mater. Med. 2007, 18, 1849-1858.
[471] Ohgaki, M., Yamashita, K. Preparation of polymethylmethacrylate-reinforced functionally graded hydroxyapatite composites. J. Am. Ceram. Soc. 2003, 86, 1440-1442.
[472] del Real, R. P., Padilla, S., Vallet-Regi, M. Gentamicin release from hydroxyapatite/poly(ethyl methacrylate)/poly(methyl methacrylate)composites. J. Biomed. Mater. Res. 2000, 52, 1-7.
[473] Saito, M., Maruoka, A., Mori, T., Sugano, N., Hino, K. Experimental studies on a new bioactive bone cement: hydroxyapatite composite resin. Biomaterials 1994, 15, 156-160.
[474] Kawagoe, K., Saito, M., Shibuya, T., Nakashima, T., Hino, K., Yoshikawa, H. Augmentation of cancellous screw fixation with hydroxyapatite composite resin (CAP) in vivo. J. Biomed. Mater. Res. 2000, 53, 678-684.
[475] Turner, A. W. L., Gillies, R. M., Svehla, M. J., Saito, M., Walsh, W. R. Hydroxyapatite composite resin cement augmentation of pedicle screw fixation. Clin. Orthop. Rel. Res. 2003, 406, 253-261.
[476] Watson, K. E., Ten Huisen, K. S., Brown, P. W. The formation of hydroxyapatite – calcium polyacrylate composites. J. Mater. Sci. Mater. Med. 1999, 10, 205-213.
[477] Reed, C. S., Ten Huisen, K. S., Brown, P. W., Allcock, H. R. Thermal stability and compressive strength of calcium-deficient hydroxyapatite – poly[bis(carboxylatophenoxy)phosphazene] composites. Chem. Mater. 1996, 8, 440-447.
[478] Peter, S. J., Kim, P., Yasko, A. W., Yaszemski, M. J., Mikos, A. G. Crosslinking characteristics of an injectable poly(propylene fumarate)/β-tricalcium phosphate paste and mechanical properties of the crosslinked composite for use as a biodegradable bone cement. J. Biomed. Mater. Res. 1999, 44, 314-321.
[479] He, S., Yaszemski, M. J., Yasko, A. W., Engel, P. S., Mikos, A. G. Injectable biodegradable polymer composites based on poly (propylene fumarate) crosslinked with poly(ethylene glycol)-dimethacrylate. Biomaterials 2000, 21, 2389-2394.
[480] Ignjatovic, N., Jovanovic, J., Suljovrujic, E., Uskokovic, D. Injectable polydimethylsiloxane/hydroxyapatite composite cement. Biomed. Mater. Eng. 2003, 13, 401-410.
[481] Fei, Z., Hu, Y., Wu, D., Wu, H., Lu, R., Bai, J., Song, H. Preparation and property of a novel bone graft composite consisting of rhBMP-2 loaded PLGA microspheres and calcium phosphate cement. J. Mater. Sci. Mater. Med. 2008, 19, 1109-1116.
[482] Link, D. P., van den Dolder, J., van den Beucken, J. J. J. P., Cuijpers, V. M., Wolke, J. G. C., Mikos, A. G., Jansen, J. A. Evaluation of the biocompatibility of calcium phosphate cement/PLGA microparticle composites. J. Biomed. Mater. Res. A 2008, 87A, 760-769.

[483] Chiang, T. Y., Ho, C. C., Chen, D. C. H., Lai, M. H., Ding, S. J. Physicochemical properties and biocompatibility of chitosan oligosaccharide/gelatin/calcium phosphate hybrid cements. Mater. Chem. Phys. 2010, 120, 282-288.

[484] Fujishiro, Y., Takahashi, K., Sato, T. Preparation and compressive strength of α-tricalcium phosphate/gelatin gel composite cement. J. Biomed. Mater. Res. 2001, 54, 525-230.

[485] Miyazaki, K., Horibe, T., Antonucci, J. M., Takagi, S., Chow, L. C. Polymeric calcium phosphate cements: analysis of reaction products and properties. Dental Mater. 1993, 9, 41-45.

[486] Miyazaki, K., Horibe, T., Antonucci, J. M., Takagi, S., Chow, L. C. Polymeric calcium phosphate cements: setting reaction modifiers. Dental Mater. 1993, 9, 46-50.

[487] dos Santos, L. A., de Oliveira, L. C., Rigo, E. C. S., Carrodeguas, R. G., Boschi, A. O., de Arruda, A. C. F. Influence of polymeric additives on the mechanical properties of α-tricalcium phosphate cement. Bone 1999, 25, 99S-102S.

[488] Greish, Y. E., Brown, P. W., Bender, J. D., Allcockm H. R., Lakshmim S., Laurencin, C. T. Hydroxyapatite-polyphosphazane composites prepared at low temperatures. J. Am. Ceram. Soc. 2007, 90, 2728-2734.

[489] Greish, Y. E., Bender, J. D., Lakshmi, S., Brown, P. W., Allcock, H. R., Laurencin, C. T. Formation of hydroxyapatite-polyphosphazene polymer composites at physiologic temperature. J. Biomed. Mater. Res. A 2006, 77A, 416-425.

[490] Greish, Y. E., Bender, J. D., Lakshmi, S., Brown, P. W., Allcock, H. R., Laurencin, C. T. Low temperature formation of hydroxyapatite-poly(alkyl oxybenzoate) phosphazene composites for biomedical applications. Biomaterials 2005, 26, 1-9.

[491] Mickiewicz, R. A., Mayes, A. M., Knaack, D. Polymer – calcium phosphate cement composites for bone substitutes. J. Biomed. Mater. Res. 2002, 61, 581-592.

[492] Carey, L. E., Xu, H. H. K., Simon, C. G., Takagi, S., Chow, L. C. Premixed rapid-setting calcium phosphate composites for bone repair. Biomaterials 2005, 26, 5002-5014.

[493] Krüger, R., Seitz, J. M., Ewald, A., Bach, F. W., Groll, J. Strong and tough magnesium wire reinforced phosphate cement composites for load-bearing bone replacement. J. Mech. Behav. Biomed. Mater. 2013, 20, 36-44.

[494] Miao, X., Tan, L. P., Tan, L. S., Huang, X. Porous calcium phosphate ceramics modified with PLGA – bioactive glass. Mater. Sci. Eng. C 2007, 27, 274-279.

[495] Lickorish, D., Guan, L., Davies, J. E. A three-phase, fully resorbable, polyester/calcium phosphate scaffold for bone tissue engineering: evolution of scaffold design. Biomaterials 2007, 28, 1495-1502.

[496] Qian, G., Fan, P., He, F., Ye, J. Novel strategy to accelerate bone regeneration of calcium phosphate cement by incorporating 3D plotted poly(lactic-co-glycolic acid) network and bioactive wollastonite. Adv. Healthcare Mater. 2019, 8, 1801325.

[497] Iwasakia, Y., Takahataa, Y., Fujii, S. Self-setting particle-stabilized emulsion for hard-tissue engineering. Colloid Surface B 2015, 126, 394-400.

[498] Xu, H. H. K., Simon, C. G. Fast setting calcium phosphate-chitosan scaffold: mechanical properties and biocompatibility. Biomaterials 2005, 26, 1337-1348.

[499] Zhang, L., Li, Y., Zhou, G., Lu, G. Y., Zuo, Y. Setting mechanism of nano-hydroxyapatite/chitosan bone cement. J. Inorg. Mater. 2006, 21, 1197-1202.

[500] Ruhe, P. Q., Hedberg, E. L., Padron, N. T., Spauwen, P. H. M., Jansen, J. A., Mikos, A. G. Biocompatibility and degradation of poly(D,L-lactic-co-glycolic acid)/calcium phosphate cement composites. J. Biomed. Mater. Res. A 2005, 74A, 533-544.

[501] Guo, D. G., Sun, H. L., Xu, K. W., Han, Y. Long-term variations in mechanical properties and in vivo degradability of CPC/PLGA composite. J. Biomed. Mater. Res. B Appl. Biomater. 2007, 82B, 533-544.

[502] Habraken, W. J. E. M., Wolke, J. G. C., Mikos, A. G., Jansen, J. A. Injectable PLGA microsphere/calcium phosphate cements: physical properties and degradation characteristics. J. Biomater. Sci. Polym. Edn. 2006, 17, 1057-1074.

[503] Ruhe, P. Q., Hedberg-Dirk, E. L., Padron, N. T., Spauwen, P. H. M., Jansen, J. A., Mikos, A. G. Porous poly (D,L-lactic-co-glycolic acid)/calcium phosphate cement composite for reconstruction of bone defects. Tissue Eng. 2006, 12, 789-800.

[504] Ruhe, P. Q., Hedberg, E. L., Padron, N. T., Spauwen, P. H. M., Jansen, J. A., Mikos, A. G. rhBMP-2 release from injectable poly (D,L-lactic-co-glycolic acid)/calcium phosphate composites. J. Bone Joint Surg. (Am.) 2003, 85A, Suppl. 3, 75-81.

[505] Plachokova, A., Link. D., van den Dolder. J., van den Beucken. J., Jansen, J. A. Bone regenerative properties of injectable PGLA–CaP composite with TGF-β1 in a rat augmentation model. J. Tissue Eng. Regen. Med. 2007, 1, 457-464.

[506] Roy, A., Jhunjhunwala, S., Bayer, E., Fedorchak, M., Little, S. R., Kumta, P. N. Porous calcium phosphate–poly (lactic-co-glycolic) acid composite bone cement: a viable tunable drug delivery system. Mater. Sci. Eng. C 2016, 59, 92-101.

[507] Bian, T., Zhao, K., Meng, Q., Jiao, H., Tang, Y., Luo, J. Fabrication and performance of calcium phosphate cement/small intestinal submucosa composite bionic bone scaffolds with different microstructures. Ceram. Int. 2018, 44, 9181-9187.

[508] Moussi, H., Weiss, P., le Bideau, J., Gautier, H., Charbonnier, B. Injectable macromolecule-based calcium phosphate bone substitutes. Mater. Adv., 2022, 3, 6125-6141.

[509] Liu, C. Collagen–hydroxyapatite composite scaffolds for tissue engineering. In: Hydroxyapatite (HAp) for biomedical applications. Mucalo, M. R. (Ed.); Woodhead publishing series in biomaterials: Number 95; Elsevier; Cambridge, UK, 2015; pp. 211-234.

[510] Hellmich, C., Ulm, F. J. Are mineralized tissues open crystal foams reinforced by crosslinked collagen? – Some energy arguments. J. Biomech. 2002, 35, 1199-1212.

[511] Veiga, A., Castro, F., Rocha, F., Oliveira, A. L. An update on hydroxyapatite/collagen composites: what is there left to say about these bioinspired materials? J. Biomed. Mater. Res. B Appl. Biomater. 2022, 110B, 1192-1205.

[512] Roveri, N., Falini, G., Sidoti, M. C., Tampieri, A., Landi, E., Sandri, M., Parma, B. Biologically inspired growth of hydroxyapatite nanocrystals inside self-assembled collagen fibers. Mater. Sci. Eng. C 2003, 23, 441-446.

[513] Tampieri, A., Celotti, G., Landi, E. From biomimetic apatites to biologically inspired composites. Anal. Bioanal. Chem. 2005, 381, 568-576.

[514] Tampieri, A., Celotti, G., Landi, E., Sandri, M., Roveri, N., Falini, G. Biologically inspired synthesis of bone-like composite: self-assembled collagen fibers/hydroxyapatite nanocrystals. J. Biomed. Mater. Res. A 2003, 67A, 618-625.

[515] Clarke, K. I., Graves, S. E., Wong, A. T. C., Triffit, J. T., Francis, M. J. O., Czernuszka, J. T. Investigation into the formation and mechanical properties of a bioactive material based on collagen and calcium phosphate. J. Mater. Sci. Mater. Med. 1993, 4, 107-110.

[516] Ten Huisen, K. S., Martin, R. I., Klimkiewicz, M., Brown, P. W. Formation and properties of a synthetic bone composite: hydroxyapatite-collagen. J. Biomed. Mater. Res. 1995, 29, 803-810.

[517] Ishikawa, H., Koshino, T., Takeuchi, R., Saito, T. Effects of collagen gel mixed with hydroxyapatite power on interface between newly formed bone and grafted Achilles tendon in rabbit femoral bone tunnel. Biomaterials 2001, 22, 1689-1694.

[518] ltoh, S., Kikuchi, M., Takakuda, K., Koyama, Y., Matsumoto, H. N., Ichinose, S., Tanaka, J., Kawauchi, T., Shinomiya, K. The biocompatibility and osteoconductive activity of a novel hydroxyapatite/collagen composite biomaterial and its function as a carrier of rhBMP-2. J. Biomed. Mater. Res. 2001, 54, 445-453.

[519] Uskokovic, V., Ignjatovic, N., Petranovic, N. Synthesis and characterization of hydroxyapatite-collagen biocomposite materials. Mater. Sci. Forum 2002, 413, 269-274.

[520] Yoon, B. H., Kim, H. W., Lee, S. H., Bae, C. J., Koh, Y. H., Kong, Y. M., Kim, H. E. Stability and cellular responses to fluorapatite-collagen composites. Biomaterials 2005, 26, 2957-2963.

[521] Wahl, D. A., Czernuszka, J. T. Collagen-hydroxyapatite composites for hard tissue repair. Eur. Cell Mater. 2006, 11, 43-56.

[522] Sachlos, E., Gotora, D., Czernuszka, J. T. Collagen scaffolds reinforced with biomimetic composite nano-sized carbonate-substituted hydroxyapatite crystals and shaped by rapid prototyping to contain internal microchannels. Tissue Eng. 2006, 12, 2479-2487.

[523] Teng, S. H., Lee, E. J., Park, C. S., Choi, W. Y., Shin, D. S., Kim, H. E. Bioactive nanocomposite coatings of collagen/hydroxyapatite on titanium substrates. J. Mater. Sci. Mater. Med. 2008, 19, 2453-2461.

[524] Song, J. H., Kim, H. E., Kim, H. W. Collagen-apatite nanocomposite membranes for guided bone regeneration. J. Biomed. Mater. Res. B Appl. Biomater. 2007, 83B, 248-257.

[525] Pek, Y. S., Gao, S., Arshad, M. S. M., Leck, K. J., Ying, J. Y. Porous collagen-apatite nanocomposite foams as bone regeneration scaffolds. Biomaterials 2008, 29, 4300-4305.

[526] Zhao, H., Huang, C., Jin, H., Cai, J. A novel route for collagen/hydroxyapatite preparation by enzymatic decomposition of urea. J. Compos. Mater. 2010, 44, 2127-2133.

[527] Kozłowska, J., Sionkowska, A. Effects of different crosslinking methods on the properties of collagen-calcium phosphate composite materials. Int. J. Biol. Macromol. 2015, 74, 397-403.

[528] Banglmaier, R. F., Sander, E. A., VandeVord, P. J. Induction and quantification of collagen fiber alignment in a three-dimensional hydroxyapatite–collagen composite scaffold. Acta Biomater. 2015, 17, 26-35.

[529] Banks, E., Nakajima, S., Shapiro, L. C., Tilevitz, O., Alonzo, J. R., Chianelli, R. R. Fibrous apatite grown on modified collagen. Science 1977, 198, 1164-1166.

[530] Hayashi, K., Yabuki, T., Tabuchi, K., Fujii, T. Repair of experimental bone defect with a collagen block containing synthesized apatite. Arch. Orthop. Traum. Surg. 1982, 99, 265-269.

[531] Mittelmeier, H., Nizzard, M. Knochenregeneration mit industriell gefertigtem Collagen Apatit Implantat ("Collapat"). In: Osteogenese und Knochenwachstum, Hackenbroch, M. H., Refior, H. J., Jäger, M. G. (Eds.); Thieme: Stuttgart, Germany, 1982; pp. 194-197.

[532] Serre, C. M., Papillard, M., Chavassieux, P., Boivin, G. In vitro induction of a calcifying matrix by biomaterials constituted of collagen and/or hydroxyapatite: an ultrastructural comparison of three types of biomaterials. Biomaterials 1993, 14, 97-106.

[533] Scabbia, A., Trombelli, L. A comparative study on the use of a HA/collagen/chondroitin sulphate biomaterial (Biostite®) and a bovine-derived HA xenograft (Bio-Oss®) in the treatment of deep intraosseous defects. J. Clin. Periodontol. 2004, 31, 348-355.

[534] Yamasaki, Y., Yoshida, Y., Okazaki, M., Shimazu, A., Kubo, T., Akagawa, Y., Uchida, T. Action of FGMgCO_3Ap-collagen composite in promoting bone formation. Biomaterials 2003, 24, 4913-4920.

[535] Wang, X., Grogan, S. P., Rieser, F., Winkelmann, V., Maquet, V., Berge, M. L. Mainil-Varlet, P. Tissue engineering of biphasic cartilage constructs using various biodegradable scaffolds: an in vitro study. Biomaterials 2004, 25, 3681-3688.

[536] Zahn, D. Multi-scale simulations of apatite–collagen composites: from molecules to materials. Front. Mater. Sci. 2017, 11, 1-12.

[537] Chang, M. C., Ikoma, T., Kikuchi, M., Tanaka, J. The cross-linkage effect of hydroxyapatite/collagen nanocomposites on a self-organization phenomenon. J. Mater. Sci. Mater. Med. 2002, 13, 993-997.

[538] Iijima, M., Moriwaki, Y., Kuboki, Y. Oriented growth of octacalcium phosphate on and inside the collagenous matrix in vitro. Connect. Tissue Res. 1996, 32, 519-524.

[539] Miyamoto, Y., Ishikawa, K., Takechi, M., Toh, T., Yuasa, T., Nagayama, M., Suzuki, K. Basic properties of calcium phosphate cement containing atelocollagen in its liquid or powder phases. Biomaterials 1998, 19, 707-715.

[540] Iijima, M., Moriwaki, Y., Kuboki, Y. In vitro crystal growth of octacalcium phosphate on type I collagen fiber. J. Cryst. Growth 1994, 137, 553-560.

[541] Iijima, M., Iijima, K., Moriwaki, Y., Kuboki, Y. Oriented growth of octacalcium phosphate crystals on type I collagen fibrils under physiological conditions. J. Cryst. Growth 1994, 140, 91-99.

[542] Lawson, A. C., Czernuszka, J. T. Collagen – calcium phosphate composites. Proc. Inst. Mech. Eng. H 1998, 212, 413-425.

[543] Du, C., Cui, F. Z., Zhang, W., Feng, Q. L., Zhu, X. D., de Groot, K. Formation of calcium phosphate/collagen composites through mineralization of collagen matrix. J. Biomed. Mater. Res. 2000, 50, 518-527.

[544] Kamakura, S., Sasaki, K., Honda, Y., Anada, T., Suzuki, O. Octacalcium phosphate combined with collagen orthotopically enhances bone regeneration. J. Biomed. Mater. Res. B Appl. Biomater. 2006, 79B, 210-217.

[545] Kawai, T., Anada, T., Honda, Y., Kamakura, S., Matsui, K., Matsui, A., Sasaki, K., Morimoto, S., Echigo, S, Suzuki, O. Synthetic octacalcium phosphate augments bone regeneration correlated with its content in collagen scaffold. Tissue Eng. A 2009, 15, 23-32.

[546] Masuda, T., Kawai, T., Anada, T., Kamakura, S., Suzuki, O. Quality of regenerated bone enhanced by implantation of octacalcium phosphate-collagen composite. Tissue Eng. C 2010, 16, 471-478.

[547] Kikuchi, M., Ikoma, T., Itoh, S., Matsumoto, H. N., Koyama, Y., Takakuda, K., Shinomiya, K., Tanaka, J. Biomimetic synthesis of bone-like nanocomposites using the self-organization mechanism of hydroxyapatite and collagen. Compos. Sci. Technol. 2004, 64, 819-825.

[548] Yunoki, S., Ikoma, T., Monkawal, A., Ohtal, K., Tanaka, J. Preparation and characterization of hydroxyapatite/collagen nanocomposite gel. J. Nanosci. Nanotechnol. 2007, 7, 818-821.

[549] Li, X., Chang, J. Preparation of bone-like apatite-collagen nanocomposites by a biomimetic process with phosphorylated collagen. J. Biomed. Mater. Res. A 2008, 85A, 293-300.

[550] Ficai, A., Andronescu, E., Voicu, G., Ghitulica, C., Vasile, B. S., Ficai, D., Trandafir, V. Self-assembled collagen/hydroxyapatite composite materials. Chem. Eng. J. 2010, 160, 794-800.

[551] Jee, S. S., Thula, T. T., Gower, L. B. Development of bone-like composites via the polymer-induced liquid-precursor (PILP) process. Part 1: Influence of polymer molecular weight. Acta Biomater. 2010, 6, 3676-3686.

[552] Kane, R. J., Weiss-Bilka, H. E., Meagher, M. J., Liu, Y., Gargac, J. A., Niebur, G. L., Wagner, D. R., Roeder, R. K. Hydroxyapatite reinforced collagen scaffolds with improved architecture and mechanical properties. Acta Biomater. 2015, 17, 16-25.

[553] Bu, H., Li, G. Comparative investigation of hydroxyapatite/collagen composites prepared by $CaCl_2$ addition at different time points in collagen self-assembly process. J. Mater. Sci. 2018, 53, 6313-6324.

[554] Bian, T., Zhao, K., Meng, Q., Tang, Y., Jiao, H., Luo, J. The construction and performance of multi-level hierarchical hydroxyapatite (HA)/collagen composite implant based on biomimetic bone Haversian motif. Mater. Des. 2019, 162, 60-69.

[555] Andronescu, E., Ficai, M., Voicu, G., Ficai, D., Maganu, M., Ficai, A. Synthesis and characterization of collagen/hydroxyapatite: magnetite composite material for bone cancer treatment. J. Mater. Sci. Mater. Med. 2010, 21, 2237-2242.

[556] Ficai, M., Andronescu, E., Ficai, D., Voicu, G., Ficai, A. Synthesis and characterization of COLL-PVA/HA hybrid materials with stratified morphology. Colloid Surface B 2010, 81, 614-619.

[557] Chacon, E. L., Bertolo, M. R. V., de Guzzi Plepis, A. M., da Conceição Amaro Martins, V., dos Santos, G. R., Pinto, C. A. L., Pelegrine, A. A., Teixeira, M. L., Buchaim, D. V., Nazari, F. M., Buchaim, R. L., Sugano, G. T., da Cunha, M. R. Collagen-chitosan-hydroxyapatite composite scaffolds for bone repair in ovariectomized rats. Sci. Rep. 2023, 13, 28.

[558] Hoshi, M., Taira, M., Sawada, T., Hachinohe, Y., Hatakeyama, W., Takafuji, K., Tekemoto, S., Kondo, H. Preparation of collagen/hydroxyapatite composites using the alternate immersion method and evaluation of the cranial bone-forming capability of composites complexed with acidic gelatin and b-FGF. Materials 2022, 15, 8802.

[559] Animut, T. Y., Ningsih, H. S., Shih, H. H., Wu, M. H., Shih, S. J. Effect of calcium silicate and β-tricalcium phosphate reinforcement on the mechanical–biological properties of freeze-dried collagen composite scaffolds for bone tissue engineering applications. Ceramics 2023, 6, 548-560.

[560] Inzana, J. A., Olvera, D., Fuller, S. M., Kelly, J. P., Graeve, O. A., Schwarz, E. M., Kates, S. L., Awad, H. A. 3D printing of composite calcium phosphate and collagen scaffolds for bone regeneration. Biomaterials 2014, 35, 4026-4034.

[561] Tamimi, F., Kumarasami, B., Doillon, C., Gbureck, U., Nihouannen, D. L., Cabarcos, E. L., Barralet, J. E. Brushite-collagen composites for bone regeneration. Acta Biomater. 2008, 4, 1315-1321.

[562] Mai, R., Reinstorf, A., Pilling, E., Hlawitschka, M., Jung, R., Gelinsky, M., Schneider, M., Loukota, R., Pompe, W., Eckelt, U., Stadlinger, B. Histologic study of incorporation and resorption of a bone cement-collagen composite: an in vivo study in the minipig. Oral Surg. Oral Med. Oral Pathol. Oral Radiol. Endod. 2008, 105, e9-e14.

[563] Moreau, J. L., Weir, M. D., Xu, H. H. K. Self-setting collagen-calcium phosphate bone cement: mechanical and cellular properties. J. Biomed. Mater. Res. A 2009, 91A, 605-613.

[564] Liu, X., Wang, X. M., Chen, Z., Cui, F. Z., Liu, H. Y., Mao, K., Wang, Y. Injectable bone cement based on mineralized collagen. J. Biomed. Mater. Res. B Appl. Biomater. 2010, 94B, 72-79.

[565] Otsuka, M., Nakagawa, H., Ito, A., Higuchi, W. I. Effect of geometrical structure on drug release rate of a three-dimensionally perforated porous apatite/collagen composite cement. J. Pharm. Sci. 2010, 99, 286-292.

[566] Cui, F. Z., Li, Y., Ge, J. Self-assembly of mineralized collagen composites. Mater. Sci. Eng. R 2007, 57, 1-27.

[567] Hirota, K., Nishihara, K., Tanaka, H. Pressure sintering of apatite-collagen composite. Biomed. Mater. Eng. 1993, 3, 147-151.

[568] Zahn, D., Hochrein, O., Kawska, A., Brickmann, J., Kniep, R. Towards an atomistic understanding of apatite-collagen biomaterials: linking molecular simulation studies of complex-, crystal- and composite-formation to experimental findings. J. Mater. Sci. 2007, 42, 8966-8973.

[569] Silva, C. C., Pinheiro, A. G., Figueiro, S. D., Goes, J. C., Sasaki, J. M., Miranda, M. A. R., Sombra, A. S. B. Piezoelectric properties of collagen-nanocrystalline hydroxyapatite composites. J. Mater. Sci. 2002, 37, 2061-2070.

[570] Yunoki, S., Ikoma, T., Tsuchiya, A., Monkawa, A., Ohta, K., Sotome, S., Shinomiya, K., Tanaka, J. Fabrication and mechanical and tissue ingrowth properties of unidirectionally porous hydroxyapatite/collagen composite. J. Biomed. Mater. Res. B Appl. Biomater. 2007, 80B, 166-173.

[571] Keeney, M., Collin, E., Pandit, A. Multi-channelled collagen-calcium phosphate scaffolds: their physical properties and human cell response. Tissue Eng. C 2009, 15, 265-273.

[572] Chapman, M. W., Bucholz, R., Cornell, C. Treatment of acute fractures with a collagen-calcium phosphate graft material: a randomized clinical trial. J. Bone Joint Surg. (Am.) 1997, 79A, 495-502.

[573] Rodrigues, C. V. M., Serricella, P., Linhares, A. B. R., Guerdes, R. M., Borojevic, R., Rossi, M. A., Duarte, M. E. L., Farina, M. Characterization of a bovine collagen-hydroxyapatite composite scaffold for bone tissue engineering. Biomaterials 2003, 24, 4987-4997.

[574] Miura, K.-I., Anada, T., Honda, Y., Shiwaku, Y., Kawai, T., Echigo, S., Takahashi, T., Suzuki, O. Characterization and bioactivity of nano-submicro octacalcium phosphate/gelatin composite. Appl. Surf. Sci. 2013, 282, 138-145.

[575] Kawai, T., Echigo, S., Matsui, K., Tanuma, Y., Takahashi, T., Suzuki, O., Kamakura, S. First clinical application of octacalcium phosphate collagen composite in human bone defect. Tissue Eng. A 2014, 20, 1336-1341.

[576] Lickorish, D., Ramshaw, J. A. M., Werkmeister, J. A., Glattauer, V., Howlett, C. R. Development of a collagen-hydroxyapatite composite biomaterial via biomimetic process. J. Biomed. Mater. Res. A 2004, 68A, 19-27.

[577] Sionkowska, A., Kozłowska, J. Characterization of collagen/hydroxyapatite composite sponges as a potential bone substitute. Int. J. Biol. Macromol. 2010, 47, 483-487.

[578] Hsu, F. Y., Chueh, S. C., Wang, J. Y. Microspheres of hydroxyapatite/reconstituted collagen as supports for osteoblast cell growth. Biomaterials 1999, 20, 1931-1936.

[579] Wu, T. J., Huang, H. H., Lan, C. W., Lin, C. H., Hsu, F. Y., Wang, Y. J. Studies on the microspheres comprised of reconstituted collagen and hydroxyapatite. Biomaterials 2004, 25, 651-658.

[580] Wei, Q., Lu, J., Wang, Q., Fan, H., Zhang, X. Novel synthesis strategy for composite hydrogel of collagen/hydroxyapatite-microsphere originating from conversion of $CaCO_3$ templates. Nanotechnology 2015, 26, 115605.

[581] Liao, S. S., Watari, F., Uo, M., Ohkawa, S., Tamura, K., Wang, W., Cui, F. Z. The preparation and characteristics of a carbonated hydroxyapatite/collagen composite at room temperature. J. Biomed. Mater. Res. B Appl. Biomater. 2005, 74B, 817-821.

[582] Yokoyama, A., Gelinsky, M., Kawasaki, T., Kohgo, T., König, U., Pompe, W., Watari, F. Biomimetic porous scaffolds with high elasticity made from mineralized collagen – an animal study. J. Biomed. Mater. Res. B Appl. Biomater. 2005, 75B, 464-472.

[583] Zou, C., Weng, W., Deng, X. J., Cheng, K., Liu, X., Du, P., Shen, G., Han, G. Preparation and characterization of porous β-tricalcium phosphate/collagen composites with an integrated structure. Biomaterials 2005, 26, 5276-5284.

[584] https://en.wikipedia.org/wiki/Bioconjugation (accessed in August 2023).

[585] Gotterbarm, T., Richter, W., Jung, M., Berardi-Vilei, S., Mainil-Varlet, P., Yamashita, T., Breusch, S. J. An in vivo study of a growth-factor enhanced, cell free, two-layered collagen-tricalcium phosphate in deep osteochondral defects. Biomaterials 2006, 27, 3387-3395.

[586] Sobczak-Kupiec, A., Drabczyk, A., Florkiewicz, W., Głąb, M., Kudłacik-Kramarczyk, S., Słota, D., Tomala, A., Tyliszczak, B. Review of the applications of biomedical compositions containing hydroxyapatite and collagen modified by bioactive components. Materials 2021, 14, 2096.

[587] Jayaraman, M., Subramanian, M. V. Preparation and characterization of two new composites: collagen-brushite and collagen-octacalcium phosphate. Medical Sci. Monitor 2002, 8, BR481-BR487.

[588] Matsui, K., Matsui, A., Handa, T., Kawai, T., Suzuki, O., Kamakura, S., Echigo, S. Bone regeneration by octacalcium phosphate collagen composites in a dog alveolar cleft model. Int. J. Oral Maxillofac. Surg. 2010, 39, 1218-1225.

[589] Iibuchi, S., Matsui, K., Kawai, T., Sasaki, K., Suzuki, O., Kamakura, S., Echigo, S. Octacalcium phosphate (OCP) collagen composites enhance bone healing in a dog tooth extraction socket model. Int. J. Oral Maxillofac. Surg. 2010, 39, 161-168.

[590] Kawai, T., Matsui, K., Ezoe, Y., Kajii, F., Suzuki, O., Takahashi, T., Kamakura, S. Efficacy of octacalcium phosphate collagen composite for titanium dental implants in dogs. Materials 2018, 11, 229.

[591] Ni, T., Zhu, Y., Hao, L., Chen, Y., Cheng, T. Preparation of photothermal-sensitive PDGF@ZIF-8-PDA@COL/PLGA-TCP composite scaffolds for bone defect repair. Mater. Des. 2022, 217, 110643.

[592] Kołodziejska, B., Kaflak, A., Kolmas, J. Biologically inspired collagen/apatite composite biomaterials for potential use in bone tissue regeneration – a review. Materials 2020, 13, 1748.

[593] Ikeda, H., Yamaza, T., Yoshinari, M., Ohsaki, Y., Ayukawa, Y., Kido, M. A., Inoue, T., Shimono, M., Koyano, K., Tanaka, T. Ultrastructural and immunoelectron microscopic studies of the peri-implant epithelium-implant (Ti-6Al-4V) interface of rat maxilla. J. Periodontol. 2000, 71, 961-973.

[594] Uchida, M., Oyane, A., Kim, H. M., Kokubo, T., Ito, A. Biomimetic coating of laminin-apatite composite on titanium metal and its excellent cell-adhesive properties. Adv. Mater. 2004, 16, 1071-1074.

[595] Oyane, A., Uchida, M., Ito, A. Laminin-apatite composite coating to enhance cell adhesion to ethylene-vinyl alcohol copolymer. J. Biomed. Mater. Res. A 2005, 72A, 168-174.

[596] Oyane, A., Uchida, M., Onuma, K., Ito, A. Spontaneous growth of a laminin-apatite nano-composite in a metastable calcium phosphate solution. Biomaterials 2006, 27, 167-175.

[597] Oyane, A., Tsurushima, H., Ito, A. Highly efficient gene transfer system using a laminin-DNA-apatite composite layer. J. Gene Med. 2010, 12, 194-206.

[598] Oyane, A., Wang, X., Sogo, Y., Ito, A., Tsurushima, H. Calcium phosphate composite layers for surface-mediated gene transfer. Acta Biomater. 2012, 8, 2034-2046.

[599] Yaylaoglu, M. B., Korkusuz, P., Ors, U., Korkusuz, F., Hasirci, V. Development of a calcium phosphate-gelatin composite as a bone substitute and its use in drug release. Biomaterials 1999, 20, 711-719.

[600] Kim, H. W., Knowles, J. C., Kim, H. E. Porous scaffolds of gelatin-hydroxyapatite nanocomposites obtained by biomimetic approach: characterization and antibiotic drug release. J. Biomed. Mater. Res. B Appl. Biomater. 2005, 74B, 686-698.

[601] Hillig, W. B., Choi, Y., Murtha, S., Natravali, N., Ajayan, P. An open-pored gelatin/hydroxyapatite composite as a potential bone substitute. J. Mater. Sci. Mater. Med. 2008, 19, 11-17.

[602] Chang, M. C., Douglas, W. H., Tanaka, J. Organic-inorganic interaction and the growth mechanism of hydroxyapatite crystals in gelatin matrices between 37 and 80°C. J. Mater. Sci. Mater. Med. 2006, 17, 387-396.

[603] Chang, M. C., Douglas, W. H. Cross-linkage of hydroxyapatite/gelatin nanocomposite using imide-based zero-length cross-linker. J. Mater. Sci. Mater. Med. 2007, 18, 2045-2051.

[604] Liu, X., Smith, L. A., Hu, J., Ma, P. X. Biomimetic nanofibrous gelatin/apatite composite scaffolds for bone tissue engineering. Biomaterials 2009, 30, 2252-2258.

[605] Lin, H. R., Yeh, Y. J. Porous alginate/hydroxyapatite composite scaffolds for bone tissue engineering: preparation, characterization and in vitro studies. J. Biomed. Mater. Res. B Appl. Biomater. 2004, 71B, 52-65.

[606] Turco, G., Marsich, E., Bellomo, F., Semeraro, S., Donati, I., Brun, F., Grandolfo, M., Accardo, A., Paoletti, S. Alginate/hydroxyapatite biocomposite for bone ingrowth: a trabecular structure with high and isotropic connectivity. Biomacromolecules 2009, 10, 1575-1583.

[607] Chae, T., Yang, H., Leung, V., Ko, F., Troczynski, T. Novel biomimetic hydroxyapatite/alginate nanocomposite fibrous scaffolds for bone tissue regeneration. J. Mater. Sci. Mater. Med. 2013, 24, 1885-1894.

[608] Li, S., Kan, B., Zhao, K., Ren, T., Lin, B., Wei, J., Chen, T. Preparation of tricalcium phosphate-calcium alginate composite flat sheet membranes and their application for protein release. Polym. Compos. 2015, 36, 1899-1906.

[609] Wulandari, W., Islami, D. M., Wellia, D. V., Emriadi, E., Sisca, V., Jamarun, N. The effect of alginate concentration on crystallinity, morphology, and thermal stability properties of hydroxyapatite/alginate composite. Polymers 2023, 15, 614.

[610] Won, J. E., El-Fiqi, A., Jegal, S. H., Han, C. M., Lee, E. J., Knowles, J. C., Kim, H. W. Gelatin-apatite bone mimetic co-precipitates incorporated within biopolymer matrix to improve mechanical and biological properties useful for hard tissue repair. J. Biomater. Appl. 2014, 28, 1213-1225.

[611] Sarkar, C., Kumari, P., Anuvrat, K., Sahu, S. K., Chakraborty, J., Garai, S. Synthesis and characterization of mechanically strong carboxymethyl cellulose–gelatin–hydroxyapatite nanocomposite for load-bearing orthopedic application. J. Mater. Sci. 2018, 53, 230-246.

[612] Yamaguchi, I., Tokuchi, K., Fukuzaki, H., Koyama, Y., Takakuda, K., Monma, H., Tanaka, J. Preparation and microstructure analysis of chitosan/hydroxyapatite nanocomposites. J. Biomed. Mater. Res. 2001, 55, 20-27.

[613] Tang, X. J., Gui, L., Lü, X. Y. Hard tissue compatibility of natural hydroxyapatite/chitosan composite. Biomed. Mater. 2008, 3, 044115.

[614] Cai, X., Tong, H., Shen, X., Chen, W., Yan, J., Hu, J. Preparation and characterization of homogeneous chitosan-polylactic acid/hydroxyapatite nanocomposite for bone tissue engineering and evaluation of its mechanical properties. Acta Biomater. 2010, 5, 2693-2703.

[615] Ge, H., Zhao, B., Lai, Y., Hu, X., Zhang, D., Hu, K. From crabshell to chitosan-hydroxyapatite composite material via a biomorphic mineralization synthesis method. J. Mater. Sci. Mater. Med. 2010, 21, 1781-1787.

[616] Onoki, T., Nakahira, A., Tago, T., Hasegawa, Y., Kuno, T. Novel low temperature processing techniques for apatite ceramics and chitosan polymer composite bulk materials and its mechanical properties. Appl. Surf. Sci. 2012, 262, 263-266.

[617] Fernández, T., Olave, G., Valencia, C. H., Arce, S., Quinn, J. M. W., Thouas, G. A., Chen, Q. Z. Effects of calcium phosphate/chitosan composite on bone healing in rats: calcium phosphate induces osteon formation. Tissue Eng. A 2014, 20, 1948-1960.

[618] Kucharska, M., Walenko, K., Lewandowska-Szumieł, M., Brynk, T., Jaroszewicz, J., Ciach, T. Chitosan and composite microsphere-based scaffold for bone tissue engineering: evaluation of tricalcium phosphate content influence on physical and biological properties. J. Mater. Sci. Mater. Med. 2015, 26, 143.

[619] Marques, C. F., Olhero, S. M., Torres, P. M. C., Abrantes, J. C. C., Fateixa, S., Nogueira, H. I. S., Ribeiro, I. A. C., Bettencourt, A., Sousa, A., Granja, P. L., Ferreira, J. M. F. Novel sintering-free scaffolds obtained by additive manufacturing for concurrent bone regeneration and drug delivery: proof of concept. Mater. Sci. Eng. C 2019, 94, 426-436.

[620] Ait Said, H., Noukrati, H., Ben Youcef, H., Mahdi, I., Oudadesse, H., Barroug, A. In situ precipitated hydroxyapatite-chitosan composite loaded with ciprofloxacin: formulation, mechanical, in vitro antibiotic uptake, release, and antibacterial properties. Mater. Chem. Phys. 2023, 294, 127008.

[621] Scalera, F., Pereira, S. I. A., Bucciarelli, A., Tobaldi, D. M., Quarta, A., Gervaso, F., Castro P. M. L., Polini, A., Piccirillo, C. Chitosan-hydroxyapatite composites made from sustainable sources: a morphology and antibacterial study. Mater. Today Sustainability 2023, 21, 100334.

[622] Wan, A. C. A., Khor, E., Hastings, G. W. Preparation of a chitin-apatite composite by in situ precipitation onto porous chitin scaffolds. J. Biomed. Mater. Res. 1998, 41, 541-548.

[623] Wan, A. C. A., Khor, E., Hastings, G. W. Hydroxyapatite modified chitin as potential hard tissue substitute material. J. Biomed. Mater. Res. 1997, 38, 235-241.

[624] Geçer, A., Yldz, N., Erol, M., Çalml, A. Synthesis of chitin calcium phosphate composite in different growth media. Polym. Composite 2008, 29, 84-91.

[625] Dong, H., Ye, J. D., Wang, X. P., Yang, J. J. Preparation of calcium phosphate cement tissue engineering scaffold reinforced with chitin fiber. J. Inorg. Mater. 2007, 22, 1007-1010.
[626] Silva, S. S., Duarte, A. R. C., Oliveira, J. M., Mano, J. F., Reis, R. L. Alternative methodology for chitin–hydroxyapatite composites using ionic liquids and supercritical fluid technology. J. Bioact. Compat. Polym. 2013, 28, 481-491.
[627] Wang, J., Sun, Q. Z., Gao, J., Liu, D. M., Meng, X. C., Li, M. Q. Preparation and properties on silk fibers reinforced hydroxyapatite/chitosan composites. Adv. Mater. Res. 2010, 105-106, 557-560.
[628] Zhang, Y., Reddy, V. J., Wong, S. Y., Li, X., Su, B., Ramakrishna, S., Lim, C. T. Enhanced biomineralization in osteoblasts on a novel electrospun biocomposite nanofibrous substrate of hydroxyapatite/collagen/chitosan. Tissue Eng. A 2010, 16, 1949-1960.
[629] Xu, Z., Li, K., Zhou, K., Li, S., Chen, H., Zeng, J., Hu, R. 3D printing silk fibroin/hydroxyapatite/sodium alginate composite scaffolds for bone tissue engineering. Fiber. Polym. 2023, 24, 275-283.
[630] Dehghani, N., Haghiralsadat, F., Yazdian, F., Sadeghian-Nodoushan, F., Ghasemi, N., Mazaheri, F., Pourmadadi, M., Naghib, S. M. Chitosan/silk fibroin/nitrogen-doped carbon quantum dot/α-tricalcium phosphate nanocomposite electrospinned as a scaffold for wound healing application: in vitro and in vivo studies. Int. J. Biol. Macromol. 2023, 238, 124078.
[631] Kousalya, G. N., Gandhi, R. M., Sundaram, S. C., Meenakshi, S. Synthesis of nano-hydroxyapatite chitin/chitosan hybrid biocomposites for the removal of Fe(III). Carbohydr. Polym. 2010, 82, 594-599.
[632] Sundaram, C. S., Viswanathan, N., Meenakshi, S. Uptake of fluoride by nano-hydroxyapatite/chitosan, a bioinorganic composite. Bioresour. Technol. 2008, 99, 8226-8230.
[633] Sundaram, C. S., Viswanathan, N., Meenakshi, S. Fluoride sorption by nano-hydroxyapatite/chitin composite. J. Hazard. Mater. 2009, 172, 147-151.
[634] Salama, A. Recent progress in preparation and applications of chitosan/calcium phosphate composite materials. Int. J. Biol. Macromol, 2021, 178, 240-252.
[635] Wen, H. B., de Wijn, J. R., van Blitterswijk, C. A., de Groot, K. Incorporation of bovine serum albumin in calcium phosphate coating on titanium. J. Biomed. Mater. Res. 1999, 46, 245-252.
[636] Liu, T. Y., Chen, S. Y., Liu, D. M., Liou, S. C. On the study of BSA-loaded calcium-deficient hydroxyapatite nano-carriers for controlled drug delivery. J. Control. Release 2005, 107, 112-121.
[637] Liu, Y., Hunziker, E., Randall, N., de Groot, K., Layrolle, P. Proteins incorporated into biomimetically prepared calcium phosphate coatings modulate their mechanical strength and dissolution rate. Biomaterials 2003, 24, 65-70.
[638] Dorozhkin, S. V., Dorozhkina, E. I. The influence of bovine serum albumin on the crystallization of calcium phosphates from a revised simulated body fluid. Colloid Surface A 2003, 215, 191-199.
[639] Fu, H. H., Hu, Y. H., McNelis, T., Hollinger, J. O. A calcium phosphate-based gene delivery system. J. Biomed. Mater. Res. A 2005, 74A, 40-48.

[640] Bisht, S., Bhakta, G., Mitra, S., Maitra, A. pDNA loaded calcium phosphate nanoparticles: highly efficient non-viral vector for gene delivery. Int. J. Pharm. 2005, 288, 157-168.
[641] Kakizawa, Y., Miyata, K., Furukawa, S., Kataoka, K. Size-controlled formation of a calcium phosphate-based organic-inorganic hybrid vector for gene delivery using poly(ethylene glycol)-block-poly(aspartic acid). Adv. Mater. 2004, 16, 699-702.
[642] Singh, R., Saxena, A., Mozumdar, S. Calcium phosphate – DNA nanocomposites: morphological studies and their bile duct infusion for liver-directed gene therapy. Int. J. Appl. Ceram. Technol. 2008, 5, 1-10.
[643] Oyane, A., Araki, H., Sogo, Y., Ito, A., Tsurushima, H. Coprecipitation of DNA and calcium phosphate using an infusion fluid mixture. Key Eng. Mater. 2013, 529-530, 465-470.
[644] Sporysh, I., Shynkaruk, E., Lysko, O., Shynkaruk, A., Dubok, V., Buzaneva, E., Ritter, U., Scharff, P. Biomimetic hydroxyapatite nanocrystals in composites with C_{60} and Au-DNA nanoparticles: IR-spectral study. Mater. Sci. Eng. B 2010, 169, 128-133.
[645] Taguchi, T., Kishida, A., Akashi, M. Hydroxyapatite formation on/in poly(vinyl alcohol) hydrogel matrices using a novel alternate soaking process. Chem. Lett. 1998, 8, 711-712.
[646] Tachaboonyakiat, W., Serizawa, T., Akashi, M. Hydroxyapatite formation on/in biodegradable chitosan hydrogels by an alternate soaking process. Polym. J. 2001, 33, 177-181.
[647] Schnepp, Z. A. C., Gonzalez-McQuire, R., Mann, S. Hybrid biocomposites based on calcium phosphate mineralization of self-assembled supramolecular hydrogels. Adv. Mater. 2006, 18, 1869-1872.
[648] Patel, M., Patel, K. J., Caccamese, J. F., Coletti, D. P., Sauk, J. J., Fisher, J. P. Characterization of cyclic acetal hydroxyapatite nanocomposites for craniofacial tissue engineering. J. Biomed. Mater. Res. A 2010, 94A, 408-418.
[649] Yu, H., Song, J., Zhang, X., Jiang, K., Fan, H., Li, Y., Zhao, Y., Liu, S., Hao, D., Li, G. Hydroxyapatite-tethered peptide hydrogel promotes osteogenesis. Gels 2022, 8, 804.
[650] Bigi, A., Boanini, E., Gazzano, M., Kojdecki, M. A., Rubini, K. Microstructural investigation of hydroxyapatite-polyelectrolyte composites. J. Mater. Chem. 2004, 14, 274-279.
[651] Bigi, A., Boanini, E., Gazzano, M., Rubini, K., Torricelli, P. Nanocrystalline hydroxyapatite-polyaspartate composites. Biomed. Mater. Eng. 2004, 14, 573-579.
[652] Boanini, E., Fini, M., Gazzano, M., Bigi, A. Hydroxyapatite nanocrystals modified with acidic amino acids. Eur. J. Inorg. Chem. 2006, 4821-4826.
[653] Boanini, E., Torricelli, P., Gazzano, M., Giardino, R., Bigi, A. Nanocomposites of hydroxyapatite with aspartic acid and glutamic acid and their interaction with osteoblast-like cells. Biomaterials 2006, 27, 4428-4433.
[654] Ikawa, N., Kimura, T., Oumi, Y., Sano, T. Amino acid containing amorphous calcium phosphates and the rapid transformation into apatite. J. Mater. Chem. 2009, 19, 4906-4913.

[655] Sánchez-Salcedo, S., Nieto, A., Vallet-Regi, M. Hydroxyapatite/β-tricalcium phosphate/agarose macroporous scaffolds for bone tissue engineering. Chem. Eng. J. 2005, 137, 62-71.

[656] Román, J., Cabañas, M. V., Peña, J., Doadrio, J. C., Vallet-Regi, M. An optimized β-tricalcium phosphate and agarose scaffold fabrication technique. J. Biomed. Mater. Res. A 2008, 84A, 99-107.

[657] Alcaide, M., Serrano, M. C., Pagani, R., Sánchez-Salcedo, S., Nieto, A., Vallet-Regí, M., Portolés, M. T. L929 fibroblast and SAOS-2 osteoblast response to hydroxyapatite-βTCP/ agarose biomaterial. J. Biomed. Mater. Res. A 2009, 89A, 539-549.

[658] Abiraman, S., Varma, H., Umashankar, P., John, A. Fibrin sealant as an osteoinductive protein in a mouse model. Biomaterials 2002, 23, 3023-3031.

[659] Wittkampf, A. Fibrin sealant as sealant for hydroxyapatite granules. J. Craniomaxillofac. Surg. 1989, 17, 179-181.

[660] d'Arc, M. B., Daculsi, G. Micro macroporous biphasic ceramics and fibrin sealant as a mouldable material for bone reconstruction in chronic otitis media surgery: a 15 years experience. J. Mater. Sci. Mater. Med. 2003, 14, 229-233.

[661] le Nihouannen, D., Guehennec, L. L., Rouillon, T., Pilet, P., Bilban, M., Layrolle, P., Daculsi, G. Micro-architecture of calcium phosphate granules and fibrin glue composites for bone tissue engineering. Biomaterials 2006, 27, 2716-2722.

[662] le Nihouannen D., Saffarzadeh A., Aguado E., Goyenvalle E., Gauthier O., Moreau F., Pilet P., Spaethe R., Daculsi, G., Layrolle, P. Osteogenic properties of calcium phosphate ceramics and fibrin glue based composites. J. Mater. Sci. Mater. Med. 2007, 18, 225-235.

[663] le Nihouannen, D., Goyenvalle, E., Aguado, E., Pilet, P., Bilban, M., Daculsi, G., Layrolle, P. Hybrid composites of calcium phosphate granules, fibrin glue, and bone marrow for skeletal repair. J. Biomed. Mater. Res. A 2007, 81A, 399-408.

[664] Yoh, R., Matsumoto, T., Sasaki, J. I., Sohmura, T. Biomimetic fabrication of fibrin/apatite composite material. J. Biomed. Mater. Res. A 2008, 87A, 222-228.

[665] Cui, G., Li, J., Lei, W., Bi, L., Tang, P., Liang, Y., Tao, S., Wang, Y. The mechanical and biological properties of an injectable calcium phosphate cement-fibrin glue composite for bone regeneration. J. Biomed. Mater. Res. B Appl. Biomater. 2010, 92B, 377-385.

[666] Boanini, E., Torricelli, P., Gazzano, M., Giardino, R., Bigi, A. Alendronate-hydroxyapatite nanocomposites and their interaction with osteoclasts and osteoblast-like cells. Biomaterials 2008, 29, 790-796.

[667] Wang, L., Nemoto, R., Senna, M. Microstructure and chemical states of hydroxyapatite/silk fibroin nanocomposites synthesized via a wet-mechanochemical route. J. Nanopart. Res. 2002, 4, 535-540.

[668] Nemoto, R., Wang, L., Ikoma, T., Tanaka, J., Senna, M. Preferential alignment of hydroxyapatite crystallites in nanocomposites with chemically disintegrated silk fibroin. J. Nanopart. Res. 2004, 6, 259-265.

[669] Wang, L., Li, C. Z., Senna, M. High-affinity integration of hydroxyapatite nanoparticles with chemically modified silk fibroin. J. Nanopart. Res. 2007, 9, 919-929.

[670] Li, L., Wei, K. M., Lin, F., Kong, X. D., Yao, J. M. Effect of silicon on the formation of silk fibroin/calcium phosphate composite. J. Mater. Sci. Mater. Med. 2008, 19, 577-582.
[671] Fan, C., Li, J., Xu, G., He, H., Ye, X., Chen, Y., Sheng, X., Fu, J., He, D. Facile fabrication of nano-hydroxyapatite/silk fibroin composite via a simplified coprecipitation route. J. Mater. Sci. 2010, 45, 5814-5819.
[672] Liu, H., Xu, G. W., Wang, Y. F., Zhao, H. S., Xiong, S., Wu, Y., Heng, B. C., An, C. R., Zhu, G. H., Xie, D. H. Composite scaffolds of nano-hydroxyapatite and silk fibroin enhance mesenchymal stem cell-based bone regeneration via the interleukin 1 alpha autocrine/paracrine signaling loop. Biomaterials 2015, 49, 103-112.
[673] Farokhi, M., Mottaghitalab, F., Samani, S., Shokrgozar, M. A., Kundu, S. C., Reis, R. L., Fatahi, Y., Kaplan, D. L. Silk fibroin/hydroxyapatite composites for bone tissue engineering. Biotechnol. Adv. 2018, 36, 68-91.
[674] Milazzo, M., Fitzpatrick, V., Owens, C. E., Carraretto, I. M., McKinley, G. H., Kaplan, D. L., Buehler, M. J. 3D printability of silk/hydroxyapatite composites for microprosthetic applications. ACS Biomater. Sci. Eng. 2023, 9, 1285-1295.
[675] Salama, A., Neumann, M., Günter, C., Taubert, A. Ionic liquid-assisted formation of cellulose/calcium phosphate hybrid materials. Beilstein J. Nanotechnol. 2014, 5, 1553-1568.
[676] Wang, L., Li, C. Z. Preparation and physicochemical properties of a novel hydroxyapatite/chitosan-silk fibroin composite. Carbohydr. Polym. 2007, 68, 740-745.
[677] Shahbazarab, Z., Teimouri, A., Chermahini, A. N., Azadi, M. Fabrication and characterization of nanobiocomposite scaffold of zein/chitosan/nanohydroxyapatite prepared by freeze-drying method for bone tissue engineering. Int. J. Biol. Macromol. 2018, 108, 1017-1027.
[678] Oliveira, J. M., Costa, S. A., Leonor, I. B., Malafaya, P. B., Mano, J. F., Reis, R. L. Novel hydroxyapatite/carboxymethylchitosan composite scaffolds prepared through an innovative "autocatalytic" electroless coprecipitation route. J. Biomed. Mater. Res. A 2009, 88, 470-480.
[679] Sogo, Y., Ito, A., Matsuno, T., Oyane, A., Tamazawa, G., Satoh, T., Yamazaki, A., Uchimura, E., Ohno, T. Fibronectin-calcium phosphate composite layer on hydroxyapatite to enhance adhesion, cell spread and osteogenic differentiation of human mesenchymal stem cells in vitro. Biomed. Mater. 2007, 2, 116-123.
[680] Lin, J., Fan, Y., Hutchinson, D. J., Malkoch, M. Soft hydroxyapatite composites based on triazine-trione systems as potential biomedical engineering frameworks. ACS Appl. Mater. Interfaces 2023, 15, 7329-7339.
[681] Rhee, S. H., Suetsugu, Y., Tanaka, J. Biomimetic configurational arrays of hydroxyapatite nanocrystals on bio-organics. Biomaterials 2001, 22, 2843-2847.
[682] Cross, K. J., Huq, N. L., Palamara, J. E., Perich, J. W., Reynolds, E. C. Physicochemical characterization of casein phosphopeptide-amorphous calcium phosphate nanocomplexes. J. Biol. Chem. 2005, 280, 15362-15369.
[683] Dimopoulou, M., Ritzoulis, C., Papastergiadis, E. S., Panayiotou, C. Composite materials based on okra hydrocolloids and hydroxyapatite. Food Hydrocolloid 2014, 42, 348-354.

[684] Nakata, R., Tachibana, A., Tanabe, T. Preparation of keratin hydrogel/hydroxyapatite composite and its evaluation as a controlled drug release carrier. Mater. Sci. Eng. C 2014, 41, 59-64.

[685] Manda, M. G., da Silva, L. P., Cerqueira, M. T., Pereira, D. R., Oliveira, M. B., Mano, J. F., Marques, A. P., Oliveira, J. M., Correlo, V. M., Reis, R. L. Gellan gum-hydroxyapatite composite spongy-like hydrogels for bone tissue engineering. J. Biomed. Mater. Res. A 2018, 106A, 479-490.

[686] Yu, P., Bao, R. Y., Shi, X. J., Yang, W., Yang, M. B. Self-assembled high-strength hydroxyapatite/graphene oxide/chitosan composite hydrogel for bone tissue engineering. Carbohyd. Polym. 2017, 155, 507-515.

[687] Li, C., Born, A. K., Schweizer, T., Zenobi-Wong, M., Cerruti, M., Mezzenga, R. Amyloid-hydroxyapatite bone biomimetic composites. Adv. Mater. 2014, 26, 3207-3212.

[688] Kolanthai, E., Colon, V. S. D., Sindu, P. A., Chandra, V. S., Karthikeyan, K. R., Babu, M. S., Sundaram, S. M., Palanichamy, M., Kalkura, S. N. Effect of solvent; enhancing the wettability and engineering the porous structure of a calcium phosphate/agarose composite for drug delivery. RSC Adv. 2015, 5, 18301-18311.

[689] Krukowski, S., Lysenko, N., Kolodziejski, W. Synthesis and characterization of nanocrystalline composites containing calcium hydroxyapatite and glycine. J. Solid State Chem. 2018, 264, 59-67.

[690] Jung, J. Y., Hong, Y. J., Choi, Y. S., Jeong, S., Lee, W. K. A new method for the preparation of bioactive calcium phosphate films hybridized with 1α,25-dihydroxyvitamin D_3. J. Mater. Sci. Mater. Med. 2010, 20, 2441-2453.

[691] Killion, J. A., Geever, L. M., Devine, D. M., Higginbotham, C. L. Fabrication and in vitro biological evaluation of photopolymerisable hydroxyapatite hydrogel composites for bone regeneration. J. Biomater. Appl. 2014, 28, 1274-1283.

[692] Shchukin, D. G., Sukhorukov, G. B., Möhwald, H. Biomimetic fabrication of nanoengineered hydroxyapatite/polyelectrolyte composite shell. Chem. Mater. 2003, 15, 3947-3950.

[693] Busch, S., Dolhaine, H., DuChesne, A., Heinz, S., Hochrein, O., Laeri, F., Podebrad, O., Vietze, U., Weiland, T., Kniep R. Biomimetic morphogenesis of fluorapatite-gelatin composites: fractal growth, the question of intrinsic electric fields, core/shell assemblies, hollow spheres and reorganization of denatured collagen. Eur. J. Inorg. Chem. 1999, 1999, 1643-1653.

[694] Simon, P., Schwarz, U., Kniep, R. Hierarchical architecture and real structure in a biomimetic nano-composite of fluorapatite with gelatine: a model system for steps in dentino- and osteogenesis? J. Mater. Chem. 2005, 15, 4992-4996.

[695] Tlatlik, H., Simon, P., Kawska, A., Zahn, D., Kniep, R. Biomimetic fluorapatite-gelatin nanocomposites: pre-structuring of gelatin matrices by ion impregnation and its effect on form development. Angew. Chem. Int. Ed. Engl. 2006, 45, 1905-1910.

[696] Simon, P., Zahn, D., Lichte, H., Kniep, R. Intrinsic electric dipole fields and the induction of hierarchical form developments in fluorapatite-gelatin nanocomposites: a general principle for morphogenesis of biominerals? Angew. Chem. Int. Ed. Engl. 2006, 45, 1911-1915.

[697] Kniep, R., Simon, P. "Hidden" hierarchy of microfibrils within 3D-periodic fluorapatite-gelatin nanocomposites: development of complexity and form in a biomimetic system. Angew. Chem. Int. Ed. Engl. 2008, 47, 1405-1409.

[698] Brickmann, J., Paparcone, R., Kokolakis, S., Zahn, D., Duchstein, P., Carrillocabrera, W., Simon, P., Kniep, R. Fluorapatite-gelatine nanocomposite superstructures: new insights into a biomimetic system of high complexity. Chem Phys Chem 2010, 11, 1851-1853.

[699] Vyalikh, A., Simon, P., Rosseeva, E., Buder, J., Kniep, R., Scheler, U. Intergrowth and interfacial structure of biomimetic fluorapatite-gelatin nanocomposite: a solid-state NMR study. J. Phys. Chem. B 2014, 118, 724-730.

[700] Gashti, M. P., Stir, M., Hulliger, J. Synthesis of bone-like micro-porous calcium phosphate/iota-carrageenan composites by gel diffusion. Colloid Surface B 2013, 110, 426-433.

[701] Jacoveila, P. F. Use of calcium hydroxylapatite (Radiesse®) for facial augmentation. Clin. Interv. Aging 2008, 3, 161-174.

[702] Lizzul, P. F., Narurkar, V. A. The role of calcium hydroxylapatite (Radiesse®) in nonsurgical aesthetic rejuvenation. J. Drugs Dermatol. 2010, 9, 446-450.

[703] Klesing, J., Chernousova, S., Kovtun, A., Neumann, S., Ruiz, L., Gonzalez-Calbet, J. M., Vallet-Regi, M., Heumann, R., Epple, M. An injectable paste of calcium phosphate nanorods, functionalized with nucleic acids, for cell transfection and gene silencing. J. Mater. Chem. 2010, 20, 6144-6148.

[704] Thai, V. V., Lee, B. T. Fabrication of calcium phosphate-calcium sulfate injectable bone substitute using hydroxy-propyl-methyl-cellulose and citric acid. J. Mater. Sci. Mater. Med. 2010, 21, 1867-1874.

[705] Low, K. L., Tan, S. H., Zein, S. H. S., Roether, J. A., Mouriño, V., Boccaccini, A. R. Calcium phosphate-based composites as injectable bone substitute materials. J. Biomed. Mater. Res. B Appl. Biomater. 2010, 94B, 273-286.

[706] D'Este, M., Eglin, D. Hydrogels in calcium phosphate moldable and injectable bone substitutes: sticky excipients or advanced 3-D carriers? Acta Biomater. 2013, 9, 5421-5430.

[707] Weiss, P., Gauthier, O., Bouler, J. M., Grimandi, G., Daculsi, G. Injectable bone substitute using a hydrophilic polymer. Bone 1999, 25, Suppl. 2, 67S-70S.

[708] Daculsi, G., Weiss, P., Bouler, J. M., Gauthier, O., Millot, F., Aguado, E. Biphasic calcium phosphate/hydrosoluble polymer composites: a new concept for bone and dental substitution biomaterials. Bone 1999, 25, Suppl. 2, 59S-61S.

[709] Turczyn, R., Weiss, P., Lapkowski, M., Daculsi, G. In situ self-hardening bioactive composite for bone and dental surgery. J. Biomater. Sci. Polym. Edn. 2000, 11, 217-223.

[710] Bennett, S., Connolly, K., Lee, D. R., Jiang, Y., Buck, D., Hollinger, J. O., Gruskin, E. A. Initial biocompatibility studies of a novel degradable polymeric bone substitute that hardens in situ. Bone 1996, 19, 101S-107S.

[711] Bongio, M., van den Beucken, J. J. J. P., Nejadnik, M. R., Leeuwenburgh, S. C. G., Kinard, L. A., Kasper, F. K., Mikos, A. G., Jansen, J. A. Biomimetic modification of synthetic hydrogels by incorporation of adhesive peptides and calcium

phosphate nanoparticles: in vitro evaluation of cell behavior. Eur. Cell. Mater. 2011, 22, 359-376.

[712] Bongio, M., van den Beucken, J. J. J. P., Nejadnik, M. R., Birgani, Z. T., Habibovic, P., Kinard, L. A., Kasper, F. K., Mikos, A. G., Leeuwenburgh, S. C. G., Jansen, J. A. Subcutaneous tissue response and osteogenic performance of calcium phosphate nanoparticle-enriched hydrogels in the tibial medullary cavity of guinea pigs. Acta Biomater. 2013, 9, 5464-5474.

[713] Chernousova, S., Klesing, J., Soklakova, N., Epple, M. A genetically active nano-calcium phosphate paste for bone substitution, encoding the formation of BMP-7 and VEGF-A. RSC Adv. 2013, 3, 11155-11161.

[714] Yu, B., Zhang, Y., Li, X., Wang, Q., Ouyang, Y., Xia, Y., Lin, B., Li, S., Fan, Y., Chen, Y. The use of injectable chitosan/nanohydroxyapatite/collagen composites with bone marrow mesenchymal stem cells to promote ectopic bone formation in vivo. J. Nanomater. 2013, 2013, art. no. 506593.

[715] Bodakhe, S., Verma, S., Garkhal, K., Samal, S. K., Sharma, S. S., Kumar, N. Injectable photocrosslinkable nanocomposite based on poly(glycerol sebacate) fumarate and hydroxyapatite: development, biocompatibility and bone regeneration in a rat calvarial bone defect model. Nanomedicine 2013, 8, 1777-1795.

[716] Fricain, J. C., Aid, R., Lanouar, S., Maurel, D. B., le Nihouannen, D., Delmond, S., Letourneur, D., Vilamitjana, J. A., Catros, S. In-vitro and in-vivo design and validation of an injectable polysaccharide-hydroxyapatite composite material for sinus floor augmentation. Dent. Mater. 2018, 34, 1024-1035.

[717] Lin, G., Cosimbescu, L., Karin, N. J., Tarasevich, B. J. Injectable and thermosensitive PLGA-g-PEG hydrogels containing hydroxyapatite: preparation, characterization and in vitro release behavior. Biomed. Mater. 2012, 7, 024107.

[718] Nejadnik, M. R., Yang, X., Bongio, M., Alghamdi, H. S., van den Beucken, J. J. J. P., Huysmans, M. C., Jansen, J. A., Hilborn, J., Ossipov, D., Leeuwenburgh, S. C. G. Self-healing hybrid nanocomposites consisting of bisphosphonated hyaluronan and calcium phosphate nanoparticles. Biomaterials 2014, 35, 6918-6929.

[719] Munarin, F., Petrini, P., Gentilini, R., Pillai, R. S., Dirè, S., Tanzi, M. C., Sglavo, V. M. Micro- and nano-hydroxyapatite as active reinforcement for soft biocomposites. Int. J. Biol. Macromol. 2015, 72, 199-209.

[720] Douglas, T. E. L., Schietse, J., Zima, A., Gorodzha, S., Parakhonskiy, B. V., Khalenkow, D., Shkarin, R., Ivanova, A., Baumbach, T., Weinhardt, V., Stevens, C. V., Vanhoorne, V., Vervaet, C., Balcaen, L., Vanhaecke, F., Slośarczyk, A., Surmeneva, M. A., Surmenev, R. A., Skirtach, A. G. Novel self-gelling injectable hydrogel/alpha-tricalcium phosphate composites for bone regeneration: physiochemical and microcomputer tomographical characterization. J. Biomed. Mater. Res. A 2018, 106A, 822-828.

[721] Daculsi, G., Rohanizadeh, R., Weiss, P., Bouler, J. M. Crystal polymer interaction with new injectable bone substitute: SEM and HrTEM study. J. Biomed. Mater. Res. 2000, 50, 1-7.

[722] Grimande, G., Weiss, P., Millot, F., Daculsi, G. In vitro evaluation of a new injectable calcium phosphate material. J. Biomed. Mater. Res. 1998, 39, 660-666.

[723] Weiss, P., Lapkowski, M., LeGeros, R. Z., Bouler, J. M., Jean, A., Daculsi, G. FTIR spectroscopic study of an organic/mineral composite for bone and dental substitute materials. J. Mater. Sci. Mater. Med. 1997, 8, 621-629.
[724] Weiss, P., Bohic, S., Lapkowski, M., Daculsi, G. Application of FTIR microspectroscopy to the study of an injectable composite for bone and dental surgery. J. Biomed. Mater. Res. 1998, 41, 167-170.
[725] Schmitt, M., Weiss, P., Bourges, X., del Valle, G. A., Daculsi, G. Crystallization at the polymer/calcium-phosphate interface in a sterilized injectable bone substitute IBS. Biomaterials 2002, 23, 2789-2794.
[726] Gauthier, O., Müller, R., von Stechow, D., Lamy, B., Weiss, P., Bouler, J. M., Aguado, E., Daculsi, G. In vivo bone regeneration with injectable calcium phosphate biomaterial: a three-dimensional micro-computed tomographic, biomechanical and SEM study. Biomaterials 2005, 26, 5444-5453.
[727] Weiss, P., Layrolle, P., Clergeau, L. P., Enckel, B., Pilet, P., Amouriq, Y., Daculsi, G., Giumelli, B. The safety and efficacy of an injectable bone substitute in dental sockets demonstrated in a human clinical trial. Biomaterials 2007, 28, 3295-3305.
[728] Fatimi, A., Tassin, J. F., Axelos, M. A. V., Weiss, P. The stability mechanisms of an injectable calcium phosphate ceramic suspension. J. Mater. Sci. Mater. Med. 2010, 21, 1799-1809.
[729] Trojani, C., Boukhechba, F., Scimeca, J. C., Vandenbos, F., Michiels, J. F., Daculsi, G., Boileau, P., Weiss, P., Carle, G. F., Rochet, N. Ectopic bone formation using an injectable biphasic calcium phosphate/Si-HPMC hydrogel composite loaded with undifferentiated bone marrow stromal cells. Biomaterials 2006, 27, 3256-3264.
[730] Zhang, S. M., Lü, G. Clinical application of compound injectable bone substitutes in bone injury repair. J. Clin. Rehabil. Tissue Eng. Res. 2009, 13, 10117-10120.
[731] Daculsi, G., Uzel, P. A., Bourgeois, N., le François, T., Rouvillain, J. L., Bourges, X., Baroth, S. New injectable bone substitute using reversible thermosensitive hydrogel and BCP granules: in vivo rabbit experiments. Key Eng. Mater. 2009, 396-398, 457-460.
[732] Iooss, P., le Ray, A. M., Grimandi, G., Daculsi, G., Merle, C. A new injectable bone substitute combining poly(ε-caprolactone) microparticles with biphasic calcium phosphate granules. Biomaterials 2001, 22, 2785-2794.
[733] Bohner, M. Design of ceramic-based cements and putties for bone graft substitution. Eur. Cell Mater. 2010, 20, 1-12.
[734] Meyers, M. A., Lin, A. Y. M., Seki, Y., Chen, P. Y., Kad, B. K., Bodde, S. Structural biological composites: an overview. JOM 2006, 58, 36-43.
[735] Polo-Corrales, L., Latorre-Esteves, M., Ramirez-Vick, J. E. Scaffold design for bone regeneration. J. Nanosci. Nanotechnol. 2014, 14, 15-56.
[736] Uskoković, V. When 1 + 1 > 2: Nanostructured composites for hard tissue engineering applications. Mater. Sci. Eng. C 2015, 57, 434-451.
[737] Mo, X., Zhang, D., Liu, K., Zhao, X., Li, X., Wang, W. Nano-hydroxyapatite composite scaffolds loaded with bioactive factors and drugs for bone tissue engineering. Int. J. Mol. Sci. 2023, 24, 1291.

[738] Shi, J., Dai, W., Gupta, A., Zhang, B., Wu, Z., Zhang, Y., Pan, L., Wang, L. Frontiers of hydroxyapatite composites in bionic bone tissue engineering. Materials 2022, 15, 8475.

Chapter 2

Boron-Substituted Biomaterials

Nader Abuhamed[1,*]
Saynur Arslan
and Kerim Emre Öksüz
Sivas Cumhuriyet University, Department of Metallurgical and Materials Engineering, Sivas, Türkey

Abstract

Biomaterial scientists are interested in boron as a dopant element because of its natural functions in human hard tissues. However, boron does not exist in its elemental form in nature. Instead, it reacts with oxygen to form boric acid (H_3BO_3). Boron occurs naturally as borates, including orthoborate BO_3^{3-}, metaborate BO_2^{-}, or tetraborate $B_4O_7^{2-}$, or as any salt containing these anions, such as sodium metaborate, $Na^+[BO_2]^-$, and borax $(Na^+)_2[B_4O_7]^{2-}$. Boron exerts a range of biological effects, including impacts on reproduction and embryogenesis, promoting wound healing and enhancing responses to injury or infection, influencing calcium and bone metabolism, and benefiting central nervous function. It also affects the presence or function of various hormones, such as thyroid hormone, insulin, estrogen, and progesterone, as well as calcium, magnesium, and vitamin D, which are involved in hard tissue metabolism. Research conducted at the molecular level has demonstrated that boron can increase RNA transcription within placental nuclei and activate mRNA translation, particularly for growth factors involved in wound repair and the formation of new blood vessels (angiogenesis). This chapter will provide a summary of the recent advances in the

* Corresponding Author's Email: abuhamednader6@gmail.com.

In: Biocomposites
Editor: James R. Bush
ISBN: 979-8-89113-267-2

development of boron biomaterials that respond to specific biomedical applications.

Keywords: biomedical applications, boron, chronic wounds, ion-substitution, responsive biomaterials

Introduction

Bioceramic materials for biomedical applications are utilized to restore and revitalize compromised, fractured, or infected segments of the musculoskeletal system within the human body. The roster of bioceramics encompasses aluminum oxide (Al_2O_3), silica (SiO_2), glass ceramics, bioactive bioglasses such as 45S5, S53P4, 13-93B1, 13-93B3, $Li_2O \times Al_2O_3 \times nSiO_2$ (LAS glass system), $MgO \times Al_2O_3 \times nSiO_2$ (MAS glass system), and the $ZnO \times Al_2O_3 \times nSiO_2$ (ZAS glass system) (Öksüz et al. 2023; Boccaccini, Brauer, and Hupa 2016), titanium dioxide (TiO_2), zirconia (ZrO_2) and calcium phosphates (Ca_3PO_4) (Hench 1991). Thanks to their favorable interactions with human tissue, these bioceramics find versatile employment in various medical contexts. They serve as substitutes for teeth, knees, hips, ligaments, and tendons, support spinal fusion, rectify periodontal disease, reconstruct maxillofacial structures, enhance and stabilize jawbones, temporarily fill bone voids post tumor removal, and furnish thrombo-resistant carbon coatings for prosthetic heart valves. This extensive range of applications is well-documented in studies by some researchers in literature (Dearnaley and Arps 2005; Eslami et al. 2018; Simsir, Akkan, and Öksüz 2021; Öksüz 2019). Bioceramics that have the potential to elicit specific tissue responses when interacting with the body's physiological system are classified as bioactive. For instance, calcium phosphate-based materials in surgical applications are prominent examples of bioactive ceramics (Goksen Tosun et al. 2023; Özer and Öksüz 2019). Bioactive glass possesses inherent osteogenic capabilities and can facilitate the regeneration of damaged or missing bone tissue. Based on their solubility, bioactive ceramics can be further categorized as either resorbable biomaterial or nonresorbable biomaterial. For instance, hydroxyapatite (HA), when used as a bone filling material, promotes the healing of bony defects by forming new bone as it gradually dissolves over time (Pina and Ferreira 2012). The solubility of HA is influenced by its level of porosity. Tricalcium phosphates (TCPs), utilized as bone fillers in either powdered or compacted form, are highly resorbable bioceramics.

Additionally, materials of animal origin like mother of pearl and coral, primarily composed of aragonite calcium carbonates, exhibit considerable resorbability (Öksüz and Şereflişan 2022; Öksüz and Şereflişan 2023).

On the contrary, the category of non-resorbable ceramics encompasses notable examples such as metal oxides, including ZrO_2 and Al_2O_3. These ceramic materials find utilization that extends beyond serving as mere bone fillers; they also play a pivotal role in the realm of dental implants, contributing to the advancement of restorative dentistry and implantology (Siddiqi, Khan, and Zafar 2017; Öksüz and Özer 2017; Öksüz, Kilinç, and Özer 2019). Formun Üstü An illustration showing the various applications of medical bioceramic implants in the human body is given in Figure 1.

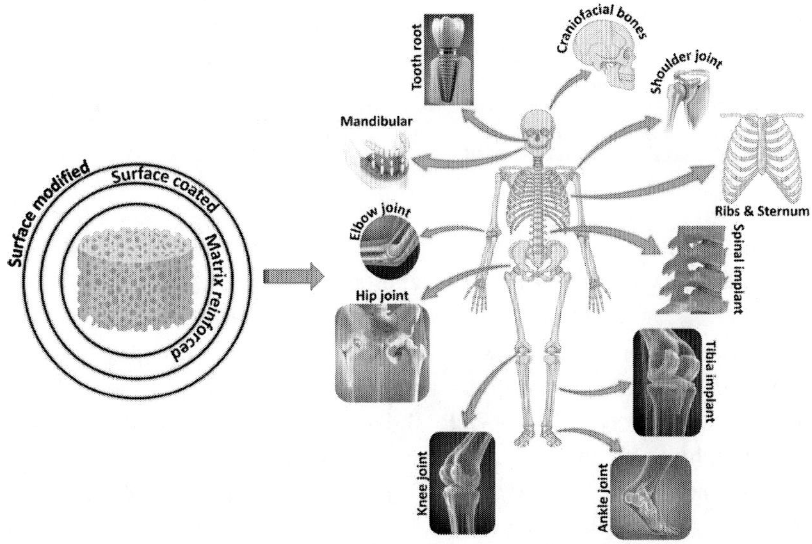

Figure 1. Visual representation showcasing diverse applications of medical bioceramic implants within the human body. Implants and orthopedic application devices play a crucial role in enhancing the quality of human life by seamlessly integrating cutting-edge medical technology with the intricate biomechanics of the human body. These remarkable innovations, meticulously designed and precision-engineered, serve as bridges between medical science and the human form, offering myriad solutions that span from regaining mobility and functionality to fostering accelerated healing processes.

The conceptualization an advancement of sophisticated bioceramic biomaterials tailored for specific applications are made attainable through manipulating ion exchange within their framework. Among the array of dopant elements available, boron emerges as a compelling focus for biomaterial researchers, forced by its inherent roles within the human hard tissues. This chapter examines recent investigations concerning the integration of boron into the architecture of bioceramic materials, with a primary emphasis on their consequential impact within biological contexts. An exploration of the existing body of research suggests that the current exploration of boron incorporation into the matrix of bioceramics displays notable potential in augmenting the performance attributes of these biomaterials. The studies conducted thus far in the scientific literature indicate promising developments in enhancing these biomaterials' functional properties and efficacy within living organisms. This book chapter overviews these research endeavors, shedding light on the evolving landscape of boron-doped bioceramics and their intricate interplay with biological systems.

Synthesis and Characterization Techniques for Boron-Substituted Bioceramics

Bioceramics has witnessed significant advancements over the years, with boron-substituted bioceramics emerging as a pivotal area of interest. These bioceramics, enriched with boron, have shown promising potential in various biomedical applications. However, the synthesis and characterization of these materials remain crucial for understanding their properties and ensuring their efficacy. This section delves into the methods employed in synthesizing boron-substituted bioceramics, the techniques used for their characterization, and the challenges faced in these areas. By exploring these aspects, a comprehensive overview of the current state of boron-substituted bioceramics is provided.

Synthesis Techniques for Boron-Substituted Bioceramics

Boron-substituted bioceramics have distinguished themselves as an important part of the broader bioceramics landscape. Their potential covers a broad range of biomedical applications, emphasizing the significance of their synthesis and characterization. Understanding these materials is critical for appreciating their distinct properties and ensuring optimal application in various biomedical settings. This section explores the various synthesis methodologies, the techniques used to characterize them, and the challenges researchers face. Doing so provides a comprehensive overview of the current state of boron-substituted bioceramics and the promising avenues for future research.

Among the synthesis methods explored for boron-substituted bioceramics, techniques such as sol-gel, melt-and-quench, etc., have been prominently employed. The sol-gel approach captures significant attention owing to its promising implications in biomedical engineering. The importance of boron in human health, particularly its influence on hard tissues, cannot be overstated. By integrating boron into bioceramics like HA, calcium phosphates, and bioglasses, there's potential to significantly enhance their role in promoting vascularization and facilitating bone regeneration. Recent studies have looked into the potential of boron-enriched sol-gel-derived mesoporous bioactive glass nanoparticles, particularly in tissue applications. When loaded with specific medicinal compounds and bioactive ions, these nanoparticles offer options for customization, allowing researchers to fine-tune their biological features to meet specific needs. Furthermore, boron-enhanced bioactive glasses have showed exceptional bioactivity. Their ability to promote bone formation and vascular development has been particularly notable, indicating a viable path for future study and applications (Ege et al. 2022).

The sol-gel synthesis process was used to study boron-substituted cordierite glass ceramics, and the results were interesting. Using low-cost inorganic salts as precursors, researchers observed that substituting Boron trioxide (B_2O_3) for Al_2O_3 significantly enhances the sintering capability of amorphous cordierite powder. This substitution further aids in promoting the crystallization of α-cordierite. A notable outcome of this boron substitution is the improved near-infrared spectral emissivity of the glass ceramics. However, challenges persist, especially concerning the control of boron content, given B_2O_3 propensity to volatilize (Wang and Kuang 2010).

Further research into the sol-gel process for creating bioactive glasses has revealed the effect of boron substitution on calcium. Furthermore, boron increases the crystallinity and bioactivity of these glasses. Because of the harmonious balance of calcium and boron, a sample containing 10% boron demonstrated maximum bioactivity. However, it is important to note that boron level that exceeds 10 mol% affects the biocompatibility of these glasses. For example, researchers discovered that a sample with 15 mol% boron had the highest toxicity levels. When their composition is improved, these glasses can be used in bone tissue engineering (Seyedmomeni et al. 2018). Another study used the sol-gel approach to coat carbon fiber with boron-doped yttria-stabilized ZrO_2 to improve its oxidation resistance, particularly for high-temperature applications. Compared to standard approaches, the boron-doped system exhibited a cubic phase, providing greater oxidation protection. However, the process was not without its difficulties. Obtaining the appropriate coating thickness and ensuring uniformity proved difficult, with parameters such as the sol's Zr content and drying conditions playing a key role. Despite these challenges, the study highlighted this coating methodology's potential to improve the thermal stability of carbon fiber (Ratnayake et al. 2018).

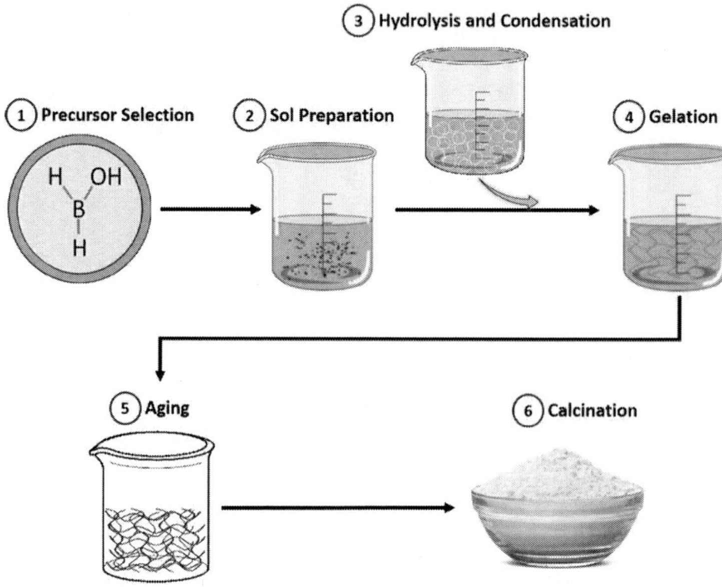

Figure 2. Schematic representation of the sol-gel method employed in the synthesis of boron-substituted biomaterials. The process involves precursor solution preparation, gel formation, and subsequent heat treatment to achieve controlled boron incorporation. The resulting biomaterials exhibit enhanced properties suitable for diverse biomedical applications.

Bioactive glasses, especially those containing borate, share similarities with their silicate-based counterparts regarding synthesis techniques. The melt-and-quench method, as seen in Figure 2, is a commonly employed technique for their synthesis. An exciting aspect of these glasses is their lower melting temperature than standard silicate-based bioactive glasses. B_2O_3

lowers the melting point, which serves as a flux during the melting phase. The integration of B_2O_3 generally bolsters the glass's formability and stability, making it more amenable to being transformed into fibers and coatings. This adaptability contrasts with the difficulties encountered by typical silicate-based bioactive glasses. The sol-gel approach, which includes borate and borosilicate variations, has also gained interest in manufacturing bioactive glasses. In terms of bioactivity, sol-gel-produced glasses frequently outperform melt-and-quench-produced glasses. This increased bioactivity is mostly due to the sol-gel-derived glasses higher surface area and porosity, resulting in faster ion release and degradation rates. Notably, bioactive glasses with a high SiO_2 concentration, normally linked with the melt-and-quench method, have shown improved bioactivity when treated using the sol-gel technique. While the sol-gel method does come with its set of challenges, such as the requirement for specific metal cation precursors, these limitations have not deterred its widespread application, especially in the context of borate bioactive glasses (Naseri et al. 2019; Lepry, Naseri, and Nazhat 2017). A notable example of the sol-gel methods efficacy is evident in a study that developed a soda-lime phosphorate bioactive glass with a distinct composition. Remarkably, this glass was converted to hydroxyl carbonate apatite in under 3h, showing better bioactivity than its counterpart produced using the melt-and-quench method (Lepry and Nazhat 2015).

A study investigated the synthesis of bioceramics, with a particular focus on a Ti-containing borate glass series based on a specific system. The melt-quenching technique was used to create this glass. After that, the mixture was ball-milled and sieved to obtain glass particles of 20 μm. A polymer foam replication approach was used to create the scaffold. Polyurethane foams with linked macroporosity were immersed in an ethanol-dispersed slurry of borate glass powder. Following drying, these coated foams were subjected to high temperatures at a particular heating rate and duration. One of the approach's major advantages is the ability to modulate the degradation rates and ion scaffolds lease parameters of the scaffolds by changing the glass system. These ions have been shown to improve angiogenesis and bone production. In addition, certain scaffold modifications demonstrated antibacterial efficacy against Gram-positive and Gram-negative bacterial strains. Nonetheless, the study had some shortcomings. The lack of research into the influence of surface roughness on cell-to-surface interactions, which might influence cell adhesion, was a severe limitation (Shafaghi et al. 2021).

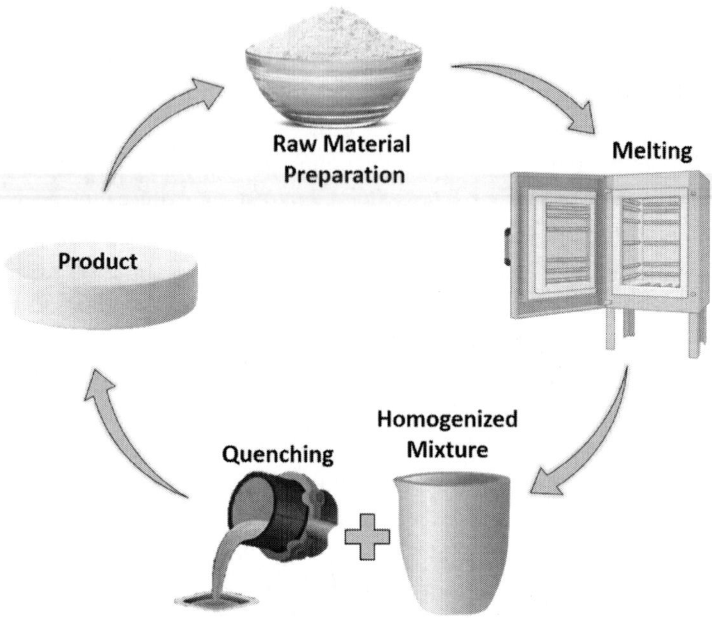

Figure 3. Schematic representation of the melt-quenching technique used for the fabrication of boron-substituted biomaterials. The method involves melting the constituents and rapid quenching. The resulting biomaterials display tailored properties suitable for a range of biomedical uses.

Furthermore, recent studies comparing these two approaches have demonstrated that the synthesis technique used substantially impacts the resultant texture and structure of the glass. As the B_2O_3 content of silicate glass increases, the melting point decreases noticeably while thermal stability improves (Gharbi, El Feki, and Oudadesse 2016). The melt-quenching method necessitates the use of covered crucibles. This precautionary measure is essential to mitigate the evaporation of specific components, including alkalis, boron, phosphorus, and fluorides (Gharbi, El Feki, and Oudadesse 2016; Hench 2013). On the other hand, glasses created using the sol-gel process have various attributes such as a more polymerized structure, increased porosity, and higher specific surface area values. These properties contribute to greater solubility. The sol-gel technique offers several advantages, making it a preferred choice for various applications. Among the benefits are low-temperature processing, better purity, homogenous component distribution, and the capacity to modify particle size likewise, the improved porosity and specific surface area values of sol-gel-derived glasses are significant.

Improving these qualities in bioactive glasses can speed up their decomposition. A study aimed at developing a biodegradable copper-borosilicate glass for wound dressings by contrasting two synthesis techniques: traditional melt casting and the sol-gel method. The sol-gel approach was better due to its improved ion release control and excellent biocompatibility, as evidenced by a more than 80% cell survivability rate. On the other hand, the melt casting method encountered difficulties, most notably a reduced cell viability of less than 10%, raising worries about biocompatibility. Furthermore, the sol-gel-derived glass passed through a crystallization phase, forming combeite, a Food and Drug Administration (FDA) approved bioactive phase. As a result, the sol-gel-derived glass may be suitable for wound healing and a larger range of tissue engineering applications. Alternatively, despite its historical dominance, the melt casting technique has limitations for this application (Saghatchi et al. 2023).

In another experimental study, researchers attempted to develop a simplified process for producing bioactive glass using minimal equipment and then analyzed the properties of the resulting glass and its glass ceramics. The glycol thermal technique created the borate bioactive glass designated 45B5. The transesterification reaction between precursors and glycol lies at the heart of this novel chemical route. The approach successfully produced a vitreous structure with a lamellar morphology after calcination. The glass-ceramics successfully integrated oxide into their network after heat treatments at appropriate temperatures, resulting in the crystallization of separate phases. *In vitro* tests revealed that the glass and glass ceramics were highly soluble and quickly transformed into HA, a bioactivity indication. Nonetheless, various samples early mineralization stages revealed the presence of $CaCO_3$ species, indicating differences in apatite production capacities. The 45B5 glass and its accompanying glass ceramics displayed remarkable bioactivity, making them viable candidates for tissue engineering applications, particularly in wound healing materials. The glycol thermal approach opens up a new avenue for synthesizing various bioactive glass compositions, perhaps ushering in a new age in their application in tissue engineering (dos Santos et al. 2023).

Characterization Methods

Glass fibers have been produced from borate bioactive glasses, identified for their potential in various biomedical applications. Typically woven, these glass fibers have been utilized in the engineering of soft tissues, such as muscles and ligaments, due to their resemblance to the fibrous nature of these tissues. Furthermore, these fibers possess a chemistry suitable for adjusting the biodegradation rate. Through research, it has been demonstrated that an open morphology is provided by scaffolds made of these glass fibers when they incorporate three-dimensional meshes. This structure has been found to facilitate the easy diffusion of nutrients and waste in and out of the scaffold. Additionally, these scaffolds offer structural support, and the growth of blood vesicles into them is allowed. As a result, high effectiveness for cell attachment and proliferation has been observed (Abou Neel et al. 2009).

In neurology, these glass fibers play a role as nerve conduits. In clinical settings, the effectiveness of borate glass fiber, especially in the form of glass wool, has been established. For the treatment of conditions like chronic wounds and soft tissue repairs, including diabetic ulcers that have not responded well to traditional methods, this material has been found beneficial (Jung et al. 2011). Bioactive glasses containing B_2O_3 have been employed as coating layers on metallic implants that bear loads (Joy-anne et al. 2019; Kuo and Du 2022). However, a challenge has been faced by these metallic implants, such as those made of Ti and its alloys. Despite their recognized strength and biocompatibility, they do not form a direct bond with the host tissue (Peddi, Brow, and Brown 2008). Bioactive coatings, such as hydroxyl apatite or bioactive glasses, are placed to the implant surface to promote osteointegration and increase stability at the implant-bone contact. Creating strong and airtight connections with metallic elements is a distinguishing feature of bioactive glasses containing B_2O_3 (Peddi, Brow, and Brown 2008). When combined with their intrinsic bioactivity and an equivalent thermal expansion coefficient attained through composition design, this bonding property reveals their potential as materials for bioactive coatings on metallic implants.

Such coatings are also utilized as corrosion inhibitors for metallic implants and bioactive layers that promote greater compatibility with host tissues. The production of metallic titanium diboride (TiB2) at the glass-metal interface has been discovered as the cause of the strong binding of the borate glass bioactive coating. Bioactive glasses using B_2O_3 have been found to maintain the bioactivity inherent in base silicate bioactive glasses while also having improved antibacterial characteristics (Datta et al. 2018). The glass coating may be applied via the viscous sintering of glass powders, also known as the enameling method. Alternative procedures, such as laser processing and various coating techniques, have been used (Kuo et al. 2019, 309). The idea of a gradient bioactive glass coating has been proposed to ease the shift from an active biological surface to a layer resistant to corrosion (Lopez-Esteban et al. 2009).

Various instruments, including solid-state NMR, X-ray, neutron scattering, Raman, and Infrared spectroscopy, have been employed for analyzing structures. Solid-state NMR is particularly useful for examining specific atomic arrangements in glass materials. Bioactive glasses containing B_2O_3 have been thoroughly researched, as boron coordination is highly sensitive to 11B solid-state NMR (Stone-Weiss et al. 2018). This technique has also been applied to study the local environments of elements like silicon, phosphorus, calcium, and strontium in bioactive glasses (Bonhomme et al. 2018; Bonhomme et al. 2012). Tests for bioactivity are usually performed through *in vitro* methods, centering on the creation of hydroxyl apatite (HA) in fluids that mimic bodily conditions, such as simulated body fluid (SBF). However, comparing bioactivity test outcomes is challenging due to the variety of protocols, which differ in the types of testing solutions, sample shapes, and experimental conditions. To address this, a round-robin test was designed to provide a uniform *in vitro* evaluation of bioactive glasses ability to form apatite. Importantly, the increase in pH following the dissolution of standard bioactive glasses is less significant compared to conventional phosphosilicate bioactive glasses, owing to the acidic characteristics of boron ions (Macon et al. 2015).

Osteogenic and Angiogenic Properties of Boron-Substituted Bioceramics for Bone Tissue Engineering

Boron, a trace element, is pivotal in maintaining bone health. Integrating boron into bioceramics began in 1990 when Brink introduced the first borosilicate glass tailored for medical applications (Brink 1997). This innovative material was further explored by Huang et al. who, by substituting SiO_2 with B_2O_3, observed a significant enhancement in the glass's ability to convert to HA when exposed to aqueous phosphate solutions (Huang et al. 2006). Borosilicate glasses are characterized by their high content of boron oxide ($B_2O_3 > 8$ wt%) and SiO_2, which form the primary structure of the glass network (Hasanuzzaman et al. 2016). While both borosilicate and silicate glasses transform to form an apatite layer, borosilicate glasses uniquely develop a borate-rich layer, in contrast to the silicate-rich layer formed by silicate glasses. Incorporating boron into silicate glass reduces its melting point, accelerates its conversion to HA, and enhances biodegradability. Borate glasses, when compared to their silica-based counterparts, demonstrate a rapid and complete conversion to HA. This can be attributed to their relatively lower chemical stability and heightened reactivity (Huang et al. 2006).

The borate 13-93B3 bioactive glass stands out for its remarkable elastic modulus and compressive strength. When borosilicate glasses are implanted, they release a substantial amount of boron at the site. This release decelerates the degradation of the biomaterial and elevates the pH of the surrounding growth medium. The polymer foam replication technique was employed to craft scaffolds from these glass ceramics, resulting in a microstructure reminiscent of natural trabecular bone. In a quest to further optimize the properties of boron-based glasses, Awad et al. employed the melt-quenching method to produce a glass with the composition $24.5Na_2O$-$xSiO_2$-$(45-x) B_2O_3$-$24.5CaO$-$6P_2O_5$. Their objective was to enhance the bonding capabilities of the glass, expedite the production of HA, and fine-tune its dissolution rate (Awad, Abdelghany, and Meikhail 2022). Boron's significance as a trace element for the human body has been underscored by a plethora of scientific evidence (Bose et al. 2013). Most research on boron's impact on biological functions has been rooted in nutritional studies. Comprehensive work has documented various animal biological responses to boron addition (Abdelnour et al. 2018). The elements supplementary effects on various organs and systems, including the brain, liver, bone metabolism, immune responses, wound healing, and hormonal activities, have highlighted its pivotal role in metabolic processes (Khaliq, Juming, and Ke-Mei 2018).

One of the most noteworthy roles boron plays is in bone development. It is particularly influential in mineralization and bone growth. Research focusing on osteogenesis has illuminated boron's significant contribution to the mineralization activities of osteoblasts. This is achieved through its influence on specific gene expressions and hormones, notably 17β-estradiol (E2) and testosterone (Khaliq, Juming, and Ke-Mei 2018). Furthermore, boron has been identified as a stimulant for mRNA, particularly those that encode growth factors pivotal for angiogenesis and wound healing (Dzondo-Gadet et al. 2002). The proteins influenced by boron span a wide range, from bone-centric ones like bone morphogenic proteins (BMP-4, BMP-6, BMP-7), collagen type 1, osteopontin, bone sialoprotein, and osteocalcin to enzymes crucial for wound healing present in fibroblasts, such as elastase, collagenase, and alkaline phosphatase (Swager and Luppino 2015).

One of boron's pivotal attributes is its controlled release during its degradation. Given that minimal boron concentrations can stimulate osteogenesis, this controlled boron release can significantly amplify the bone repair process. The 13-93B3 fibers, a variant of boron substitutes, have been identified to play a dual role in bone regeneration. They initiate bone formation and promote bone resorption, a phenomenon attributed to their specific degradation patterns and ion release rates (Liu, Rahaman, and Day 2013). The inherent ability of boron-based bioglasses to modulate the release of calcium and other ions positions them as potential regulators of osteogenic gene expression. This controlled ion release, combined with BBGs capacity to facilitate the delivery of essential nutrients, underscores their importance in ensuring the growth and stability of the newly formed bone. Further attesting to the bioactivity of boron-based biomaterials is the formation of crystalline HA coating on their surface (Yao et al. 2007; Cho and Kang 2009).

In one study, radiographic imaging highlighted the bone-forming and therapeutic capabilities of 13-93B3 scaffolds. Compared to silicate-based bioglasses, boron-based bioglasses scaffolds displayed better biocompatibility and bioactivity, a distinction largely credited to their controlled degradation behavior. Another noteworthy advantage of Boron-substituted biomaterials is their ability to foster the proliferation and differentiation of mesenchymal stem cells. Specifically, bone marrow-derived stromal cells can be rejuvenated with boron substitutes, emphasizing their potential in stem cell-based bone regeneration therapies (Xie et al. 2009). In a comparative study, while silicate 13–93B3 scaffolds underwent only partial conversion, the 13-93B3 scaffolds were entirely transformed into HA. This observation underscores the heightened potential of 13-93B3 scaffolds in bone regenerative applications (Gu et al. 2013). Recent advancements in bone tissue engineering have underscored the potential of boron-based bioglasses in promoting angiogenesis and osteogenesis. A scaffold composed of boron-based bioglass and zoledronic acid, with a specific composition of 20.0CaO-4.6Na_2O-3.8P_2O_5-53.8B_2O_3-12.1K_2O-4.6 MgO (wt%), has been found to significantly enhance the angiogenic and osteogenic gene expressions in human endothelial cells and bone marrow mesenchymal stem cells (Su et al. 2020). The inherent flexibility in the composition of boron-based bioglasses allows for incorporating a diverse array of physiologically active ions. This adaptability positions boron-based bioglasses as a promising material for applications in bone development and osteoblast differentiation (Cui et al. 2017; Zhang et al. 2015). Zhang et al. highlighted the potential of doping boron-based bioglasses with strontium, revealing that osteogenic differentiation of BMSC was notably enhanced in 9 mol% Sr doped 13-93B3 cements when compared to their undoped counterparts (Zhang et al. 2015).

In a study by Rahaman et al. rat calvarial defects were addressed using 4% CuO-13-93B3 scaffolds (Bi et al. 2013). These scaffolds, enriched with angiogenic copper ions, show enhanced bone growth compared to the pure 13-93B3 scaffolds, emphasizing the potential of metal ion doping in boron-based bioglasses. Further innovations in boron-based bioglass scaffolds include the addition of 5% graphene, which resulted in an increase in their electrical conductivity to 0.06 S/cm. This suggests that these graphene-infused boron-based bioglass scaffolds could be particularly effective in bone regeneration applications (Turk and Deliormanlı 2017). Boron-substituted bioceramics are poised to revolutionize bone tissue engineering by harnessing their unique properties to enhance osteogenesis, angiogenesis, and overall bone regeneration.

Mechanical and Biological Properties of Boron-Substituted Bioceramics

Boron-substituted bioceramics inherently exhibit enhanced physicochemical stability, an essential feature for preserving long-term functionality within biological settings (Li et al. 2023). Concurrently, these biomaterials demonstrate significant biocompatibility, suggesting optimal integration with human tissues and a diminished potential for inducing adverse reactions. Specifically, boron-substituted bioceramics, resembling boron-substituted HA, display remarkable efficacy in bolstering bone regeneration and amplifying osteogenic processes (Yılmaz and Evis 2016; Xia et al. 2019). Notably, their potential to counteract bone loss is particularly significant, especially in post-menopausal stages where bone-related issues are prevalent (Yılmaz and Evis 2016). The deliberate incorporation of boron into the crystal lattice of bioceramics, like HA, offers a mechanism to enhance their mechanical attributes, encompassing hardness and strength (Yılmaz and Evis 2016). Including ions through HA substitutions modulates its structural, mechanical, and biological properties. A specific study analyzed the impact of boron doping in HA on its compressive strength and Young's modulus in the context of bacterial cellulose, boron-doped HA, and gelatin scaffolds. Following boron addition, the compressive strength increased from 75.3 MPa to 94.9 MPa, albeit with a slight decline in Young's modulus. At temperatures of 1000°C and 1200°C respectively, the compressive strength of 0.5, 1, and 2 wt.% boron-doped HA decreased from 39.5 MPa to 29.1 MPa and then remarkably rise to 161.2 MPa. Vicker's microhardness tests highlighted a pronounced decrease in microhardness for the boron-doped samples. However, elevating the sintering temperature from 1000°C to 1200°C resulted in a substantial 16-fold increase in microhardness values (Uysal, Yılmaz, and Evis 2020). Another experimental study identified that introducing graphene oxide to boron-doped HA scaffolds augmented mechanical properties such as compressive strength and elastic modulus (Li et al. 2023). Further research revealed that boron-substituted hydroxyapatite (BHAp) degrades faster than HA or β-tricalcium phosphate in *in vitro* biodegradation settings, highlighting boron's potential role in influencing bio-ceramic degradation kinetics (Barheine et al. 2011). Incorporating borate glasses facilitated the rapid formation of an HA layer and fostered bonding with bone tissue (Massera et al. 2015). Other study aimed to expound the implications of boron substitution on the structural properties of fluoro-aluminosilicate glasses and their resultant ceramics. The density of both boron-substituted glasses and glass-ceramics diminished with increasing boron substitution for aluminum. Boron atoms manifested as BO_4 triangles and BO_3 tetrahedra within the glass structure. Remarkably, a pronounced degree of boron substitution prompted shifts in the crystalline growth behavior of fluorapatite, inducing the emergence of highly oriented clusters (Zhang, Stamboulis, and Ni 2018). Due to their biocompatibility, silica-based glasses like 45S5 (Mitruţ et al. 2022) and S53P4 (Lee et al. 2020) have emerged as viable candidates for diverse biomedical applications, particularly in bone grafting (Massera et al. 2015). In biomedical applications, phosphate glasses are increasingly prominent as sought-after alternatives to silicate glasses, especially concerning bone repair and reconstruction. Unlike 45S5 silicate glasses, borate and phosphate glasses present consistent dissolution behaviors. Importantly, when immersed in phosphate-rich solutions, borate glasses respond more swiftly than 45S5 silicate glasses, culminating in the formation of HA (Gharbi et al. 2023). Boron's inclusion in calcium phosphate glasses bolstered their bioactivity, an effect attributed to boron's ability to alter coordination and introduce hydroxyl groups onto the glass surface (Saranti, Koutselas, and Karakassides 2006). Incorporating boron into specific glass systems, exemplified by P_2O_5–CaO–Na_2O–MgO, has yielded positive outcomes in cellular metabolic activity, growth, and structure (Massera et al. 2015). A study conducted by Alqahtani (2020) showcased an enhancement in cold-cured PMMA's mechanical properties following the introduction of hexagonal boron nitride (h-BN) nanopowder and zirconia oxide (ZrO_2). This material combination amplified stiffness and strength, rendering it apt for durable denture construction

(Abdulai and Cengiz 2023). Prior research into the physico-chemical properties of calcium borophosphate glasses, characterized by the formula (1-x)$Ca(PO_3)_2$–x(B_2O_3) where x=B/B+P ≤6, highlighted that BO_4 integration raised the glass transition temperature, crystallization temperature, density, and microhardness, while diminishing water solubility and the UV spectrums cut-off wavelength (Yılmaz and Evis 2016). In another research effort, a unique borophosphate glass state was synthesized by integrating selenium (Se) in varying mol% of SeO_2, resulting in the composition xSeO_2 (100−x) ($48P_2O_5$–50CaO–$2B_2O_3$). Predominantly, Se ions supplanted the B units, and SeO_2 content influenced shifts in the electronic density of B, P, and O atoms bonding states. Furthermore, the utility of copper (Cu) in tandem with B-enriched bioceramics has been explored. Utilizing Cu's angiogenic properties, B-enriched bioactive glass scaffolds and borosilicate glasses were complemented with Cu (Yılmaz and Evis 2016). Lee et al. (2020) assessed the feasibility of Boron Nitride Nanoplates (BNNPs) as reinforcing agents in dental materials. Integrating BNNPs into a ZrO_2 framework led to discernible enhancements in strength and fracture toughness (Abdulai and Cengiz 2023).

Boron-Containing Bioactive Glasses: Drug Delivery and Tissue Engineering Applications

The field of nanobiotechnology has ushered in groundbreaking anticancer strategies, leveraging a variety of nanosystems. These include nanosheets, nanospheres, nanotubes, nanocarriers, microneedles, nanocapsules, dendrimer-based lipid nanoparticles, and micelles. While each system offers unique advantages, there is a growing emphasis on the potential of boron-substituted biomaterials. Carbon and, more prominently, boron-based biomaterials have emerged as central elements in the drug delivery landscape. Their adaptability and physicochemical properties make them suitable for targeted and efficient drug delivery mechanisms. The recent surge in their application underscores their significance in reshaping cancer therapeutics, focusing on enhancing the precision and effectiveness of drug delivery (Das et al. 2023). As research continues, boron-substituted biomaterials are poised to be at the forefront of innovative drug delivery solutions.

Researchers developed a novel drug delivery system using hydroxy boron nitride (BN) nanosheets combined with Pd nanoparticles. This system was created through a thermal substitution method, enhancing the biocompatibility of BN nanomaterials. The Pd nanoparticles showed high efficiency in loading the anticancer drug, doxorubicin, and its release was responsive to changes in pH, glutathione concentration, and near-infrared irradiation. Cellular studies confirmed faster drug release under near-infrared light. When tested in mice, combining Pd nanoparticles with doxorubicin significantly inhibited tumor growth, showcasing the benefits of combined chemotherapy and photothermal therapy. This innovative drug delivery system displayed low toxicity, stability, and multi-stimuli responsiveness, making it a promising tool for cancer treatment (Zhang et al. 2019).

Utilizing molecular dynamics simulation, a study explored the potential of BN nanotubes as a delivery system for platinum-based anticancer drugs, specifically carboplatin. Initially, the study compared the penetration rates of water molecules and carboplatin in functionalized and pristine BN nanotubes. Consequently, it was found that the functionalized BN nanotubes, due to their water-mediated hydrogen bonding, had a notably higher penetration rate than their pristine counterparts. Furthermore, when encapsulating multiple carboplatin drugs within the functionalized BN nanotubes, the study revealed that the hydrogen bond formation between adjacent drug molecules played a significant role. Additionally, the van der Waals interaction between the drug and the BN nanotubes inner surface became particularly influential in drug positioning at higher drug concentrations. Therefore, this research suggests that functionalized BN nanotubes might offer enhanced capabilities for drug delivery, especially for platinum-based anticancer treatments (Khatti and Hashemianzadeh 2016).

The therapeutic potential of BN nanomaterials, especially BN nanospheres (BNNS), as vehicles for targeted anticancer drug delivery has shown much interest. Nonetheless, certain impediments, such as suboptimal dispersity in biological milieus and limited drug loading proficiency, have constrained their widespread utilization. To overcome these limitations, scholars have pioneered an innovative drug delivery paradigm: folate-anchored mesoporous silica (MS)-enriched BNNS, denoted as BNMS-FA. This strategic alteration markedly augmented the dispersity and drug accommodation capability of BNNS. Crucially, these BNMS-FA assemblies exhibited biocompatibility up to 100 μg/mL concentrations and were adeptly assimilated by HeLa and MCF-7 cells via folate receptor-guided endocytosis. The oncology drug, Doxorubicin (DOX), was integrated into these assemblies with commendable efficacy, revealing a protracted release trajectory in diverse pH scenarios. Significantly, BNMS-FA/DOX assemblies manifested enhanced drug assimilation and tumor-inhibitory prowess relative to unbound

DOX, BNNS/DOX, and BNMS/DOX configurations. In summation, the FA-modified BNMS structures, endowed with their enhanced drug retention and precision-targeted delivery attributes, present an auspicious avenue for refining oncological interventions, particularly when factoring in the prospective utility of BNNS in boron neutron capture therapy (Feng et al. 2019).

The domain of drug delivery has recently witnessed a burgeoning interest in boron-infused polymeric nanomaterials. These polymeric nanomaterials offer numerous advantages for drug delivery, including extended circulation in the bloodstream, facile surface modification, and precise targeting, thereby enhancing therapeutic efficacy. To optimize drug accumulation at the target site, drug delivery mechanisms have primarily adopted two strategies: (i) enhancing the affinity between the delivery system and the target site and (ii) modulating drug release in response to specific triggers such as pH, ROS, or particular proteins/carbohydrates (Liu and He 2017; Brooks and Sumerlin 2016; Kamaly et al. 2016).

While some drug delivery mechanisms are constructed using naturally occurring biomolecules, which can respond to specific enzymes, these systems often present challenges. They may elicit immunogenic reactions, exhibit diminished stability, and pose storage and processing complications. Conversely, mechanisms crafted from synthetic chemicals often circumvent these limitations, albeit sometimes at the cost of selectivity. Despite potential synthetic hurdles, aryl boronic acid-infused materials have become central to research endeavors focused on selectivity, stimuli-responsiveness, and drug delivery, as this segment seeks to elucidate (Li, Chen, and Liu 2015; Lu et al. 2013). A particular emphasis is placed on insulin delivery mechanisms, leveraging the unique covalent reversible chemistry of aryl boronic acid-infused materials, targeting diabetes mellitus treatment.

The delivery of therapeutic proteins is pivotal for numerous biomedical endeavors. However, ensuring protein stability during both delivery and prolonged storage poses challenges. One study introduces a trehalose-based hydrogel, synthesized using a trehalose polymer and a phenylboronic acid end-functionalized 8-arm poly (ethylene glycol) (PEG), which effectively stabilizes insulin against high temperatures and facilitates its glucose-triggered release. The hydrogel's formation hinges on boronate ester linkages between the trehalose side chains and the PEG boronic acid cross-linker. The hydrogel dissolves upon glucose introduction, given glucose's stronger affinity to boronic acid than trehalose, enabling glucose-responsive insulin release. Remarkably, this hydrogel offers significant thermal protection to insulin, retaining 74% of its activity post-heating, in contrast to a mere 2% without additives. The study presents a glucose-responsive hydrogel derived from a trehalose glycopolymer tailored for insulin delivery and thermal stabilization. The hydrogel's formation is physiologically compatible, and its dissolution rate is influenced by glucose concentration, leading to controlled insulin release. The trehalose hydrogel's ability to thermally shield insulin suggests its potential to prevent the need for specialized refrigeration, enhancing patient convenience. Given boronic acid is known pH-responsive properties, these trehalose-boronic acid hydrogels could be further explored for delivering a broader spectrum of protein therapeutics, including the targeted release of anticancer drugs in tumor environments (Lee et al. 2018).

Moreover, for diabetes treatment, there is a pressing need for self-regulated, painless insulin delivery that can precisely manage glycemic levels. Phenylboronic acid (PBA), recognized for its synthetic nature and reversible glucose-binding capability, has been incorporated into hydrogels to address this. Although previous PBA hydrogels demonstrated glucose sensitivity, their temperature-dependent release profiles limited clinical applications. This study introduces an optimized, temperature-stable, boronate-infused hydrogel, which has been fashioned into a microneedle (MN) patch for on-demand insulin delivery. This MN patch, distinct from many glucose-responsive patches that rely on glucose oxidase and nanoparticles, offers temperature-independent, rapid, and sustained glucose-responsive insulin release. Its entirely synthetic, protein-free, and nanoparticle-free composition ensures safety and supports scalability for mass production. The hydrogel's unique structure and boronate content make it ideal for drug delivery, potentially revolutionizing diabetic care by providing consistent glycemic control and enhancing patient quality of life (Chen et al. 2020).

Boron-substituted bioglasses have garnered significant attention for their potential in tissue regeneration, particularly in facilitating extended wound healing. Wound recovery is a multifaceted process encompassing homeostasis, inflammation, cell proliferation, and tissue maturation. Central to this is angiogenesis, the formation of new blood vessels, which ensures the delivery of essential growth factors, oxygen, and nutrients to the wound site. Boron plays a pivotal role in this context, promoting angiogenesis through its involvement in ionic dissolution. It further augments protein and collagen secretion, vital for extracellular matrix production, and accelerates wound healing by fostering keratinocyte migration (Aslam et al. 2023). Notably, 13-93B3 micro fibrous scaffolds, when exposed to SBF, have demonstrated stable and constant calcium ion release compared to 45S5 bioglass, underscoring their enhanced wound healing capabilities (Lepry and Nazhat 2020). Calcium ions are instrumental in the latter stages of wound healing, aiding epidermal cell migration and regeneration. While the necessity of a HA layer in wound healing remains a topic of debate, some studies suggest its potential to expedite the healing trajectory. The orchestration of new tissue formation is further facilitated by activating specific healing factors, reinforcing the significance of boron-substituted materials in drug delivery for tissue regeneration (Tang, Pang, and Wang 2017; Balasubramanian et al. 2018).

Boron-substituted bio glasses, particularly the 13-93B3 BG, have shown promise in tissue engineering, especially for wound healing applications. This bioglass was incorporated into a polyvinyl acetate hydrogel with 5 mol% SrO, enhancing the polymers mechanical properties by serving as a cross-linking agent and a filler. Importantly, these scaffolds ensured safety by releasing boron levels below the critical 100 ppm threshold (Tang, Pang, and Wang 2017). In another endeavor, 13-93B3 particles were embedded into a methylcellulose hydrogel, cross-linked using manuka honey, a recognized biocompatible and antibacterial agent (Schuhladen et al. 2021). The biocompatibility of these 3D-printed scaffolds was confirmed through *in vitro* tests with human dermal fibroblasts. Incorporating boron-substituted bioglass into polymers has been observed to fortify their mechanical strength and modulate boron ion release rates. While doping these scaffolds can enhance their wound-healing efficacy, it's essential to be cautious of potential toxicity (Naseri et al. 2022). Adding copper ions to boron-substituted bioglasses has promoted endothelial cell proliferation and rapid HA layer development (Wang et al. 2014). Furthermore, Cu-doped 13-93B3 microfiber dressings created a conducive environment for wound healing by enhancing VEGF production and stimulating angiogenesis. This was corroborated by a significant increase in vascular development in wounds treated with Cu-infused 13-93B3 scaffolds (Zhao et al. 2015).

The development of boron-substituted biomaterials for tissue engineering and drug delivery demonstrates their critical contribution to advancing therapeutic interventions. These biomaterials can change cancer treatment, drug release, and wound healing through creative methods. Boron-substituted materials show their prowess in precise drug administration and efficient tissue regeneration by utilizing their biocompatibility, stimuli-responsive behaviors, and remarkable binding capabilities. These biomaterials are expected to influence the future of medical interventions as research continues to find their diverse applications, providing treatments that are safer, more efficient, and more accessible.

Boron-Substituted Bioceramics for Dental Applications: Implants and Restorative

Calcium phosphate (CaP) compounds, particularly HA, hold prominence in bone defect repair, joint/tissue replacement, and surface coatings for metal implants to enhance biocompatibility. The CaP family includes various members with distinct characteristics. For instance, HA is highly stable and biocompatible, while others like monocalcium phosphate monohydrate (MCPM) and dicalcium phosphate (DCP) exhibit varying solubilities and stabilities. HA and TCP are the most widely used CaP compounds in medical areas (Yılmaz and Evis 2016). In literature, boron-doped HA applications predominantly encompass its utilization as a component in bone tissue engineering scaffolds and its auxiliary role in phosphors for fluorescent lamps (Uysal, Yılmaz, and Evis 2020). Boron is used in other bioceramics, particularly glasses, and is a dopant or component in composites made of CaPs. Using dental tissue engineering and stem cell therapy, emerging therapies aim to restore teeth's essential functions (Rad et al. 2018).

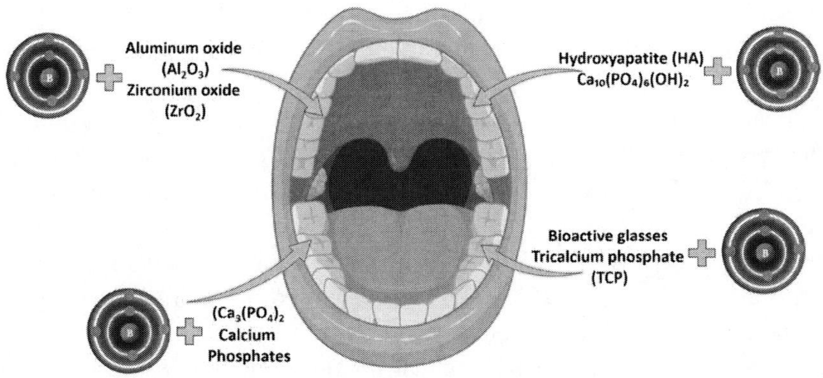

Figure 4. Illustration of well-known bioceramics with boron for dental applications. The illustration shows boron-substituted variations of well-known bioceramic materials, emphasizing their potential for improving mechanical strength, biocompatibility, and therapeutic properties of dental biomaterials.

Bioactive glasses (BGs) are central to this domain, forming bonds with hard tissues via a HA layer, though silicate-based BGs face conversion limitations. Modifying silicate-based BGs by incorporating boron enhances their biodegradability, aligning with the ideal degradation pace for tissue regeneration (Rad et al. 2018). Beyond traditional bioceramics, advancements in nanotechnology have introduced novel boron-based materials with potential in dentistry. Boron has emerged as a promising candidate due to its demonstrated capacity to adhere to dental materials and promote the growth and differentiation of bone-derived cells. Boron nitride nanotubes (BNNTs) have enhanced bioactivity and physicochemical properties, positioning them as potential reinforcements for dental adhesives (Mitruţ et al. 2022).

Recent studies have focused on composite biomaterials containing boron. Ali et al. employed BNNTs as a HA and -TCP reinforcing additive. These BNNTs hold promise as innovative fillers for dental adhesives. Their inclusion has been observed to enhance the chemical and mechanical properties of the adhesive resin (Abdulai and Cengiz 2023). Dental implants fabricated from BNNPs present biocompatible reinforcing materials, exhibiting characteristics akin to graphene. The substance known as white graphene, hexagonal boron nitride (hBN), is another promising candidate. In contrast to graphene, hBN preserves the aesthetics of color for dental applications. hBN materials are used in various dental products, enhancing mineral deposition and antibacterial activity because of their exceptional biocompatibility and mechanical qualities. Furthermore, their chemical stability facilitates the effective anchoring of metal nanoparticles (Lee et al. 2020).

In light of existing research explored the synthesis of composites by amalgamating HA with hBN at the nanoscale. The study, which involved implanting these composites in rats, found that blood boron levels in the test subjects were significantly elevated compared to a control group (Yılmaz and Evis 2016). The widespread adoption of Ti alloys in various industries is curtailed by their elevated costs. However, strategies might be devised to reduce the processing costs of Ti alloys, potentially broadening their industrial applications. Zhu et al. (2003) observed enhanced microstructure and tensile properties in Ti alloys upon boron addition across different alloy types. This study aimed to determine the optimal boron concentration (0.57 wt%) for dental cast Ti alloys. It was revealed that the boron addition significantly refined the microstructure of Ti alloys, necessitating further exploration of its implications. In a separate investigation, Sopcak et al. (2023) assessed the impact of boron addition on dental cement using a biphasic calcium magnesium silicate cement (CaMgSi). The addition of boron, specifically at 0.25 and 0.5 mol, enhanced the cement's hydration, self-setting capabilities, and mechanical properties, with the degree of enhancement contingent on the boron concentration. This improvement in mechanical strength might be attributed to the presence of borate clusters fostering a denser microstructure (Rad et al. 2018).

Efforts to enhance the fracture toughness of bioceramic materials have frequently involved the incorporation of reinforcements to create composite materials, with SiC, Al_2O_3, and ZrO_2 serving as common reinforcement agents. A study evaluated the suitability of boron nitride nanoplates (BNNP) as a reinforcement for dental ceramics. A composite of 3 mol.% Y_2O_3 and ZrO_2, termed tetragonal ZrO_2 polycrystal (3Y-TZP), acted as the ceramic matrix. To protect BNNP from oxidative damage and unforeseen liquid phase sintering, 3 wt% of Al_2O_3 powder was incorporated. This introduced aluminum borate ($Al_{18}B_4O_{33}$) on the BNNP surface, providing a robust layer against thermal and oxidative degradation. Results showcased that ZrO_2 with dispersed BNNPs delivered enhanced strength (up to 27.3%) and increased fracture toughness (up to 37.5%) upon adding 1–1.5 vol.% BNNPs. Furthermore, cytotoxicity assays affirmed the non-toxic nature of BNNPs, while barrier attributes of BNNPs almost entirely impeded the degradation of the ZrO_2 matrix in moist conditions (Lee et al. 2020). Based on the research results, it can be inferred that incorporating boron into the formulation of next-generation dental materials could potentially position them as the favored choice for dental restorations, thereby advancing them to the forefront of dental therapeutic interventions.

In-Vitro and *In-Vivo* Studies of Boron Substituted Biomaterials

Despite its status as an essential nutrient for plants, the role of boron in biological systems is a relatively recent discovery. Its significance lies in being a foundational element, as it serves a pivotal structural function within the intricate framework of plant cell walls. This newfound awareness has led to a deeper understanding of boron's dynamic role within the broader context of biological processes, shedding light on its implications for not only plant health but also the intricate interplay between elements in the natural world (Nielsen 2008; Ateş, Aydemir, and Öksüz 2023). Boron, also recognized as an indispensable trace element crucial to human well-being, assumes a multifaceted role, particularly in wound healing functions and the maintenance of robust bone health. Its acceptance as an essential component in the human systems from its influence on fundamental physiological processes, showcasing its ability to expedite the intricate process of wound healing, thereby facilitating the body's regenerative capacities. Moreover, boron's impact extends to bone health, where it contributes to optimizing bone density and strength, ensuring the structural integrity of the skeletal system. This understanding of boron's significance underscores its intricate involvement in enhancing human health and its position as a pivotal micronutrient in the spectrum of biological functionality (Bose et al. 2013).

Boron element has many biological responses/effects, including (i) influences on reproduction and embryogenesis, (ii) facilitation of wound healing and response to injury or infection, (iii) modifications to calcium and bone metabolisms, (iv) favorable effects on central nervous system function, and (v) interactions with the presence and function of essential components like vitamin D and hormones—such as thyroid hormone, insulin, estrogen, and progesterone (Nielsen 2008; Bose et al. 2013; Dzondo-Gadet et al. 2002). This element's dynamic interplay with Ca, vitamin D, and Mg, integral to bone metabolism, further underscores its intricate role (Dzondo-Gadet et al. 2002). Operating at the molecular stratum, studies have unveiled that boron augments RNA transcription within isolated placental nuclei, concurrently stimulating mRNA translation, particularly those coding for growth factors critical in angiogenesis and wound healing (Ateş, Aydemir, and Öksüz 2023; Lakhkar et al. 2013).

Boron accumulation within bone surpasses the levels found in both blood and soft tissues, with concentrations hinging on intake variability (Chapin et al. 1998; Moseman 1994). A singular instance demonstrates a human bone tissue boron level of 0.90 ppm (Moseman 1994). While definitive estimated average requirements or adequate intake levels for boron remain absent, relying on animal-derived data, a tolerable upper intake level of 20 mg/day has been established for adults (Shenkin 2003). Remarkably, an investigation employing a daily regimen of 3 mg/day boron supplement or a placebo revealed even this modest intake elicited a slight elevation in bone mineral density, as showcased in a particular study (Meacham, Taper, and Volpe 1994). A study examining boron deficiency in rats (0 vs. 3 mg/kg) showed that the lack of boron resulted in reduced weight gain and femoral strength, along with diminished mineral concentrations (Cu, Fe, Mg) attributed to the organic matrix. Furthermore, boron supplementation induced changes in the vertebral microarchitecture, leading to an augmentation in the trabecular thickness of rats receiving boron supplementation compared to those deficient in boron (Nielsen and Stoecker 2006). Another study exploring the impact of boron on mineral element concentrations within the bone's organic matrix revealed an elevation in the levels of Zn and K in the tibia following boron supplementation (Nielsen 2004). A separate study uncovered that administering boron supplementation (at a rate of 50 mg/kg body weight of boron within 96 hours) led to noteworthy enhancements in bone mineral density, as well as in the maximum breaking force of the femur and the compression strength of the tibia among rabbits that were fed a high-energy diet (Hakki et al. 2013).

In vitro cell studies also revealed boron dose-dependent regulation of osteoblastic cells. The cell viability tests on MC3T3-E1 cells using varying boron concentrations in cell culture media and they found that adding boron at 1000 ng/ml or higher reduced cell survival within a short time (24 hours), but over a longer term, there were no significant differences compared to the untreated control group (Hakki, Bozkurt, and Hakki 2010). The experimental study also observed positive regulation in osteoblastic function, evidenced by enhanced expressions of collagen type I, osteopontin, bone sialoprotein, osteocalcin, and Runx2 mRNA in the Boron treated groups. Bone morphogenetic protein (BMP) levels increased at 0.1, 1, 10, and 100 ng/ml B concentrations. Likewise, Gümüşderelioğlu et al. (2015) demonstrated that boron released from chitosan nanoparticles doped with H3BO3 (approximately 175 nm in diameter) impacted the proliferation and osteogenic differentiation of MC3T3-E1 cells. Their study revealed elevated alkaline phosphatase (ALP) activity—a marker of early osteogenic differentiation—on scaffolds containing boron-encapsulated nanoparticles.

Conclusion

Recently, a diverse range of ceramic biomaterials has emerged, each crafted to achieve heightened bioactivity and enhanced biocompatibility. The continuous surge in research within this domain paints a promising picture, suggesting a remarkable advancement trajectory in this arena. This book chapter aims to offer a concise yet comprehensive exploration into the realm of boron-substituted biomaterials. The data presented herein underscores the escalating intrigue surrounding the latest boron substitution breakthroughs within ceramic, glassy, or hybrid biomaterial systems, primarily driven by its intrinsic connection with living tissues. Looking ahead, integrating boron into bioceramic materials holds promising potential and a set of challenges in biomaterial research. As the investigation into these composite materials progresses, a clearer understanding of their enhanced advantages in medical applications is likely to emerge. However, several challenges lie ahead, including optimizing the boron incorporation process to achieve consistent and controlled concentrations within the biomaterials. Achieving a delicate balance between improved biocompatibility, enhanced mechanical properties, and controlled degradation rate will be crucial for achieving the desired clinical outcomes. Furthermore, ensuring long-term stability, compatibility with host tissues, and minimizing potential adverse effects will be critical considerations in developing and utilizing boron-substituted bioceramic materials. Addressing these challenges will pave the way for realizing safer and more effective biomaterials that can revolutionize various medical applications, ranging from orthopedics to tissue engineering.

References

Abdulai, Dalndushe, and Seda Cengiz. 2023. "The Effect of Boron Addition on the Prosthetic Dental Materials: A Review." *International Journal of Applied Dental Sciences* 9 (2): 31–37. doi:10.22271/oral.2023.v9.i2a.1710.

Abdelnour, Sameh A., Mohamed E. Abd El-Hack, Ayman A. Swelum, Antonella Perillo, and Caterina Losacco. 2018. "The vital roles of boron in animal health and production: A comprehensive review." *Journal of Trace Elements in Medicine and Biology* 50: 296-304. https://doi.org/10.1016/j.jtemb.2018.07.018.

Abou Neel, Ensanya A., David M. Pickup, Sabeel P. Valappil, Robert J. Newport, and Jonathan C. Knowles. 2009. "Bioactive functional materials: a perspective on phosphate-based glasses." *Journal of Materials Chemistry* 19 (6): 690-701. https://doi.org/10.1039/c6ra07819b.

Alipour S., Jannat F., Hosseini L. Teaching Breast Cancer Screening via Text Messages as Part of Continuing Education for Working Nurses: A Case-control Study. *Asian Pac J Cancer P* (2014) 15(14):5607-5609. https://doi.org/10.7314/apjcp.2014.15.14.5607.

Aslam, Awais Ali, Javeria Akram, Rana Adeel Mehmood, Arifa Mubarak, Amna Khatoon, Uzma Akbar, Sheikh Asrar Ahmad, and Muhammad Atif. 2023. "Boron-based bioactive glasses: Properties, processing, characterization and applications." *Ceramics International*. https://doi.org/10.1016/j.ceramint.2023.03.164.

Ates, Ayten, Burçak Aydemir, and Kerim Emre Öksüz. 2023. "Investigation of Physicochemical and Biological Properties of Boron-Doped Biochar." *Biomass Conversion and Biorefinery*, July. Springer Nature. https://doi.org/10.1007/s13399-023-04567-1.

Awad, Walla M., Amr M. Abdelghany, and Mahrous S. Meikhail. 2022. "Preparation and Characterization of Borosilicate Bioglass." *Egyptian Journal of Chemistry* 65 (7): 363-369. doi: 10.21608/ejchem.2021.104384.4818.

Balasubramanian, Preethi, Teresa Buettner, Valentina Miguez Pacheco, and Aldo R. Boccaccini. 2018. "Boron-containing bioactive glasses in bone and soft tissue engineering." *Journal of the European Ceramic Society* 38 (3): 855-869. https://doi.org/10.1016/j.jeurceramsoc.2017.11.001.

Barheine, Sabrina, Satoshi Hayakawa, Christian Jäger, Yuki Shirosaki, and Akiyoshi Osaka. 2011. "Effect of Disordered Structure of Boron-Containing Calcium Phosphates on Their *In Vitro* Biodegradability." *Journal of the American Ceramic Society* 94 (8). Wiley-Blackwell: 2656–62. doi:10.1111/j.1551-2916.2011.04400.x.

Bi, Lianxiang, Mohamed N. Rahaman, Delbert E. Day, Zackary Brown, Christopher Samujh, Xin Liu, Ali Mohammadkhah, Vladimir Dusevich, J. David Eick, and Lynda F. Bonewald. 2013. "Effect of bioactive borate glass microstructure on bone regeneration, angiogenesis, and hydroxyapatite conversion in a rat calvarial defect model." *Acta Biomaterialia* 9 (8): 8015-8026. https://doi.org/10.1016/j.actbio.2013.04.043.

Boccaccini, Aldo R., Delia S. Brauer, and Leena Hupa. 2016. *Bioactive Glasses. RSC Smart Materials*. https://doi.org/10.1039/9781782622017.

Bonhomme, Christian, Christel Gervais, Nicolas Folliet, Frédérique Pourpoint, Cristina Coelho Diogo, Jonathan Lao, Edouard Jallot, Joséphine Lacroix, Jean-Marie Nedelec, Dinu Iuga, John V. Hanna, Mark E. Smith, Ye Xiang, Jincheng Du and Danielle Laurencin. 2012. "87Sr solid-state NMR as a structurally sensitive tool for the investigation of materials: antiosteoporotic pharmaceuticals and bioactive glasses." *Journal of the American Chemical Society* 134 (30): 12611-12628. https://doi.org/10.1021/ja303505g.

Bonhomme, Christian, Xiaoling Wang, Ivan Hung, Zhehong Gan, Christel Gervais, Capucine Sassoye, Jessica Rimsza, Jincheng Du, Mark E. Smith, John V. Hanna, Stéphanie Sarda, Pierre Gras, Christèle Combesg and Danielle Laurencin. 2018. "Pushing the limits of sensitivity and resolution for natural abundance 43 Ca NMR using ultra-high magnetic field (35.2 T)." *Chemical Communications* 54 (69): 9591-9594. https://doi.org/10.1039/C8CC05193C.

Bose, Susmita, Gary Fielding, Solaiman Tarafder, and Amit Bandyopadhyay. 2013. "Understanding of dopant-induced osteogenesis and angiogenesis in calcium phosphate ceramics." *Trends in Biotechnology* 31 (10): 594-605. https://doi.org/10.1016/j.tibtech.2013.06.005.

Brink, Maria. 1997. "The influence of alkali and alkaline earths on the working range for bioactive glasses." *Journal of Biomedical Materials Research: An Official Journal of The Society for Biomaterials and The Japanese Society for Biomaterials* 36 (1): 109-117. https://doi.org/10.1002/(sici)1097-4636(199707)36:1.

Brooks, William L. A., and Brent S. Sumerlin. 2016. "Synthesis and applications of boronic acid-containing polymers: from materials to medicine." *Chemical Reviews* 116 (3): 1375-1397. https://doi.org/10.1021/acs.chemrev.5b00300.

Chapin, Robert E., Warren W. Ku, Mary Alice Kenney, and Harriet McCoy. 1998. "The Effects of Dietary Boric Acid on Bone Strength in Rats." *Biological Trace Element Research* 66 (1–3). Springer Science+Business Media: 395–99. doi:10.1007/bf02783150.

Chen, Siyuan, Takuya Miyazaki, Michiko Itoh, Hiroko Matsumoto, Yuki Moro-oka, Miyako Tanaka, Yuji Miyahara, Takayoshi Suganami, and Akira Matsumoto. 2020. "Temperature-stable boronate gel-based microneedle technology for self-regulated insulin delivery." *ACS Applied Polymer Materials* 2 (7): 2781-2790. https://doi.org/10.1021/acsapm.0c00341.

Cho, Jung Sang, and Yun Chan Kang. 2009. "Synthesis of spherical shape borate-based bioactive glass powders prepared by ultrasonic spray pyrolysis." *Ceramics International* 35 (6): 2103-2109. https://doi.org/10.1016/j.ceramint.2008.11.010.

Cui, Xu, Chengcheng Huang, Meng Zhang, Changshun Ruan, Songlin Peng, Li Li, Wenlong Liu, Ting Wang, Bing Li, Wenhai Huang, M. N. Rahaman, William Weijia Lu. Haobo Pan. 2017. "Enhanced osteointegration of poly (methylmethacrylate) bone cements by incorporating strontium-containing borate bioactive glass." *Journal of the Royal Society Interface* 14 (131): 20161057. https://doi.org/10.1098/rsif.2016.1057.

Das, Bhaskar C., Parthiban Chokkalingam, Pavithra Masilamani, Srushti Shukla, and Sasmita Das. 2023. "Stimuli-responsive boron-based materials in drug delivery." *International Journal of Molecular Sciences* 24 (3): 2757. https://doi.org/10.3390/ijms24032757.

Datta, Susmit, Tarun Adarsh, Prerna Diwan, K. Annapurna, Biswanath Kundu, and Kaushik Biswas. 2018. "Effect of boron oxide addition on structural, thermal, *in vitro* bioactivity and antibacterial properties of bioactive glasses in the base S53P4 composition." *Journal of Non-Crystalline Solids* 498: 204-215. https://doi.org/10.1016/j.jnoncrysol.2018.06.027.

Dearnaley, G., and James H. Arps. 2005. "Biomedical Applications of Diamond-like Carbon (DLC) Coatings: A Review." *Surface & Coatings Technology* 200 (7). Elsevier BV: 2518–24. doi:10.1016/j.surfcoat.2005.07.077.

dos Santos, V. R., T. M. B. Campos, G. P. Thim, A. L. S. Borges, and E. de Sousa Trichês. 2023. "Glycol thermal synthesis of the 45B5 bioactive borate glass: Structural, physical, and apatite mineralization *in vitro*." Ceramics International 49 (7): 11236-11248. https://doi.org/10.1016/j.ceramint.2022.11.321.

Dzondo-Gadet, M., R. Mayap-Nzietchueng, K. Hess, P. Nabet, F. Belleville, and B. Dousset. 2002. "Action of boron at the molecular level: effects on transcription and translation in an acellular system." *Biological Trace Element Research* 85: 23-33. https://doi.org/10.1385/bter:85:1:23.

Ege, Duygu, Qaisar Nawaz, Ana M. Beltrán, and Aldo R. Boccaccini. 2022. "Effect of Boron-Doped Mesoporous Bioactive Glass Nanoparticles on C2C12 Cell Viability and Differentiation: Potential for Muscle Tissue Application." *ACS Biomaterials Science & Engineering* 8 (12): 5273-5283. https://doi.org/10.1021/acsbiomaterials.2c00876.

Eslami, Hossein, Mohammadreza Tahriri, Fathollah Moztarzadeh, Rizwan Bader, and Lobat Tayebi. 2018. "Nanostructured Hydroxyapatite for Biomedical Applications: From Powder to Bioceramic." *Journal of the Korean Ceramic Society* 55 (6). Springer Science+Business Media: 597–607. doi:10.4191/kcers.2018.55.6.10.

Feng, Shini, Huijie Zhang, Sha Xu, Chunyi Zhi, Hideki Nakanishi, and Xiao-Dong Gao. 2019. "Folate-conjugated, mesoporous silica functionalized boron nitride nanospheres for targeted delivery of doxorubicin." *Materials Science and Engineering: C* 96: 552-560. https://doi.org/10.1016/j.msec.2018.11.063.

Gharbi, Amina, Hafed El Feki, and Hassane Oudadesse. 2016. "Novel alkali borosilicate glasses: Preparation, structural investigation and thermal study." *Korean Journal of Chemical Engineering* 33: 1456-1461. https://doi.org/10.1007/s11814-015-0287-4.

Gharbi, A., Hassane Oudadesse, Hafedh Trabelsi, W. Cheikhrouhou-Koubaa, Xanthippi Chatzistavrou, Julietta V. Rau, Jyrki Heinämäki, Iulian Vasile Antoniac, Nureddin Ashammakhi, and Nabil Derbel. 2023. "High Boron Content Enhancement on Bioactive Glass Biodegradation." *Journal of Functional Biomaterials* 14 (7). Multidisciplinary Digital Publishing Institute: 364. doi:10.3390/jfb14070364.

Gu, Yifei, Wenhai Huang, Mohamed N. Rahaman, and Delbert E. Day. 2013. "Bone regeneration in rat calvarial defects implanted with fibrous scaffolds composed of silicate and borate bioactive glasses." *Acta Biomaterialia* 9 (11): 9126-9136. https://doi.org/10.1016/j.actbio.2013.06.039.

Gümüşderelioğlu, Menemşe, Ekin Ö. Tunçay, Gökçe Kaynak, Tuğrul Tolga Demirtaş, Seda Tığlı Aydın, and Sema S. Hakki. 2015. "Encapsulated Boron as an Osteoinductive Agent for Bone Scaffolds." *Journal of Trace Elements in Medicine and Biology* 31 (July). Elsevier BV: 120–28. doi:10.1016/j.jtemb.2015.03.008.

Hakki, Sema S., Buket Bozkurt, and Erdogan E. Hakki. 2010. "Boron Regulates Mineralized Tissue-Associated Proteins in Osteoblasts (MC3T3-E1)." *Journal of Trace Elements in Medicine and Biology* 24 (4). Elsevier BV: 243–50. doi:10.1016/j.jtemb.2010.03.003.

Hakki, Sema S., Niyazi Dundar, Seyit Ali Kayis, Erdogan E. Hakki, Mehmet Hamurcu, Ulku Kerimoglu, Nuri Baspinar, Abdullah Basoglu, and Forrest H. Nielsen. 2013. "Boron Enhances Strength and Alters Mineral Composition of Bone in Rabbits Fed a High Energy Diet." *Journal of Trace Elements in Medicine and Biology*27 (2). Elsevier BV: 148–53. doi:10.1016/j.jtemb.2012.07.001.

Huang, Wenhai, Delbert E. Day, Kanisa Kittiratanapiboon, and Mohamed N. Rahaman. 2006. "Kinetics and mechanisms of the conversion of silicate (45S5), borate, and borosilicate glasses to hydroxyapatite in dilute phosphate solutions." *Journal of Materials Science: Materials in Medicine* 17: 583-596. https://doi.org/10.1007/s10856-006-9220-z.

Hasanuzzaman, Muhammad, Aran Rafferty, Mustafa Sajjia, and A. G. Olabi. 2016. "*Properties of glass materials.*" 647-657. https://doi.org/10.1016/b978-0-12-803581-8.03998-9.

Hench, Larry L. 1991. "Bioceramics: From Concept to Clinic." *Journal of the American Ceramic Society* 74 (7). Wiley-Blackwell: 1487–1510. doi:10.1111/j.1151-2916.1991.tb07132.x.

Joy-anne, N. Oliver, Yingchao Su, Xiaonan Lu, Po-Hsuen Kuo, Jincheng Du, and Donghui Zhu. 2019. "Bioactive glass coatings on metallic implants for biomedical applications." *Bioactive Materials* 4: 261-270. https://doi.org/10.1016/j.bioactmat.2019.09.002.

Jung, S. B., D. E. Day, T. Day, W. Stoecker, and P. Taylor. 2011. "Treatment of non-healing diabetic venous stasis ulcers with bioactive glass nanofibers." In *Wound Repair and Regeneration* 19 (2): A30-A30. Malden, MA, USA: WILEY-BLACKWELL.

Kamaly, Nazila, Basit Yameen, Jun Wu, and Omid C. Farokhzad. 2016. "Degradable controlled-release polymers and polymeric nanoparticles: mechanisms of controlling drug release." *Chemical Reviews* 116 (4): 2602-2663. https://doi.org/10.1021/acs.chemrev.5b00346.

Kaur, Gurbinder, Gary Pickrell, Nammalwar Sriranganathan, Vishal Kumar, and Daniel Homa. 2016. "Review and the state of the art: sol–gel and melt quenched bioactive glasses for tissue engineering." *Journal of Biomedical Materials Research Part B: Applied Biomaterials* 104 (6): 1248-1275. https://doi.org/10.1002/jbm.b.33443.

Khaliq, Haseeb, Zhong Juming, and Peng Ke-Mei. 2018. "The physiological role of boron on health." *Biological Trace Element Research* 186: 31-51. https://doi.org/10.1007/s12011-018-1284-3.

Khatti, Zahra, and Seyed Majid Hashemianzadeh. 2016. "Boron nitride nanotube as a delivery system for platinum drugs: Drug encapsulation and diffusion coefficient prediction." *European Journal of Pharmaceutical Sciences* 88: 291-297. https://doi.org/10.1016/j.ejps.2016.04.011.

Kuo, Po-Hsuen, Sameehan S. Joshi, Xiaonan Lu, Yee-Hsien Ho, Ye Xiang, Narendra B. Dahotre, and Jincheng Du. 2019. "Laser coating of bioactive glasses on bioimplant titanium alloys." *International Journal of Applied Glass Science* 10 (3): 307-320. https://doi.org/10.1111/ijag.12642.

Kuo, Po-Hsuen, and Jincheng Du. 2022. "Effect of boron oxide on mechanical and thermal properties of bioactive glass coatings for biomedical applications." *Journal of the American Ceramic Society* 105 (6): 3986-4008. https://doi.org/10.1111/jace.18391.

Lakhkar, Nilay J., In-Ho Lee, Hae-Won Kim, Vehid Salih, Ivan Wall, and Jonathan C. Knowles. 2013. "Bone Formation Controlled by Biologically Relevant Inorganic Ions: Role and Controlled Delivery from Phosphate-Based Glasses." *Advanced Drug Delivery Reviews* 65 (4). Elsevier BV: 405–20. doi:10.1016/j.addr.2012.05.015.

Lee, Bin, Jae-Sung Kwon, Muhammad Waqas Khalid, Kwang Mahn Kim, Joonhui Kim, Kyoung Mook Lim, and Soon Hyung Hong. 2020. "Boron Nitride Nanoplatelets as Reinforcement Material for Dental Ceramics." *Dental Materials* 36 (6). Elsevier BV: 744–54. doi:10.1016/j.dental.2020.03.002.

Lee, Juneyoung, Jeong Hoon Ko, Kathryn M. Mansfield, Peter C. Nauka, Erhan Bat, and Heather D. Maynard. 2018. "Glucose-responsive trehalose hydrogel for insulin stabilization and delivery." *Macromolecular Bioscience* 18 (5): 1700372. https://doi.org/10.1002/mabi.201700372.

Lepry, William C., and Showan N. Nazhat. 2015. "Highly bioactive sol-gel-derived borate glasses." *Chemistry of Materials* 27 (13): 4821-4831. https://doi.org/10.1021/acs.chemmater.5b01697.

Lepry, William C., Shiva Naseri, and Showan N. Nazhat. 2017. "Effect of processing parameters on textural and bioactive properties of sol–gel-derived borate glasses." Journal of Materials Science 52: 8973-8985. https://doi.org/10.1007/s10853-017-0968-y.

Lepry, William C., and Showan N. Nazhat. 2020. "The anomaly in bioactive sol–gel borate glasses." *Materials Advances* 1 (5): 1371-1381. https://doi.org/10.1039/d0ma00360c.

Li, Daojin, Yang Chen, and Zhen Liu. 2015. "Boronate affinity materials for separation and molecular recognition: structure, properties and applications." *Chemical Society Reviews* 44 (22): 8097-8123. https://doi.org/10.1039/c5cs00013k.

Li, O., C. Feng, Q. Cao, W. Wang, Z. Ma, Y. Wu, T. He, Y. Jing, W. Tan, T. Liao, J. Xing, X. Li, Y. Wang, Y. Xiao, X. Zhu, and X. Zhang. 2023. "Strategies of Strengthening Mechanical Properties in the Osteoinductive Calcium Phosphate Bioceramics." *Regenerative Biomaterials*. https://doi.org/10.1093/rb/rbad013.

Liu, X., M. N. Rahaman, and D. E. Day. 2013. "Conversion of melt-derived microfibrous borate (13-93B3) and silicate (45S5) bioactive glass in a simulated body fluid." *Journal of Materials Science: Materials in Medicine* 24: 583-595. https://doi.org/ 10.1007/s10856-012-4831-z.

Liu, Zhen, and Hui He. 2017. "Synthesis and applications of boronate affinity materials: from class selectivity to biomimetic specificity." *Accounts of Chemical Research* 50 (9): 2185-2193. https://doi.org/10.1021/acs.accounts.7b00179.

Lopez-Esteban, S., C. F. Gutierrez-Gonzalez, Laurent Gremillard, E. Saiz, and A. P. Tomsia. 2009. "Interfaces in graded coatings on titanium-based implants." *Journal of Biomedical Materials Research Part A: An Official Journal of The Society for Biomaterials, The Japanese Society for Biomaterials, and The Australian Society for Biomaterials and the Korean Society for Biomaterials* 88 (4): 1010-1021. https://doi.org/10.1002/jbm.a.31935.

Lu, Chenchen, Hengye Li, Heye Wang, and Zhen Liu. 2013. "Probing the interactions between boronic acids and cis-diol-containing biomolecules by affinity capillary electrophoresis." *Analytical Chemistry* 85 (4): 2361-2369. https://doi.org/ 10.1021/ac3033917.

Macon, Anthony L. B., Taek B. Kim, Esther M. Valliant, Kathryn Goetschius, Richard K. Brow, Delbert E. Day, Alexander Hoppe, Aldo R Boccaccini, Ill Yong Kim, Chikara Ohtsuki, Tadashi Kokubo, Akiyoshi Osaka, Maria Vallet-Regí, Daniel Arcos, Leandro Fraile, Antonio J. Salinas, Alexandra V. Teixeira, Yuliya Vueva, Rui M. Almeida, Marta Miola, Chiara Vitale-Brovarone, Enrica Verné, Wolfram Höland, Julian R. Jones. 2015. "A unified *in vitro* evaluation for apatite-forming ability of bioactive glasses and their variants." *Journal of Materials Science: Materials in Medicine* 26: 1-10. https://doi.org/10.1007/s10856-015-5403-9.

Massera, Jonathan, Yaroslav Shpotyuk, F. Sabatier, Thierry Jouan, Catherine Boussard-Plédel, Claire Roiland, Bruno Bureau, Laeticia Petit, Nadia G. Boetti, Daniel Milanese, Leena Hupa. 2015. "Processing and Characterization of Novel Borophosphate Glasses and Fibers for Medical Applications." *Journal of Non-Crystalline Solids* 425 (October). Elsevier BV: 52–60. doi:10.1016/j.jnoncrysol. 2015.05.028.

Meacham, Susan, L. Janette Taper, and Stella L. Volpe. 1994. "Effects of Boron Supplementation on Bone Mineral Density and Dietary, Blood, and Urinary Calcium, Phosphorus, Magnesium, and Boron in Female Athletes." *Environmental Health Perspectives* 102 (November). National Institute of Environmental Health Sciences: 79. doi:10.2307/3431967.

Mitruţ, Ioana, Ion Romulus Scorei, Horia Octavian Manolea, Andrei Biţă, Laurenţiu Mogoantă, Johny Neamtu, Ludovic Everard Bejenaru, Ciocîlteu M. V., Bejenaru C., Rău G., Mogoşanu G. D. 2022. "Boron-Containing Compounds in Dentistry: A Narrative Review." *Romanian Journal of Morphology and Embryology* 63 (3). Romanian Academy's Publishing House: 477–83. doi:10.47162/rjme.63.3.01.

Moseman, Robert F. 1994. "Chemical Disposition of Boron in Animals and Humans." *Environmental Health Perspectives* 102 (suppl 7). National Institute of Environmental Health Sciences: 113–17. doi:10.1289/ehp.94102s7113.

Naseri, Shiva, William C. Lepry, Vimal B. Maisuria, Nathalie Tufenkji, and Showan N. Nazhat. 2019. "Development and characterization of silver-doped sol-gel-derived borate glasses with antibacterial activity." *Journal of Non-Crystalline Solids* 505: 438-446. https://doi.org/10.1016/j.jnoncrysol.2018.11.026.

Naseri, Shiva, Gabriele Griffanti, William C. Lepry, Vimal B. Maisuria, Nathalie Tufenkji, and Showan N. Nazhat. 2022. "Silver-doped sol-gel borate glasses: Dose-dependent effect on Pseudomonas aeruginosa biofilms and keratinocyte function." *Journal of the American Ceramic Society* 105 (3): 1711-1722. https://doi.org/10.1111/jace.17802.

Nielsen, Forrest H. 2004. "Dietary Fat Composition Modifies the Effect of Boron on Bone Characteristics and Plasma Lipids in Rats." *Biofactors* 20 (3). Wiley-Blackwell: 161–71. doi:10.1002/biof.5520200305.

Nielsen, F. H. 2008. "Is Boron Nutritionally Relevant?" *Nutrition Reviews* 66 (4): 183–191. https://doi.org/10.1111/j.1753-4887.2008.00023.x.

Nielsen, Forrest H., and Barbara Stoecker. 2006. "Dietary Boron and Fish Oil Have Desirable Effects on Vertebral Microarchitecture and Strength." *The FASEB Journal* 20 (4). Federation of American Societies for Experimental Biology. doi:10.1096/fasebj.20.4.a561-a.

Nielsen, Forrest H., and Susan L. Meacham. 2011. "Growing evidence for human health benefits of boron." *Journal of Evidence-Based Complementary & Alternative Medicine* 16 (3): 169-180. https://doi.org/10.1177/2156587211407638.

Ozer, U., Kerim Emre Öksüz, and Ali Özer. 2017. "Synthesis and Characterization of CeO_2 Doped ZrO_2 Ceramics by a Simple Sol-Gel Route." *Materials Science Forum*, September. doi:10.4028/www.scientific.net/msf.907.56.

Öksüz, K. E. 2019. "Effect of Composition and Sintering Temperature on the Structure and Properties of Porous Bioactive Glass Scaffolds." *Chiang Mai Journal of Science* 46 (3): 568–578. http://cmuir.cmu.ac.th/jspui/handle/6653943832/66038.

Öksüz, Kerim Emre, and Ali Özer. 2017. "Effect of Yttria on the Phase Formation and Sintering of HA-Al_2O_3Biocomposites." *Acta Physica Polonica A* 131 (3). Polish Academy of Sciences: 576–80. doi:10.12693/aphyspola.131.576.

Öksüz, Kerim Emre, and Hülya Şereflişan. 2022. "Microstructure of Eobania Vermiculata (Müller, 1774): SEM, F-TIR and XRD Methods." *Journal of Agricultural Production*, November. doi:10.56430/japro.1128026.

Öksüz, Kerim Emre, and H. Şereflişan. 2023."Investigation of the Structure and Hardness Properties of Anodonta Anatina Mussel Shells." *Ege Journal of Fisheries and Aquatic Sciences* 40 (2). Ege University: 132–39. doi:10.12714/egejfas.40.2.07.

Öksüz, Kerim Emre, Begüm Kurt, Zeynep Deniz Şahin İnan, and Ceylan Hepokur. 2023. "Novel Bioactive Glass/Graphene Oxide-Coated Surgical Sutures for Soft Tissue Regeneration." *ACS Omega* 8 (24). American Chemical Society: 21628–41. doi:10.1021/acsomega.3c00978.

Özer, Ali, and Kerim Emre Öksüz. 2019. "The Effect of Yttrium Oxide in Hydroxyapatite/Aluminum Oxide Hybrid Biocomposite Materials: Phase, Mechanical and Morphological Evaluation." *Materialwissenschaft Und Werkstofftechnik* 50 (11). Wiley-Blackwell: 1382–90. doi:10.1002/mawe.201800141.

Peddi, Laxmikanth, Richard K. Brow, and Roger F. Brown. 2008. "Bioactive borate glass coatings for titanium alloys." Journal of Materials Science: Materials in Medicine 19: 3145-3152. https://doi.org/10.1007/s10856-008-3419-0.

Pina, Sandra, and J. M. F. Ferreira. 2012. "Bioresorbable Plates and Screws for Clinical Applications: A Review." *Journal of Healthcare Engineering* 3 (2). Hindawi Publishing Corporation: 243–60. doi:10.1260/2040-2295.3.2.243.

Rad, Rezai Moonesi, Ammar Z. Alshemary, Zafer Evis, Dilek Keskin, Korhan Altunbaş, and Ayşen Tezcaner. 2018. "Structural and Biological Assessment of Boron Doped Bioactive Glass Nanoparticles for Dental Tissue Applications." *Ceramics International* 44 (8). Elsevier BV: 9854–64. doi:10.1016/j.ceramint.2018.02.230.

Ratnayake, S. P., N. De Silva, T. K. Kaupuge, T. S. Weerasekera, MD Senarath Yapa, V. Karunaratne, and G. A. J. Amaratunga. 2018. "Oxidation protection of carbon fiber by sol-gel derived boron doped yttria-stabilized zirconia coatings." Materials Science and Engineering: B 229: 59-64. https://doi.org/10.1016/j.mseb.2017.12.011.

Saghatchi, Mahshid, Sorosh Abdollahi, Bijan Eftekhari Yekta, and S. Mohammad Mirkazemi. 2023. "Synthesis of copper bearing borosilicate glass for soft tissue wound healing." *Ceramics International* 49 (4): 5657-5666. https://doi.org/10.1016/j.ceramint.2022.09.304.

Saranti, A., Ioannis Koutselas, and Michael A. Karakassides. 2006. "Bioactive Glasses in the System $CaO–B_2O_3–P_2O_5$: Preparation, Structural Study and *in Vitro* Evaluation." *Journal of Non-Crystalline Solids* 352 (5). Elsevier BV: 390–98. doi:10.1016/j.jnoncrysol.2006.01.042.

Schuhladen, Katharina, Vera Bednarzig, Nadine Rembold, and Aldo R. Boccaccini. 2021. "The effect of borate bioactive glass on the printability of methylcellulose-manuka honey hydrogels." *Journal of Materials Research* 36 (19): 3843-3850. https://doi.org/10.1557/s43578-021-00256-9.

Seyedmomeni, Sh S., M. Naeimi, M. Raz, J. Aghazadeh Mohandesi, F. Moztarzadeh, F. Baghbani, and M. Tahriri. 2018. "Synthesis, characterization and biological evaluation of a new sol-gel derived B and Zn-containing bioactive glass: *in vitro* study." *Silicon* 10: 197-203. https://doi.org/10.1007/s12633-016-9414-z.

Shafaghi, Romina, Omar Rodriguez, A. W. Wren, Loraine L. Y. Chiu, Emil H. Schemitsch, Paul Zalzal, Stephen D. Waldman, M. Papini, and Mark R. Towler. 2021. "*In Vitro* Evaluation of Novel Titania-containing Borate Bioactive Glass Scaffolds." *Journal of Biomedical Materials Research Part A* 109 (2). Wiley: 146–58. doi:10.1002/jbm.a.37012.

Shenkin, Alan. 2003. "Dietary Reference Values for Vitamin A, Vitamin K, Arsenic, Boron, Chromium, Copper, Iodine, Iron, Manganese, Molybdenum, Nickel, Silicon, Vanadium and Zinc." *Journal of Human Nutrition and Dietetics* 16 (3). Wiley-Blackwell: 199–200. doi:10.1046/j.1365-277x.2003.00439.x.

Siddiqi, Allauddin, Abdul Samad Khan, and Sobia Zafar. 2017. "Thirty Years of Translational Research in Zirconia Dental Implants: A Systematic Review of the Literature." *Journal of Oral Implantology* 43 (4). Allen Press: 314–25. doi:10.1563/aaid-joi-d-17-00016.

Stone-Weiss, Nicholas, Eric M. Pierce, Randall E. Youngman, Ozgur Gulbiten, Nicholas J. Smith, Jincheng Du, and Ashutosh Goel. 2018. "Understanding the structural drivers governing glass–water interactions in borosilicate based model bioactive glasses." *Acta Biomaterialia* 65: 436-449. https://doi.org/10.1016/j.actbio.2017.11.006.

Su, Zhifei, Jiehang Li, Xuan Bai, Franklin R. Tay, Min Zhang, Kunneng Liang, Libang He, He Yuan, and Jiyao Li. 2020. "Borate bioactive glass prevents zoledronate-induced osteonecrosis of the jaw by restoring osteogenesis and angiogenesis." *Oral Diseases* 26 (8): 1706-1717. https://doi.org/10.1111/odi.13436.

Swager, Timothy M., and Sarah P. Luppino. 2015. "Nothing boring about this borylation." *Synfacts* 11 (03): 0266-0266. https://doi.org/10.1055/s-0034-1380103.

Şimşir, Mehmet, H. Akkan, and Kerim Emre Öksüz. 2021. "Processing and Characterization of Porous SiC/NiTi Alloys for Biomedical Applications." *Kovové Materiály* 57 (05): 363–69. doi:10.4149/km_2019_5_363.

Tang, Yue, Libin Pang, and Deping Wang. 2017. "Preparation and characterization of borate bioactive glass cross-linked PVA hydrogel." *Journal of Non-Crystalline Solids* 476: 25-29. https://doi.org/10.1016/j.jnoncrysol.2017.07.017.

Tosun, Nazan Goksen, Ali Ozer, Tugba Bektas, Kerim Emre Oksuz, Secil Erden Tayhan, and Tugba Ozdemir. 2023. "Silk Sericin-Hydroxyapatite Nanoribbons toward Structurally Stable Osteogenic Scaffolds." *Journal of the Australian Ceramic Society*, June. Springer Science+Business Media. doi:10.1007/s41779-023-00909-4.

Turk, Mert, and Aylin M. Deliormanlı. 2017. "Electrically conductive borate-based bioactive glass scaffolds for bone tissue engineering applications." *Journal of Biomaterials Applications* 32 (1): 28-39. https://doi.org/10.1177/08853282 17709608.

Uysal, Idil, Bengi Yilmaz, and Zafer Evis. 2020. "Boron Doped Hydroxapatites in Biomedical Applications." *Bor Dergisi*, October. doi:10.30728/boron.734804.

Wang, Shuming, and Fenghua Kuang. 2010. "Sol-gel preparation and infrared radiation property of boron-substituted cordierite glass-ceramics." *Journal of Materials Science & Technology* 26 (5): 445-448. https://doi.org/10.1016/s1005-0302(10)60070-9.

Wang, Hui, Shichang Zhao, Jie Zhou, Youqu Shen, Wenhai Huang, Changqing Zhang, Mohamed N. Rahaman, and Deping Wang. 2014. "Evaluation of borate bioactive glass scaffolds as a controlled delivery system for copper ions in stimulating osteogenesis and angiogenesis in bone healing." *Journal of Materials Chemistry B* 2 (48): 8547-8557. https://doi.org/10.1039/C4TB01355G.

Xia, Lunguo, Wudi Ma, Yifeng Zhou, Zhipeng Gui, Aihua Yao, Deping Wang, Akimichi Takemura, Mamoru Uemura, Kailin Lin, and Yinghong Xu. 2019. "Stimulatory Effects of Boron Containing Bioactive Glass on Osteogenesis and Angiogenesis of Polycaprolactone: *In Vitro* Study." *BioMed Research International* 2019 (March). Hindawi Publishing Corporation: 1–12. doi:10.1155/2019/8961409.

Xie, Zongping, Xin Liu, Weitao Jia, Changqing Zhang, Wenhai Huang, and Jianqiang Wang. 2009. "Treatment of osteomyelitis and repair of bone defect by degradable bioactive borate glass releasing vancomycin." *Journal of Controlled Release* 139 (2): 118-126. https://doi.org/10.1016/j.jconrel.2009.06.012.

Yao, Aihua, Deping Wang, Wenhai Huang, Qiang Fu, Mohamed N. Rahaman, and Delbert E. Day. 2007. "*In vitro* bioactive characteristics of borate-based glasses with controllable degradation behavior." *Journal of the American Ceramic Society* 90 (1): 303-306. https://doi.org/10.1111/j.1551-2916.2006.01358.

Yılmaz, Bengi. 2016. "*Boron-Substituted Bioceramics: A Review.*" March 24. https://dergipark.org.tr/en/pub/boron/issue/18612/196468.

Zhang, Siqi, Artemis Stamboulis, and Ning Wen. 2019. "The Effect of Boron Substitution for Aluminium on the Microstructure of Calcium Fluoro-Aluminosilicate Glasses and Glass-Ceramics." *Journal of the European Ceramic Society* 39 (5). Elsevier BV: 1918–24. doi:10.1016/j.jeurceramsoc.2018.11.023.

Zhang, Yadong, Xu Cui, Shichang Zhao, Hui Wang, Mohamed N. Rahaman, Zhongtang Liu, Wenhai Huang, and Changqing Zhang. 2015. "Evaluation of injectable strontium-containing borate bioactive glass cement with enhanced osteogenic capacity in a critical-sized rabbit femoral condyle defect model." *ACS Applied Materials & Interfaces* 7 (4): 2393-2403. https://doi.org/10.1021/am507008z.

Zhao, Shichang, Le Li, Hui Wang, Yadong Zhang, Xiangguo Cheng, Nai Zhou, Mohamed N. Rahaman, Zhongtang Liu, Wenhai Huang, and Changqing Zhang. 2015. "Wound dressings composed of copper-doped borate bioactive glass microfibers stimulate angiogenesis and heal full-thickness skin defects in a rodent model." *Biomaterials* 53: 379-391. https://doi.org/10.1016/j.biomaterials.2015.02.112v.

Chapter 3

Biocomposite Materials in Orthodontics

Orhan Cicek[1]
Nurhat Ozkalayci[2]
and Mehmet Yetmez[3,*]
[1]Department of Orthodontics, Zonguldak Bulent Ecevit University, Zonguldak, Türkey
[2]Department of Healthcare Management, Sinop University, Boyabat, Sinop, Türkey
[3]Department of Mechanical Engineering, Zonguldak Bulent Ecevit University, Zonguldak, Türkey

Abstract

Biomaterials are called non-vital materials intended to interact with biological habitats. Any material that emerges as a bioengineering product can be considered as a biomaterial in the repair and reconstruction of lost, debilitated, or infarcted tissues. They differ from other materials in that they have chemical, mechanical, physical, and biological properties that make them suitable for use in a physiological medium. In this context, these biomaterials are applied in a wide range of fields such as medicine, dentistry, veterinary and pharmacology, which are in close and continuous contact with body tissues. The most fundamental feature that should be in a biomaterial planned to be used in medical applications is biocompatibility and it is seen that single-phase homogeneous, isotropic and so called isotropic materials such as metal, polymer and ceramic are used. Mechanical weakness, liquid absorption, and leaking in polymers, a tendency to create image artifacts in three-dimensional imaging methods in metals, and low mechanical reliability, brittleness, fabrication difficulties, and lack of elasticity in ceramics are main points of all disadvantages for these materials. These disadvantages have led to the development of composite biomaterials, which exhibit

* Corresponding Author's Email: yetmez@beun.edu.tr.

In: Biocomposites
Editor: James R. Bush
ISBN: 979-8-89113-267-2

multi-phase material properties without losing their properties, which are consist of two or more materials with different chemical structures. Composites, which were called biocomposites for the first time in 1991 with their use in the human body, are a mixture made by maintaining the superior properties of two different materials and losing their inferior properties. For the sake of biocompatibility, there are natural biocomposites such as bones, teeth, skin, cartilage, tendons and ligaments, as well as resin-based composite filling materials in dental applications. Biocompatible and biodegradable biocomposites with superior mechanical properties for use in oral and dental tissue engineering have been developed mainly by using multiple biocompatible material mixtures. Consequently, in the design of biocomposite materials used in humans, the suitable selection of matrix and materials, appropriate production and processing methods, and the internal-external design of the device must be taken into account correctly. In this chapter, not only biocomposite materials used in orthodontic practices are discussed but also dental implant and prosthetic bridge applications. In fixed orthodontic treatment, orthodontic brackets and arch wires placed in the slots of these brackets are used for the leveling and proper alignment of the teeth. In general, both brackets and arch wires are manufactured from metal alloys. Composite arch wires, however, offer the benefits of aesthetics, ease of forming, and the ability to alter stiffness without changing component dimensions. Again, polymer composite orthodontic brackets need to be developed due to disadvantages such as the abrasion of tooth enamel of ceramic orthodontic brackets and the unaesthetic appearance of metallic brackets and the risk of allergic reaction. The aim of this chapter is (i) to reveal the current findings on biocomposite materials used in the field of orthodontics and (ii) to discuss new and futher results for behaviors of biocomposite materials under various orthodontic force conditions.

Keywords: biomaterial, biocomposite, orthodontics, dentistry

1. Introduction

As a result of significant advances in understanding the interactions between tissues and materials over many years, it has been observed that there are huge differences between vital and non-vital materials [1]. For this reason, the terms "biomaterial" and "biocompatibility" have been used to understand the biological performance of materials [2]. In this chapter, biocompatible materials can be considered as biomaterials, and biocompatibility is defined as the ability of a material to exhibit performance against the appropriate reaction of the host during its specific application to a host [3]. This definition may be further expanded as the chemical, biological and physical compatibility of the surface of the biomaterial in the host tissue, and the structural compatibility of the biomaterial in the form of providing the most appropriate adaptation to the mechanical behavior of the host tissues [4].

Generally known that "bioengineering" is the application of the concepts and methods of science and mathematics for the repair or reconstruction of lost, damaged or dead tissues in an engineering approach, and the materials used for this purpose are called biomaterials [1]. Biomaterial was originally defined by Williams as "an inanimate material used in a medical device intended to interact with biological systems" [5]. He later updated this definition as "a material intended to interface with biological systems to evaluate, treat, augment or modify any tissue, organ or function of the body" [6]. Here, the feature of a biomaterial comes into prominence when it comes into contact with the body and/or body fluid, that it is compatible without producing a toxic or immunological response. At this point, one can mention about "composite material" briefly. Composite material is obtained by combining two or more components that do not dissolve or mix with each other [7]. The term "biocomposite" is used for those used in bioengineering of composites, a material containing two or more different constituent phases on a larger than atomic scale. In other words, while the components in a biocomposite material act in harmony while preserving their properties, they do not dissolve or fuse with each other at the same time. The mechanical properties of composites, which are structurally anisotropic in nature, exhibit different properties in different directions. In composite materials, the constituent components can be physically identified and exhibit an interface between each other. That is, properties such as the elastic modulus of a composite material may differ significantly from the properties of each component alone, but vary significantly, influenced by the structure and content of the components [1].

Additionally, biocomposites were originally developed to optimize the structure, performance, and interactions of biomedical devices with surrounding tissues, resulting in large-scale mechanical and biological properties by varying the type and distribution of reinforcement phases [1]. For example, weak and soft polymers become stronger if mixed with ceramic particles to stop or deflect crack propagation. Thus, the use of this new material in the human body, in which the good properties are preserved and the bad properties are lost as a result of mixing two different materials, is now called "biocomposite" [8] for the first time by Bernard and Picha [9].

Since traditional single-component ceramic or polymer-based materials for use in rapidly developing biomedical technologies such as tissue engineering, regenerative medicine, gene therapy and controlled drug delivery cannot meet the need, new biomaterials need to be developed. In the composite materials developed for this purpose, ceramic modifiers such as silica and wollastonite are used as well as components such as bioactive glass and calcium phosphate [10].

Bone and teeth are composite materials mainly composed of an organic matrix of collagen and hydroxyapatite [(HA), $(Ca_{10}(PO4)_6(OH)_2)$] minerals. Bioactivation has been achieved by incorporating different bioactive fillers such as bioactive glass, hydroxyapatite and wollastonite glass-ceramics into the biocomposites that have emerged with the development of polymer and inorganic fillers. Thus, biocomposites with superior biocompatibility, biodegradability and mechanical properties have taken their place in many applications in both medicine and dentistry [1, 7, 11].

Choosing biocomposite materials suitable for the physiological and mechanical characteristics of the area where it will be placed in the body facilitates the adaptation of the biomaterial. The fact that biocomposites have properties such as high strength and low modulus of elasticity that one of its components cannot have alone increases its advantage over other biomaterials [12]. In addition, the superior properties of biocomposites such as high corrosion resistance, no metal fatigue and metal ion release, and low brittleness enable these materials to be used as soft tissue implants in addition to dentistry and orthopedics [13].

Every year, millions of people have been exposed to bone defects caused by trauma, tumor or bone-related diseases and unfortunately die as a result of insufficient ideal bone tissue [14]. Although autografts and allografts are used in the treatment of these bone defects, there are limitations such as donor shortage and donor site morbidity in autografts, and disease transmission or immune reaction development in allografts. To address these limitations and the problems of bone grafting needed for surgical procedures, research has been conducted in bone tissue engineering to enable a natural tissue regeneration or replacement [15, 16]. Bone fabrication in bone tissue engineering involves the combination of biomaterials and cells that mimic the structure and composition of bone tissue at the nanoscale. For this, natural or biopolymer- based composites containing chitin, chitosan or collagen have the necessary biocompatibility and biodegradation properties [16].

With bone tissue engineering, it is aimed to develop three-dimensional scaffolds that are strong and have porosity that can naturally organize tissue regeneration or replacement when necessary. The scaffold obtained in this way must have an interconnected porous structure in order to guide the new tissue in growth and regeneration [17]. Again, biomimetic matrices are used to support osteoblast proliferation and osteogenesis required for bone regeneration [18].

Biocomposites developed for oral and dental tissue engineering have basically been developed by using multi-material mixtures [19]. Today, it is seen that many biocomposite materials are used for tissue repair, reconstruction and replacement in humans. For this purpose, the suitability of interior and exterior design in a composite structure, selection of appropriate matrix and dispersed materials, and selection of the right production and processing method are very important. Biocomposites produced with superior mechanical properties and biodegradability are also biocompatible [11]. Examples of natural composites found in plants and animals are wood, bone, dentin, collagen, cartilage and skin [1, 7, 11]. Apart from these, resin-based composites used for tooth filling in dentistry are also among biocomposites [11]. Synthetic composites, on the other hand, are essentially the combination of two constituent phases consisting of a fiber or particle reinforcement phase and a continuous phase namely matrix [20].

2. Natural Biocomposites

Composite materials naturally found in animals and plants must be properly studied and understood in order to properly design synthetic biocomposites produced for use in various applications [21]. Cellulose (fibers) in combination with lignin is a common example of polymer natural composites of wood structure [22]. Again, cotton is a material that consists of cellulose fibers alone and has a weaker and more regular structure than wood [23]. As a result of the combination and structural arrangement of these components, which turn into weak materials when separated, a strong and effective strength emerges [22, 24]. As a dynamic component of the human body, bone is also an example of natural biocomposites [25, 26]. Its main components are hydroxyapatite with its hard and brittle mineral properties, soft and flexible protein collagen, non-collagenous proteins and mucopolysaccharides. The presence of these substances in certain proportions and the way they are arranged give the bone the appropriate structure and properties to fulfill its function in the body. Collagen is also found in other natural biocomposites such as cartilage and nails, but exhibits completely different properties. This diversity in appearance and properties results from their combination with other materials and their arrangement patterns in each natural composite material.

The enamel, dentin and cementum tissues in the tooth structure are also three different natural biocomposites, and their main components are hydroxyapatite, collagen, specialized cells, other proteins and minerals. Enamel has a brittle and hard structure compared to the softer dentin and forms the outer surface of the tooth. Enamel is translucent and cannot be regenerated when damaged, but can be restored with other synthetic biocomposites (dental resin materials, nanocomposite or ceramic). Dentin, on the other hand, has a color tone that reflects from the enamel and provides darkness to the tooth. Dentin tissue has the ability to renew itself, albeit limited, and is subject to changes due to aging, physiological and pathological stimuli. It should be noted that although the main components of both biocomposites (enamel and dentin) are similar, they exhibit different properties, textures and appearances. This is because the percentages and arrangement of these components are different in each [11]. These arrangements vary greatly on the properties of each biocomposite at nano, micro and macro levels [27]. The hydroxyapatite in enamel is arranged in a special keyhole pattern that creates locked enamel prisms. Dentin and cementum tissues are natural composites with a matrix made of live, hydrated and type I collagen filled with the calcium phosphate mineral apatite [28]. Dentin forms the main part of the dental crown, which is covered with enamel in the oral cavity, while cementum covers the dentin on the surface of the tooth root, which acts as an intermediate layer to connect the root to the periodontal ligament. The periodontal ligament attaches the cementum to the socket in the alveolar bone and helps to suspend the tooth and hold it in position. As a result of the higher percentage of organic components, dentin tissue is less brittle than enamel, making it a suitable substrate to support enamel against functional forces. The higher hardness enamel protects the dentin against abrasive wear [29].

3. Synthetic Biocomposites

In synthesizing synthetic biocomposites for tissue engineering, advanced production techniques and materials science and a deeper understanding of biological reactions play an important role. Composite scaffolds are generally composed of two or more various scaffold materials such as synthetic polymer and inorganic materials. The purpose of this combination in selected biomaterials is to minimize the disadvantages by ensuring that the tissue engineered product benefits from each material [11]. Synthetic polymers such as bisphenol A-glycidyl methacrylate (Bis GMA) or urethane dimethacrylate (UDMA) are mixed with inorganic filler particles or fibers to achieve the required properties. These polymer- based composites have become increasingly popular in the restoration of lost/damaged enamel and dentin. Polymethylmethacrylate (PMMA) is the most popular polymer that has been used successfully in dentistry and medicine for a long time in plain or composite form [7, 11].

Ceramic or glass based biocomposites are currently widely used as scaffolding for dental tissue regeneration. The filler particles are added to overcome the inherent weakness of polymers such as lack of stiffness. Polymers have advantages of easy fabrication over hard and brittle materials such as ceramic scaffolds with poor processing ability to form high porosity structures. Polymers often make the environment more acidic during the degradation process. The resorption products of calcium phosphate scaffolds neutralize this acidic environment, resulting in less inflammation associated with degradation of polymers. In recent years, significant advances have been made in bone and dental tissue engineering by producing high porosity polymer/ceramic composite scaffolds with osteoconductivity and required mechanical properties. The synthetic composite materials used are biodegradable polymers as follows: (i) polyesters as polylactics, (ii) polycyanoacrylates as isobutyl cyanoacrylate polymer, (iii) polypeptides as poly (L-glutamic acid) and poly(L-lysine), and (iv) polyanhydrides as polyphoshazenes and poly(ortho esters) [11, 30].

In addition to the sufficient biocompatibility and high bioactivity of the material for hard tissue repair, significant improvements have been made in cell connectivity and viability in the polymer-bioglass-composite. When dentin is damaged, one of the hardest natural materials, it needs to be properly repaired with the right biomaterials, ideally with biocomposites. It has recently been shown that there is massive cell proliferation and differentiation when pure polylactic glycolic acid scaffold is coupled with hydroxyapatite, tricalcium phosphate, and calcium-deficient hydroxyapatite. However, it has been reported that calcium phosphate composite scaffolds effectively support the regeneration of dental tissue and are the most suitable among polylactic glycolic acid/tricalcium phosphate scaffolds for dentin pulp regeneration [11, 31]. In addition, zirconia hydroxyapatite composites are another synthetic porous material used in tissue engineering of large bone defects with adequate biocompatibility and mechanical properties, as well as excellent bone reconstruction and regeneration ability [11, 32].

4. Properties and Compatibility of Biocomposites

Biocomposites are becoming increasingly popular in dentistry due to their various desirable properties such as mechanical properties, biocompatibility, bioactivity, antibacterial activities, anti-caries and regenerative activities, adhesion to tooth structure, ease of use and high aesthetic values [11]. The interactions between the different phases of biocomposites determine their mechanical, physical or biological behavior in the living environment. The size, type and weight fraction of inorganic filler particles have a significant effect on the flexural strength, toughness and surface hardness of polymethylmethacrylate [33]. This effect is seen in polymethylmethacrylate-based bone cements, where bone particles increase the fatigue life and hardness of the cement [31]. This is similarly seen in fiber composite bone plates and femoral bodies, which have higher flexibility than their metal counterparts with better healing [34].

Although biocompatibility is one of the key properties of biocomposites, this term is not easy to define. The definition of biocompatibility has evolved over the years with the use of biomaterials for various purposes in different parts of the human body. In addition, interactions between materials and biological environments can cause a wide variety of local and systemic responses that can be considered curative, neutral, or toxic in a given situation. Therefore, a great deal of uncertainty remains about the definition of biocompatibility [11].

Bioactivation is the ability of materials to induce a specific biological response, and biomaterials such as hydroxyapatite, calcium phosphate (β-tricalcium phosphate and tetracalcium phosphate) wollastonite glass-ceramics and bioactive glasses can induce bioactivation and bone-binding ability in neutral ceramics or titanium alloys [11, 35, 36]. In addition to its osteoconductive properties, hydroxyapatite is non-toxic and non-inflammatory, making it an important component of restorative and regenerative materials [37]. Osteoblast proliferation and osteogenesis are stimulated by bioglasses through gene expression and the release of calcium, phosphorus and silicon ions [11].

The osteoconductive and osteoinductive properties of hydroxyapatite enable it to serve as a scaffold for existing osteoblasts and the formation of new osteoblasts [38]. Resin (polymer)-based restorative and protective composites made of polymeric matrix such as UDMA, Bis-GMA and PMMA mixed with non-biodegradable filler particles and widely used in dentistry are examples of non-biodegradable biocomposites. Resin composites retain their shape, size and appearance, along with their mechanical and physical properties, during their service, thanks to the polymerization shrinkage and water absorption reducing properties of the filler particles and their effects to increase wear resistance and strength [39]. Although these polymers, which can be cytotoxic or allergenic in their unreacted form, may cause adverse reactions, especially among dental personnel [40], they are neutral and safe after polymerization. Biocomposites with polymeric matrix have higher aesthetic appearance and corrosion resistance than metal alloys. In addition, biocomposites, which exhibit better fatigue strength and fracture toughness than ceramics, are not as radio-opaque as metal alloys and ceramics [41, 42].

A group of bioactive dental composites has been developed to reduce dental caries, either by suppressing the harmful activity of oral bacteria or by increasing the acid resistance of the tooth structure. Due to its favorable caries prevention activity and mechanical and physical properties, bioactive glasses containing 45S5 BAG fillers have been developed to cover pits and fissures of teeth [43]. Glass ionomers, resin modified glass ionomers and compomers are other restorative materials with composite structure that can store and release fluoride in the oral environment [44]. The fluoride ion replaces the hydroxide in the hydroxyapatite crystal to form more acid-resistant fluoroapatite and facilitates enamel remineralization by inhibiting the metabolism of cariogenic bacteria such as Streptococcus mutans [45]. Bone cements may contain antibacterial agents in the form of antibiotics such as gentamicin, tobramycin, vancomycin and cefazolin, or in the form of filler particles such as silver nanoparticles or calcium hydroxide particles [11, 46, 47].

It can be clearly pointed out that the main focus of biomaterial sciences has always been the healing and regeneration of soft and hard tissues. Calcium hydroxide, which is included as the main component of pulp capping and root canal sealers to induce dentinogenesis, is the most popular material with this capacity. These materials are used to cover and protect the exposed vital pulpal tissue in apexogenesis and apexification treatments applied to mature the tooth root or to provide closure of the apex. The alkaline PH environment caused by the hydroxyl ions released from this cement suppresses bacterial activity and creates liquefied necrosis in the superficial part of the pulp, while the deeper region of the pulp maintains neutral pH, which stimulates hard tissue formation [11]. Apart from this, Mineral Trioxide Aggregate biomaterial was developed and introduced in 1993 by Torabinejad et al. for pulp closure, root canal filling, perforation repair, apexification, apical plugging and revascularization [48, 49]. In addition, bioactive biocomposite materials such as bioglass or bioceramic are used as coating materials to improve the osteointegration of titanium and titanium- based implants [50]. It should be noted that not only the composition of biomaterials, but also their structural and surface properties can affect tissue response. For example, only porous materials with a porosity size greater than 150 μm allow tissue ingrowth when used in implants [51, 52].

Micro-retention and chemical bonding to Ca ions in the tooth structure [53] is one of the desired properties in restorative materials. Glass ionomer and resin modified glass ionomers are biomaterials with this capacity. However, conventional resin composites require a bonding agent to bond to the tooth. Therefore, although newly developed self-adhesive resin composites offer promising in-vivo results, more clinical evidence is needed [54]. Compomers that are poorly bonded to the tooth structure are indicated in low stress bearing areas that need small volume restorations. In general, less invasive restorations have been paved with improved adhesion to the teeth and there is no need for destructive cavity preparations that cause excessive loss of material for retention. Due to their improved aesthetics, durability and mechanical properties, ceramics and polymer- based composites are becoming the most popular restorative materials. These materials offer the possibility to easily use different shadow and translucency properties to mimic natural tooth color or to mask discolored teeth. Polymer composites are cured by direct chemical, light or a combination of both, depending on the production method. Although polymerization shrinkage places stress on the restoration and tooth attachment surface leading to microleakage and recurrent caries, in most cases it can be reliably controlled with the right case selection and application technique [11].

5. Mineralized Biocomposites

5.1. General Perspectives of Bone

As a natural biocomposite that provides mechanical support and flexibility to the body, remodeling of bone tissue is coordinated with the balance of activity between osteoblast and osteoclast cells. While the tensile strength against mechanical forces is provided by collagen fibers, the inorganic hydroxyapatite component increases the toughness and hardness of the tissue [55]. The bone-forming cells responsible for the formation of the organic extracellular matrix are osteoblasts. The mature embedded bone cells that protect and repair bone tissue are osteocytes. Osteoclasts, on the other hand, selectively resorb bone in a specific area in response to a biochemical or biomechanical stimulus [56].

Human bone tissue is divided into compact (cortical) and cancellous (trabecular). Compact bones have central canals (Haversian canals) and penetrating canals (Volkmann's canals) surrounded by concentric rings of matrix. At the center of the osteon or Haversian system, which consists of central canals, including small blood vessels and nerves, surrounded by concentric rings of matrix, is a large capillary blood vessel oriented parallel to the long axis of the bone. The main vessel in each osteonic canal forms a network interconnected with a larger blood vessel and pierces the periosteum covering the surface of the bone through Volkmann's canals. Cancellous bone, also called spongy (trabecular), is less dense and consists of plates and rods of bone surrounding spaces filled with red and white blood cells and differentiated platelets. These spaces are irregular and are also known as bone marrow [7, 57].

The main inorganic component of natural bone is hydroxyapatite, which is used as a column filling material and industrial catalyst for affinity chromatography to separate various proteins in orthopedic and dental materials [58]. Nano-hydroxyapatites with excellent bioactive (ability to form direct chemical bonds with surrounding tissues) and biocompatible properties are widely applied as bone repair material. In this sense, hydroxyapatites have biocompatible, bioactive and osteoconductive properties and do not cause toxic, inflammatory or immune reactions [16].

In the body, the crystallinity, porosity and composition of natural bone have the ability to adapt to biological and biomechanical environments. Thus, the mechanical properties of bone change depending on its biological location. Rigid and flexible collagen fibers reinforced with hydroxyapatite crystals provide bone with high tensile strength and fracture toughness [59].

Chitosan and its derivatives are very attractive because of their properties that allow degradation without inflammatory reaction or toxic degradation at the final stage when new tissues are formed in scaffold composites [16]. Tang et al. [60] reported that natural hydroxyapatite and chitosan composites have good hard tissue biocompatibility and excellent osteoconductivity for artificial bone implants and framework materials of tissue engineering. Jayabalan et al.

[61] reported that the nanopomposite containing calcined hydroxyapatite nanoparticles is both biocompatible and osteocompatible. From a biomimetic perspective, hydroxyapatite composites can potentially improve the biocompatibility and mechanical properties of bone graft materials [16]. In addition, electrospun nanocomposite nanofibers of hydroxyapatite/chitosans with composition and structural properties close to their natural mineralized nanofibril counterparts were prepared by Zhang et al. [62]. Chitosan/montmorillonite hydroxyapatite, a new biopolymer-based nanocomposite with improved mechanical properties, has been reported for biomedical applications [63]. The amount of nano-hydroxyapatite in biocomposites appears to have a greater effect on the early stages of osteoblast behavior (cell attachment and proliferation) than the immediate/late stages (proliferation and differentiation) [64]. Addition of nano-hydroxyapatite to chitosan scaffolds improved cell attachment and well-dispersed morphology, and higher proliferation was achieved compared to chitosan scaffolds alone [65]. It has been reported that osteoblastic cell attachment and proliferation are increased in gelatin scaffolds containing nano-hydroxyapatite [16]. In addition, membranes containing chitosan hydrogel and hydroxyapatite and membranes containing β-chitin and hydroxyapatite have been reported to be suitable for osteoblastic cell attachment and proliferation [66, 67].

Kashiwazaki et al. [68] produced porous chitosan/hydroxyapatite nanocomposites, which they stated to be biocompatible and biodegradable, by co-precipitation and porogen leaching method. Chitosan biocomponent nanofibers with an average diameter of 50-100 nanometers were successfully prepared by electrospinning a mixture solution of chitosan and polyvinyl alcohol [69]. Chitosan-hydroxyapatite nanostructured composite films have also been prepared by solvent casting of hybrid suspensions for tissue engineering applications [70]. A chitosan/hydroxyapatite nanohybrid scaffold with high porosity and homogeneous nanostructure was fabricated by a bionic process with thermally induced phase separation. These nano-hydroxyapatite particles positively guided apatite crystallization in the scaffold and provided good bioactivity of the nanohybrid scaffold [71].

Chitosan, together with other natural polymers, can be mixed with bioceramic material to prepare biomimetic scaffolds. Zhao et al. [72] demonstrated a biodegradable hydroxyapatite/chitosan-gelatin mesh composite with composition similar to normal human bone, prepared by phase separation as a three-dimensional biomimetic scaffold for bone tissue engineering. After 3 weeks, the researchers reported that the cell/scaffold structures had good biomineralization effect in culture. In yet another study,

Zhao et al. [73] produced two types of biomimetic composite materials, chitosan-gelatin and hydroxyapatite/chitosan-gelatin, and demonstrated the enhanced protein and calcium ion adsorption properties of hydroxyapatite in the chitosan-gelatin polymer network. This enhanced initial cell adhesion and long-term growth, promoting osteogenic differentiation upon induction. It also preserved the progenicity (progeny) of three- dimensional human mesenchymal stem cell structures. A new artificial bone matrix was also prepared with collagen and Ca5(PO4)3(OH) hydroxyapatite, which are the main components of natural bone. In order to improve the properties of this artificial bone matrix, chitosan has been cross-linked to scaffolds and it has been reported that this matrix can be used instead of bone [74]. The weight ratios of the components in a composite material prepared to support bone tissue growth are critical. The mechanical properties of the new composites containing nano-hydroxyapatite/chitosan/carboxymethyl cellulose prepared by a co-solution method and prepared in the ratios of 70/10/20, 70/15/15 and 70/20/10 by weight are good, biodegradation rate is adjustable and simulated body. It has been shown to be promising for bone repair materials in terms of bioactivity in the fluid [75]. Liuyun et al. [18] investigated the physicochemical and biological properties of a new biodegradable composite scaffold prepared from nanohydroxyapatite and chitosan and carboxymethyl cellulose polymers by freeze drying method. This scaffold, which is non-toxic and has good cell biocompatibility, has been reported to have good tissue biocompatibility for in-vivo implantation and thus has great potential as a bone tissue engineering material. In another study, it was reported that the addition of hydroxyapatite to chitosan-gelatin scaffolds would promote osteoblastic cell attachment, proliferation and proliferation [76]. It has also been shown that porous scaffolds composed of alginate and chitosan are produced by combining the formation of the polyelectrolyte complex with freeze drying and exhibit higher mechanical strength and better thermal stability [77].

Yoshida et al. [78] showed that in-situ composites produced from cellulose and carbonate hydroxyapatite by mechanochemical reaction have the potential to be bioactive by forming calcium phosphate on simulated body fluid surfaces. Hutchens et al. [79] synthesized a novel composite material consisting of calcium-deficient hydroxyapatite that biomimeticly deposited in a bacterial cellulose hydrogel, imitating the natural biomineralization of bone. Researchers, who stated that bacterial cellulose can be used as a template for biomimetic apatite formation, reported that this composite material has the potential to be used as an orthopedic biomaterial. In another study [80], a nanocrystalline hydroxyapatite cement containing type I collagen and its modifications with sodium citrate, calcium carbonate, phosphoserine and phosphoserine/RGD-peptide was reported to increase bone remodeling and ultimately bone formation in the early stages of bone healing. Tampieri et al. [81] noted that (i) a sublayer of enhanced biomineralized collagen corresponding to subchondral bone, (ii) an upper layer of hyaluronic acid-laden collagen that mimics the cartilage region, and (iii) the same structure as biomineralized collagen, but in a lower amount. They developed a composite osteochondral scaffold that they organized into different integrated layers consisting of a mineral-containing interlayer. Researchers reported that these composite scaffolds, a biomimetic of osteochondral tissue, have the capacity to differentially support cartilage and bone tissue formation. As can be understood from these studies, a significant amount of work is being carried out in bone tissue engineering for the development of biocomposites.

5.2. General Perspectives of Tooth

Tooth, the hardest biological material in the human body, contain three types of tissue: enamel, dentin, and pulp. It is the hardest and highest mineral material in the body, produced by the enamel ameloblast cells, found in the outer layer of the tooth. Enamel tissue consists of 97% mineral, 1% organic matter and 2% water by weight. Organic material includes soluble and insoluble proteins, enamelin, and small amounts of carbohydrates and lipids [7].

Dentin tissue, which is synthesized by odontoblasts in the tooth crown and root, is located between the enamel and the pulp, and together they form the functional unit called human tooth. Primary dentin tissue, which contains 70% hydroxyapatite, 20% organic material and 10% water by weight, is less mineralized than enamel. Primary dentin, which is synthesized before the teeth reach occlusion, forms the main part of crowns and roots and is mainly composed of type I collagen and collagen-like compounds. Primary dentin tissue has a matrix structure in which hydroxyapatite crystals are irregularly integrated. Primary dentin, which has two layers as inner and outer layers, is less mineralized than the inner layer because the outer layer has larger collagen tubes [82]. After the dental crown reaches occlusion, secondary dentin forms under the primary dentin, and while the synthesis process of this secondary dentin continues throughout life, the pulp chamber also gets smaller. Tertiary dentin, which is formed due to injury when exposed to any trauma, also has an irregular structure [83].

6. Applications of Biocomposite Materials in Oral and Dental Tissue Engineering

Tissue engineering is the use of cells, biomaterials and various chemical and biological factors together with engineering techniques to repair, heal or replace living tissue [84]. In recent years, many biomaterials have been used in combination with cells and growth factors to support the regeneration of diseased or damaged tissues in the oral region. Thus, soft and hard tissues damaged as a result of various pathological and traumatic processes in the oral cavity such as dental caries that cause demineralization and cavitation in the tooth structure, various types of periodontal diseases that cause alveolar bone resorption, and infections and /or traumas that cause irritation or ulceration in the mucosa can be healed, can be repaired or replaced [85].

Considering the metal and wood used by the ancient Egyptians and Mayans to place donor teeth in edentulous areas, it seems that the use of various materials to restore teeth dates back thousands of years. However, the beginnings in the 19th and 20th centuries for the restoration of dental caries with gold leaf form the basis of restorative dentistry. Only resin composites are popular as composites in the field of dentistry. However, many dental materials have a composite structure, such as dental amalgam, which is an alloy of silver, mercury and other metal elements and is used to fill cavitation caused by caries. In addition, amalgam is not used in anterior teeth due to its lack of tooth color despite its excellent mechanical properties. In this sense, the emergence of aesthetic resin composites has created a very important innovation in dentistry [86]. Second surgery required to remove guided tissue regeneration membranes made of non-resorbable polytetrafluoroethylene, originally used to serve as a barrier membrane between the gingival epithelium and the underlying bone for regeneration of periodontal bone, increased the likelihood of infection and patient discomfort [11, 87]. In order to overcome these problems, resorbable membranes made of synthetic polymers such as poly-L-lactic acid, polylactic-co-glycolic acid or natural materials such as collagen have been introduced [88]. These are limitations, poor mechanical properties and use, and allergic potential [89]. Biocomposites produced by incorporating nano-sized crystals such as hydroxyapatite into polymers have increased both the bioactivity and mechanical properties of directed tissue regeneration materials [90]. However, nano-sized bioactive bioceramic and biocomposites containing electrospun nanofibers that can be used for tissue regeneration have been investigated, but it has been reported that significantly more research is needed to predict its current status [11, 91].

As a recollection, it is to be known that resin based biocomposites consist of a polymer matrix (bis-phenol A glycidyl methacrylate, triethylene glycol dimethacrylate) that binds the filler particles (silica) together through a binder, glass ionomers used in primary tooth restorations consist of bioactive glass and acids [92, 93]. However, partial prosthetic devices are produced from fillers and acrylic composites to provide certain properties. In addition, current studies are directed towards making materials such as dental implants by incorporating bioactive ceramics into polymers such as polyetheretherketone [94].

Current research has focused on using stem cells carried by various scaffolds made of biodegradable biocomposites to initiate regeneration [95]. Here, the idea that stem cells planted on dental pulp may have the potential to regenerate damaged pulp, dentin and enamel is at the forefront. Essentially, natural polymers such as collagen have been reported to have the potential to provide recombinant growth factors to stimulate the regeneration of dental tissues [96]. With the combined use of growth factors, cells and biocomposites, a new paradigm of possibilities has been opened, thus, it has been reported that the repair of oral tissues will be possible only with synthetic materials, but future studies are needed for the true regeneration of living tissue [11].

7. Biocomposite Materials in Orthodontics Practices

Dental treatments, one of the most frequently applied medical treatments on humans, are quite comprehensive, from filling cavities due to caries to prosthetic applications due to lost or missing tooth(s), periodontium treatments, and orthodontic treatments of disorders in teeth and jaws. As with a wide variety of biomaterials used in all these applications, the ability of the biomaterial used in orthodontic applications to resemble and adapt to the physical, mechanical and aesthetic properties of the natural tooth structure is very important. Fixed orthodontic treatments consist of orthodontic brackets and arch wires applied to move the teeth. Correction of teeth with orthodontic treatment is provided by arch wires placed on brackets attached to the teeth. Tension changes in the archwire as a result of holding the orthodontic archwires in the bracket slot with wire or elastic ligatures ensure the alignment of the teeth. Orthodontic brackets serve as an important focal point in the transfer of forces generated from archwires to teeth. For this reason, it is very important for the brackets to have high strength and rigidity in order to avoid bending during tooth movement [1]. These materials are mostly made of metal alloys, and the order of use of these metal arch wires during the treatment is nickel-titanium and beta-titanium with less hardness at the beginning, and stainless steel and cobalt-chrome arch wires with harder corners are used, although it changes according to the biomechanical performance needed in the final stages [97].

Transparent polycarbon (polymer) and ceramic-based archwires and brackets, which appear due to the unaesthetic appearance of orthodontic appliances produced from metallic alloys, are invisible thanks to their natural tooth color and translucency properties. However, ceramic brackets made of monocrystalline or polycrystalline alumina have been introduced to the market due to both better aesthetics and higher strength due to the breaks seen in polymer brackets. Although the aesthetic properties of ceramic brackets are superior, there are also disadvantages of brittle breakage and high production cost. Teflon-coated stainless steel wires are also available as aesthetic orthodontic arch wires [1].

Traditional archwires, which are widely used in fixed orthodontic treatments, are produced from metal alloys such as stainless steel, cobalt-chromium-nickel, beta-titanium and nickel-titanium [98, 99]. Although these orthodontic archwires have superior mechanical properties compared to polymer, composite and ceramic- based materials, the unaesthetic appearance of the metallic alloy reduces the motivation for orthodontic treatment [100]. For this, materials developed from alumina, zirconia, polycarbonate and their composites have been produced as aesthetic orthodontic materials [99]. For this purpose, glass fiber/polycarbonate (GF/PC), glass fiber/nylon (GF/Nylon), glass fiber/polypropylene (GF/PP) and glass fiber/polymethylmethacrylate (GF/PMMA) composite materials have been proposed for orthodontic archwire applications [101, 102]. These composite archwires have been reported to have the advantages of aesthetics, easy shaping in the clinic, and changing stiffness without changing component dimensions [103].

The fact that aesthetic and cosmetic factors have become more important in dentistry has led the transformation of metallic alloys, which are also widely used in orthodontics, towards tooth-colored and transparent devices [104]. For this purpose, studies have focused on coating orthodontic archwires with tooth-colored Teflon (polytetrafluoroethylene), epoxy resin and polyethylene [104-107]. It has been reported that there are disadvantages of exfoliation of the coating layers in aesthetic orthodontic archwires produced by coating metal alloys with polymers such as epoxy resin and polyethylene to provide an aesthetic appearance [108, 109]. It was also reported in a study conducted by Silvia et al. [108] investigating the coating stability of four different types of aesthetically coated archwires, that these coatings have low oral strength and therefore offer low esthetics. This has led to the need to manufacture both non-coated and non-metallic arc wires [99]. For this purpose, it has been reported that orthodontic archwires produced using glass-fiber reinforced plastics are aesthetically superior due to their transparent or translucent appearance and have similar flexural (bending) properties with commercially available Ni-Ti wires [110]. In addition, color stability is an important factor in aesthetic evaluation, there are many studies on the color stability of orthodontic materials in the literature [111-114] and it has been reported that measuring the color stability of aesthetic orthodontic wires is difficult due to the shape and size of the archwires [115]. In this regard, it has been reported by Inami et al. [99] that aesthetic orthodontic archwires made of glass-fiber reinforced plastics consisting of polycarbonate and glass fibers used during orthodontic treatment have high color stability.

Continuous fiber reinforced plastics have high hardness and strength properties per unit weight. These properties increase the use of materials in areas such as the aerospace and automotive industries, as well as the reduced cost, versatile design possibilities, control of material behavior, direct bonding capabilities, translucent and aesthetic production properties of these materials also potential to be used as a structural component for dental applications [116]. Fiber-reinforced composites have low fiber loading, stress transfer, and difficulties in clinical adaptation. However, studies have reported that the use of pultruded thermoplastic prepregs can eliminate these problems [117, 118].

Fiber-reinforced archwires, which perform in the same range as metallic archwires, can be used in the initial and intermediate stages of orthodontic treatment as their hardness can be easily changed by changing the matrix composition and fiber reinforcement. In this regard, Ohtenon et al. [104] compared the force levels obtained from glass fiber reinforced composite arcwires, and it was found that polymer matrix, hardening method and water absorption had significant effects on the bending behavior of fiber reinforced archwires, and therefore the strength level and operating range were determined. It has been reported that these factors are affected by these factors and that more studies are needed on this subject.

The main material of composite archwires produced from continuous glass fiber reinforced composite by pultrusion method is epoxy resin, polyethylene-terephthalate glycol and poly(1,4)cyclohexylene dimethylene terephthalate glycol and poly-methyl-methacrylate reinforced with S-glass fiber, which has higher strength than E-glass fiber. However, the use of light-cured resins such as urethanedimethacrylate, bisphenol-A glycidyl methacrylate is quite common in the restoration of dental caries cavities. With the developments in adhesive technology, orthodontic brackets previously welded to gold or stainless steel bands in clinical practice have also begun to be applied to teeth with composite resins [119]. Although the composite archwires produced with glass fiber and light-cured resins have biocompatibility [120], the necessary properties for the control of their mechanical properties have been neglected. For this purpose, silane binding agents were used to form the chemical bond between the fiber and the matrix. Here, it is emphasized that the concentration of silane and the processing conditions have important effects on the mechanical properties of the composite archwire [121].

When producing continuous fiber-reinforced plastics for dental applications, it is important that the bending properties such as modulus, strength and tensile are created accurately and reliably. Although the length of an orthodontic wire is much greater than its cross-section, the anisotropic nature of fiber-reinforced archwires also makes the wire's transverse properties important in its mechanical performance. The longitudinal properties of the arc wire control the tensile strength and stiffness of the wire, while the transverse properties control shear strength and long-term stability [116]. The surfaces of glass fibers produced in different grades depending on their chemical composition are normally treated with sizing materials such as silane agent immediately after processing. Despite the advantages of glass fibers such as being transparent and inexpensive, they are rarely used because they are less suitable for use in the body than carbon fibers. However, when mixed with appropriate matrix materials, they can be best used in the development of orthodontic brackets and archwires for orthodontic applications where esthetics is important and stiffness is not a critical issue [1]. Jancar et al. [116], in their study investigating the analytical relationships between the load deflection data of unidirectional long fiber composite archwires and the intrinsic mechanical properties of the composite, showed that the intrinsic stiffness of continuous glass fiber reinforced orthodontic wires is not fully utilized, but it is useful in the design of dental devices made of continuous glass fiber reinforced plastics. reported that they would.

Orthodontic brackets are mostly produced from stainless steel, polycrystalline alumina and monocrystalline alumina materials. Despite the high strength and hardness of the brackets produced with metal alloys, ceramic brackets with superior aesthetics have been developed due to their poor aesthetic properties. However, ceramic brackets are more larger than metal brackets and the risk of damage due to their abrasive effects on the enamel has limited their use. Also, some patients may be hypersensitive to nickel, chromium, and cobalt metals, and severe foreign body allergic reactions that cause dermatitis in the immune system may be triggered. Since orthodontic materials produced from metallic alloys cannot be used in these cases, there is a need to develop suitable polymer composite orthodontic brackets and arch wires. Although chopped glass fiber composites are used in the production of orthodontic brackets for this purpose, these composite brackets have limited clinical use due to their lower mechanical properties than metal brackets. Further development requires the use of continuous fiber preforms as reinforcement [1].

Biomechanically, friction occurs in sliding mechanics due to the contact between the bracket and the archwire during orthodontic tooth movement [122]. Therefore, surface treatments such as electrolytic treatment and ion implantation applied to orthodontic metal brackets and archwires, and polytetrafluoroethylene (Teflon) coatings are intended to reduce friction. However, these applications have been reported to have disadvantages such as low resistance to intraoral conditions, high friction wear, and delamination or wear of the coating [32]. For this purpose, Arıcı et al. [123] investigated the friction changes between orthodontic brackets and archwires coated with aluminum oxide, titanium nitride or chromium nitride, and reported that these coatings can reduce the friction coefficient.

It is aimed to eliminate these disadvantages in clinical applications by focusing on the production of composite brackets containing interrupted chopped glass fiber reinforced polycarbonate produced by injection molding technique. Although biomechanical properties such as friction between the chopped glass fiber reinforced polycarbonate brackets and the archwire and adhesion to enamel have not been sufficiently improved, it has been reported that polycarbonate brackets with metal slots are used more than composite brackets [124].

Conclusion

Biomaterials used in oral and dental tissue bioengineering support the regeneration and healing of diseased or damaged tissues in the oral area. These are quite comprehensive, from the treatment of dental caries to prosthetic applications, periodontium treatments and orthodontic treatments of disorders in teeth and jaws. The ability of the biomaterial used in all these applications to provide the physical, mechanical and aesthetic properties of the natural tooth structure is very important. The fact that aesthetic and cosmetic approaches are at the forefront in dentistry has led to the change of brackets and arch wires, which are used in orthodontic treatments and are mostly produced from metallic alloys, towards tooth-colored transparent devices. Although aesthetic orthodontic materials developed from alumina, zirconia, polycarbonate and their composites have been developed for this purpose, some disadvantages are observed in the oral environment. Composite orthodontic brackets and arch wires made of glass fiber reinforced plastic stand out among these materials for their positive qualities in terms of both aesthetics and biomechanics. However, these aesthetic orthodontic brackets have limitations such as friction and arch wires are not effective in the initial stages of treatment. For this reason, further studies are needed to produce thick and rectangular arch wires, which are used such a way that parallel movement and root torque are needed in the later stages of orthodontic treatment, from aesthetic orthodontic biomaterials that have the least biomechanical friction and provide sufficient strength.

Disclaimer

None

References

[1] Ramakrishna, S., Huang Z.M., Kumar, A.W., and Mayer, J., *An introduction to biocomposites,* ed. A Batchelor. Vol. 1. 2004, Shelton Street, Covent Garden, London, UK: Imperial College Press.

[2] Williams, D.F., Consensus and definitions in biomaterials. *Advances in Biomaterials.* 1988, Amsterdam: Elsevier Science.

[3] Black, J. and Hastings, G., *Handbook of biomaterial properties.* 1998, Chapman and Hall, London, UK.

[4] Wintermantel, E. and Mayer, J., Anisotropic biomaterials: strategies and developments for bone implants, in *Encyclopedic Handbook of Biomaterials and Bioengineering.* 1995: New York, Marcel Dekker. p. 3-42.

[5] Williams, D.F., *Definitions in biomaterials: proceedings of a consensus conference of the European Society for Biomaterials,* Chester, England, March 3-5, 1986.
[6] Williams, D.F., *The Williams dictionary of biomaterials.* 1999: Liverpool University Press.
[7] Furtos, G., Silaghi-Dumitrescu L., Lewandowska, K., Sionkowska, A., Pascuta, P., Biocomposites for orthopedic and dental application. *Key Engineering Materials, 2016*. 672: p. 261-275.
[8] Mohanty, A.K., Misra, M., Hinrichsen, G., Biofibres, biodegradable polymers and biocomposites: An overview. *Macromolecular materials and Engineering,* 2000. 276 (1): p. 1-24.
[9] Bernard, S. and Picha, G., The use of coralline hydroxyapatite in a "biocomposite" free flap. *Plastic and reconstructive surgery,* 1991. 87 (1): p. 96-105.
[10] Dziadek, M., Stodolak-Zych, E., and Cholewa-Kowalska, K., Biodegradable ceramic-polymer composites for biomedical applications: A review. *Materials Science and Engineering: C,* 2017. 71 : p. 1175-1191.
[11] Nejatian, T., Khurshid, Z., Zafar, M.S., Najeeb, S., Zohaib, S., Mazafari, M., Hopkinson, L., and Sefat, F., *Dental biocomposites, in Biomaterials for oral and dental tissue engineering.* 2017, Elsevier. p. 65-84.
[12] Christian, S., and Billington, S., Sustainable biocomposites for construction. *Composites and polycon,* 2009: p. 1-6.
[13] Ratner, B.D., Hoffman, A.S., Schoen, F.J., and Lemons, J.E., *Biomaterials science: An introduction to materials in medicine.* 2012: Academic Press.
[14] Murugan, R. and Ramakrishna, S., Bioresorbable composite bone paste using polysaccharide based nano hydroxyapatite. *Biomaterials,* 2004. 25 (17): p. 3829-3835.
[15] Kwok, S., Partridge, N.C., Srinivasan, N., Nair, S.V., and Selvamurugan, N., Mitogen activated protein kinase-dependent inhibition of osteocalcin gene expression by transforming growth factor-β1. *Journal of cellular biochemistry,* 2009. 106 (1): p. 161-169.
[16] Swetha, M., -- Swetha, M., Sahithi, K, Moorthi, A., Srinivasan, N., Ramasamy, K., and Selvamurugan, N., Biocomposites containing natural polymers and hydroxyapatite for bone tissue engineering. *International journal of biological macromolecules,* 2010. 47 (1): p. 1-4.
[17] Manjubala, I., Scheler, S., Bössert, J., and Jandt, K.D., Mineralisation of chitosan scaffolds with nano-apatite formation by double diffusion technique. *Acta biomaterialia,* 2006. 2 (1): p. 75-84.
[18] Liuyun, J., Yubao, L., and Chengdong, X., Preparation and biological properties of a novel composite scaffold of nano-hydroxyapatite/chitosan/carboxymethyl cellulose for bone tissue engineering. *Journal of biomedical science,* 2009. 16, 65.
[19] Mohanty, A.K., Misra, M., and Drzal, L.T., *Natural fibers, biopolymers, and biocomposites.* 2005: CRC press.
[20] Ahmad, J., Introduction to polymer composites. *Machining of Polymer Composites,* 2009: p. 1-35.
[21] Sommerfeldt, D.W. and Rubin, C.T., Biology of bone and how it orchestrates the form and function of the skeleton. *European Spine Journal,* 2001. 10 : p. S86-S95.

[22] Novaes, E., Kirst, M., Chiang, V., Winter-Sederoff, H., and Sederoff, R., Lignin and biomass: a negative correlation for wood formation and lignin content in trees. *Plant physiology,* 2010. 154 (2): p. 555-561.

[23] Cao, J., Sun, X., Lu, C., Zhou, Z., Zhang, X., and Yuan, G., Water-soluble cellulose acetate from waste cotton fabrics and the aqueous processing of all-cellulose composites. *Carbohydrate polymers,* 2016. 149 : p. 60-67.

[24] Santos, R.B., Capanema, E.A., Balakshin, M.Y., Chang, H., and Jameel, H., Lignin structural variation in hardwood species. *Journal of agricultural and food chemistry,* 2012. 60 (19): p. 4923-4930.

[25] Sefat, F., Khaghani S.A., Nejatian, T., Genedy, M., Abdeldayem, A., Moghaddam, Z.S., Denyer, M.C.T., and Youseffi, M., Transforming growth factor beta (TGF-β) isomers influence cell detachment of MG-63 bone cells. *Tissue and Cell,* 2015. 47 (6): p. 567-574.

[26] Nejatian, T., Sefat, F., and Johnson, T., Impact of packing and processing technique on mechanical properties of acrylic denture base materials. *Materials,* 2015. 8 (5): p. 2093-2109.

[27] Khurshid, Z., Zafar, M., Qasim, S., Shahab, S., Naseem, M., and AbuReqaiba, A., Advances in nanotechnology for restorative dentistry. *Materials,* 2015. 8 (2): p. 717-731.

[28] Dorozhkin, S.V., A hierarchical structure for apatite crystals. *Journal of Materials Science: Materials in Medicine,* 2007. 18 (2): p. 363-366.

[29] Chun, K.J., Choi, H.H., and Lee, J.Y., Comparison of mechanical property and role between enamel and dentin in the human teeth. *Journal of dental biomechanics,* 2014, 5, 1758736014520809.

[30] Reis, R.L. and Cohn, D., *Polymer based systems on tissue engineering, replacement and regeneration.* Vol. 86. 2002: Springer Science & Business Media.

[31] Park, H.C., Liu, Y.K., and Lakes, R.S., *The material properties of bone-particle impregnated PMMA.* 1986. 108 (2): p. 141-148.

[32] Park, J.B. and Bronzino, J.D., *Biomaterials: principles and applications.* 1st ed. 2002: CRC Press.

[33] Nejatian, T., Johnson, A., and Van Noort, R., Reinforcement of denture base resin. *Advances in Science and Technology,* 2006. 49 : p. 124-129.

[34] Jockisch, K.A., Brown, S.A., Bauer, T.W., and Merritt, K., Biological response to chopped-carbon-fiber-reinforced peek. *Journal of biomedical materials research,* 1992. 26 (2): p. 133-146.

[35] Saadaldin, S.A. and Rizkalla, A.S., Synthesis and characterization of wollastonite glass–ceramics for dental implant applications. *Dental materials, 2014.* 30 (3): p. 364-371.

[36] Tanzer, M., Gollish, J., Leighton, R., Orrell, K., Giacchino, A., Welsh, P., Shea, B., and Wells, G., The effect of adjuvant calcium phosphate coating on a porous-coated femoral stem. *Clinical Orthopedics and Related Research* (1976-2007), 2004. 424 : p. 153-160.

[37] LeGeros, R.Z., Calcium phosphates in oral biology and medicine. *Monogr Oral Sci.* Vol. 15. 1991.

[38] Kumar, P., Vinitha, B., and Fathima, G., Bone grafts in dentistry. *Journal of pharmacy & bioallied sciences,* 2013. 5 (Suppl 1): p. S125-S127.

[39] Lewandrowski, K.U., Bondre, S.P., Shea, M., Untch, C.M., Hayes, W.C., Hile, D.D., Wise, D.L., and Trantolo, D.J., Composite resorbable polymer/ hydroxylapatite composite screws for fixation of osteochondral osteotomies. *Biomedical materials and engineering,* 2002. 12 (4): p. 423-438.

[40] Moharamzadeh, K., Van Noort, R., Brook, I.M., and Scutt, A.M., Cytotoxicity of resin monomers on human gingival fibroblasts and HaCaT keratinocytes. *Dental materials,* 2007. 23 (1): p. 40-44.

[41] Furtos, G., Tomoaia-Cotisel, M., and Prejmerean, C., Resin composites reinforced by glass fibers with potential biomedical applications. Structure and mechanical properties. *Particulate Science and Technology,* 2013. 31 (4): p. 332-339.

[42] Furtos, G., Silaghi-Dumitrescu, L., Moldovan, M., Baldea, B., Trusca, R., and Prejmerean, C., Influence of filler/reinforcing agent and post-curing on the flexural properties of woven and unidirectional glass fiber-reinforced composites. *Journal of Materials Science,* 2012. 47 : p. 3305-3314.

[43] Yang, S.Y., Piao, Y.Z., Kim, S.M., Lee, Y.K., Kim, K.N., and Kim, K.M., Acid neutralizing, mechanical and physical properties of pit and fissure sealants containing melt-derived 45S5 bioactive glass. *Dental Materials,* 2013. 29 (12): p. 1228-1235.

[44] Wiegand, A., Buchalla, W., and Attin, T., Review on fluoride-releasing restorative materials—fluoride release and uptake characteristics, antibacterial activity and influence on caries formation. *Dental materials,* 2007. 23 (3): p. 343-362.

[45] Buzalaf, M.A.R., Pessan, J.P., Honório, H.M., and Cate J.M.T., Mechanisms of action of fluoride for caries control. *Fluoride and the oral environment,* 2011. 22 : p. 97-114.

[46] Bistolfi, A., Massazza, G., Verne, E., Masse, A., Deledda, D., Ferraris, S., Miola, M., Galetto, F., and Crova, M., Antibiotic-loaded cement in orthopedic surgery: a review. *International Scholarly Research Notices,* 2011, 290851.

[47] Mocanu, A., Furtos, G., Rapuntean, S., Horovitz, O., Flore, C., Garbo, C., Danisteanu, A., Rapuntean, G., Prejmerean, C., and Tomoaia-Cotisel, M., Synthesis; characterization and antimicrobial effects of composites based on multi-substituted hydroxyapatite and silver nanoparticles. *Applied surface science,* 2014. 298 : p. 225-235.

[48] Torabinejad, M., Watson, T.F., and Ford, T.R.P., Sealing ability of a mineral trioxide aggregate when used as a root end filling material. *Journal of endodontics,* 1993. 19 (12): p. 591-595.

[49] Nagy, M.M., Tawfik, H.E., Hashem, A.A.R., and Abu-Seida, A.M., Regenerative potential of immature permanent teeth with necrotic pulps after different regenerative protocols. *Journal of endodontics,* 2014. 40 (2): p. 192-198.

[50] Chu, C., Xue, X., Zhu, J., and Yin, Z., Fabrication and characterization of titanium-matrix composite with 20 vol% hydroxyapatite for use as heavy load-bearing hard tissue replacement. *Journal of Materials Science: Materials in Medicine,* 2006. 17 : p. 245-251.

[51] Li, P., Ohtsuki, C., Kokubo, T., Nakanishi, K., Soga, N., and Groot, K.D., The role of hydrated silica, titania, and alumina inducing apatite on implants. *Journal of biomedical materials research,* 1994. 28 (1): p. 7-15.

[52] Simmons, C.A., Valiquette, N., and Pilliar R.M., Osseointegration of sintered porous-surfaced and plasma spray–coated implants: An animal model study of early postimplantation healing response and mechanical stability. *Journal of Biomedical Materials Research, 1999.* 47 (2): p. 127-138.

[53] Almuhaiza, M., Glass-ionomer cements in restorative dentistry: a critical appraisal. *J Contemp Dent Pract,* 2016. 17 (4): p. 331-336.

[54] Makishi, P., Pacheco, R.R., Sadr, A., Shimada, Y., Sumi, Y., Tagami, J., and Giannini, M., Assessment of self-adhesive resin composites: nondestructive imaging of resin–dentin interfacial adaptation and shear bond strength. *Microscopy and Microanalysis,* 2015. 21 (6): p. 1523-1529.

[55] Athanasiou, K.A., Zhu, C.F., Lanctot, D.R., Agrawal, C.M., and Wang, X., Fundamentals of biomechanics in tissue engineering of bone. *Tissue engineering, 2000.* 6 (4): p. 361-381.

[56] Bronner, F., Farach-Carson, M.C., and Mikos, A.G., in *Engineering of functional skeletal tissues.* 2007, Springer-Verlag London Llc.

[57] Levine, M., *Topics in dental biochemistry.* 2010: Springer Science & Business Media.

[58] Fujii, E., Ohkubo, M., Tsuru, K., Hayakawa, S., Osaka, A., Kawabata, K., Bonhomme, C., and Babonneau, F., Selective protein adsorption property and characterization of nano-crystalline zinc-containing hydroxyapatite. *Acta biomaterialia,* 2006. 2 (1): p. 69-74.

[59] Rezwan, K., Chen, Q.Z., Blaker, J.J., and Boccaccini, A.R., Biodegradable and bioactive porous polymer/inorganic composite scaffolds for bone tissue engineering. *Biomaterials,* 2006. 27 (18): p. 3413-3431.

[60] Tang, X.J., Gui, L., and Lü, X.Y., Hard tissue compatibility of natural hydroxyapatite/chitosan composite. *Biomedical Materials,* 2008. 3 (4): 044115.

[61] Jayabalan, M., Shalumon, K.T., Mitha, M.K., Ganesan, K., and Epple, M., Effect of hydroxyapatite on the biodegradation and biomechanical stability of polyester nanocomposites for orthopedic applications. *Acta Biomaterialia,* 2010. 6 (3): p. 763-775.

[62] Zhang, Y., Venugopal, J.R., El-Turki, A., Ramakrishna, S., Su, B., and Lim, C.T., Electrospun biomimetic nanocomposite nanofibers of hydroxyapatite/chitosan for bone tissue engineering. *Biomaterials,* 2008. 29 (32): p. 4314-4322.

[63] Katti, K.S., Katti, D.R., and Dash, R., Synthesis and characterization of a novel chitosan/montmorillonite/hydroxyapatite nanocomposite for bone tissue engineering. *Biomedical Materials,* 2008. 3 (3): 034122.

[64] Ngiam, M., Liao, S., Patil, A.J., Cheng, Z., Chan, C.K., and Ramakrishna, S., The fabrication of nano-hydroxyapatite on PLGA and PLGA/collagen nanofibrous composite scaffolds and their effects in osteoblastic behavior for bone tissue engineering. *Bone,* 2009. 45 (1): p. 4-16.

[65] Thein-Han, W.W. and Misra, R.D.K., Biomimetic chitosan–nanohydroxyapatite composite scaffolds for bone tissue engineering. *Acta biomaterialia,* 2009. 5 (4): p. 1182-1197.

[66] Madhumathi, K., Shalumon, K.T., Rani, V.V.D., Tamura, H., Furuike, T., Selvamurugan, N., Nair, S.V., and Jayakumar, R., Wet chemical synthesis of chitosan hydrogel–hydroxyapatite composite membranes for tissue engineering applications. *International journal of biological macromolecules,* 2009. 45 (1): p. 12-15.

[67] Madhumathi, K., Binulal, N.S., Nagahama, H., Tamura, H., Shalumon, K.T., Selvamurugan, N., Nair, S.V., Jayakumar, R., Preparation and characterization of novel β-chitin–hydroxyapatite composite membranes for tissue engineering applications. *International journal of biological macromolecules,* 2009. 44 (1): p. 1-5.

[68] Kashiwazaki, H., Kishiya, Y., Matsuda, A., Yamaguchi, K., Iizuka, T., Tanaka, J., and Inoue, N., Fabrication of porous chitosan/hydroxyapatite nanocomposites: their mechanical and biological properties. *Bio-medical materials and engineering,* 2009. 19 (2-3): p. 133-140.

[69] Shen, K., Hu, Q., Chen, L., and Shen, J., Preparation of chitosan bicomponent nanofibers filled with hydroxyapatite nanoparticles via electrospinning. *Journal of applied polymer science,* 2010. 115 (5): p. 2683-2690.

[70] Kithva, P., Grøndahl, L., Martina, D., and Trau, M., Biomimetic synthesis and tensile properties of nanostructured high volume fraction hydroxyapatite and chitosan biocomposite films. *Journal of Materials Chemistry,* 2010. 20 (2): p. 381-389.

[71] Chen, J.D., Wang, Y., and Chen, X., In situ fabrication of nano-hydroxyapatite in a macroporous chitosan scaffold for tissue engineering. *Journal of Biomaterials Science,* Polymer Edition, 2009. 20 (11): p. 1555-1565.

[72] Zhao, F., Yin, Y., Lu, W.W., Leong, J.C., Zhang, W., Zhang, J., Zhang, M., and Yao, K., Preparation and histological evaluation of biomimetic three-dimensional hydroxyapatite/chitosan-gelatin network composite scaffolds. *Biomaterials,* 2002. 23 (15): p. 3227-3234.

[73] Zhao, F., Grayson, W.L., Ma, T., Bunnell, B., and Lu, W.W., Effects of hydroxyapatite in 3-D chitosan–gelatin polymer network on human mesenchymal stem cell construct development. *Biomaterials,* 2006. 27 (9): p. 1859-1867.

[74] Wang, Y., Zhang, L., Hu, M., Liu, H., Wen, W., Xiao, H., and Niu, Y., Synthesis and characterization of collagen-chitosan-hydroxyapatite artificial bone matrix. Journal of Biomedical Materials Research Part A: *An Official Journal of The Society for Biomaterials, The Japanese Society for Biomaterials, and The Australian Society for Biomaterials and the Korean Society for Biomaterials,* 2008. 86 (1): p. 244-252.

[75] Liuyun, J., Yubao, L., Li, Z., and Jianguo, L., Preparation and properties of a novel bone repair composite: nano-hydroxyapatite/chitosan/carboxymethyl cellulose. *Journal of Materials Science: Materials in Medicine,* 2008. 19 : p. 981-987.

[76] Peter, M., Ganesh, N., Selvamurugan, N., Nair, S.V., Furuike, T., Tamura, H., and Jayakumar, R., Preparation and characterization of chitosan–gelatin/

nanohydroxyapatite composite scaffolds for tissue engineering applications. *Carbohydrate polymers,* 2010. 80 (3): p. 687-694.

[77] Han, J., Zhou, Z., Yin, R., Yang, D., and Nie, J., Alginate–chitosan/ hydroxyapatite polyelectrolyte complex porous scaffolds: Preparation and characterization. *International journal of biological macromolecules,* 2010. 46 (2): p. 199-205.

[78] Yoshida, A., Miyazaki, T., Ashizuka, M., and Ishida, E., Bioactivity and mechanical properties of cellulose/carbonate hydroxyapatite composites prepared in situ through mechanochemical reaction. *Journal of biomaterials applications,* 2006. 21 (2): p. 179-194.

[79] Hutchens, S.A., Benson, R.S., Evans, B.R., O'Neill, H.M., Rawn, C.J., Biomimetic synthesis of calcium-deficient hydroxyapatite in a natural hydrogel. *Biomaterials,* 2006. 27 (26): p. 4661-4670.

[80] Schneiders, W., Reinstorf, A., Pompe, W., Grass, R., Biewener, A., Holch, M., Zwipp, H., and Rammelt, S., Effect of modification of hydroxyapatite/collagen composites with sodium citrate, phosphoserine, phosphoserine/RGD-peptide and calcium carbonate on bone remodelling. *Bone,* 2007. 40 (4): p. 1048-1059.

[81] Tampieri, A., Sandri, M., Landi, E., Pressato, D., Francioli, S., Quarto, R., and Martin, I., Design of graded biomimetic osteochondral composite scaffolds. *Biomaterials,* 2008. 29 (26): p. 3539-3546.

[82] Hargreaves, K.M. and Cohen, S., *Cohen's pathways of the pulp.* 10th ed. 2010: Mosby.

[83] Shalaby, S.W. and Salz, U. *Polymers for dental and orthopedic applications.* 1st ed. 2006: ORC Press.

[84] Hutmacher, D.W., Scaffolds in tissue engineering bone and cartilage. *Biomaterials,* 2000. 21 (24): p. 2529-2543.

[85] Newman, G.M., Takei, H.H., and Carranza, F.A., Carranza's clinical periodontology. 2002, Philadelphia, PA, London: Saunders Elsevier.

[86] Rizvi, A., Zafar, M.S., Al-Wasifi, Y., Fareed, W., and Khurshid, Z., Role of enamel demineralization and remineralization on microtensile bond strength of resin composite. *European Journal of Dentistry,* 2016. 10 (03): p. 376-380.

[87] Caffesse, R.G., Smith, B.A., Duff, B., Morrison, E.C., Merrill, D., Becker., W., Class II furcations treated by guided tissue regeneration in humans. *Journal of Periodontology,* 1990. 61 (8): p. 510-514.

[88] Sheikh, Z., Najeeb, S., Khurshid, Z., Verma, V., Rashid, H., and Glogauer, M., Biodegradable materials for bone repair and tissue engineering applications. *Materials,* 2015. 8 (9): p. 5744-5794.

[89] Bottino, M.C. and Thomas, V., Membranes for periodontal regeneration-a materials perspective. *Biomaterials for Oral and Craniomaxillofacial Applications,* 2015. 17 : p. 90-100.

[90] Hong, Z., Zhang, P., He, C., Qiu, X., Liu, A., Chen, L., Chen, X., and Jing, X., Nano-composite of poly (L-lactide) and surface grafted hydroxyapatite: mechanical properties and biocompatibility. *Biomaterials,* 2005. 26 (32): p. 6296-6304.

[91] Zafar, M., Najeeb, S., Khurshid, Z., Vazirzadeh, M., Zohaib, S., Najeeb, B., and Sefat, F., Potential of electrospun nanofibers for biomedical and dental applications. *Materials,* 2016. 9 (2): 73.

[92] Ferracane, J.L., Resin composite—state of the art. *Dental materials,* 2011. 27(1): p. 29-38

[93] Najeeb, S., Khurshid, Z., Zafar, M.S., Khan, A.S., Zohaib, S., Martí, J.M.N., Sauro, S., Matinlinna, J.P., and Rehman, I.U., Modifications in glass ionomer cements: nano-sized fillers and bioactive nanoceramics. *International journal of molecular sciences,* 2016. 17 (7): 1134.

[94] Najeeb, S., Zafar, M.S., Khurshid, Z., and Siddiqui, F., Applications of polyetheretherketone (PEEK) in oral implantology and prosthodontics. *Journal of prosthodontic research,* 2016. 60 (1): p. 12-19.

[95] Cordeiro, M.M., Dong, Z., Kaneko, T., Zhang, Z., Miyazawa, M., Shi, S., Smith, A.J., and Nör, J.E., Dental pulp tissue engineering with stem cells from exfoliated deciduous teeth. *Journal of endodontics,* 2008. 34 (8): p. 962-969.

[96] Prescott, R.S., Alsanea, R., Fayad, M.I., Johnson, B.R., Wenckus, C.S., Hao, J., John, A.S., and George, A., In vivo generation of dental pulp-like tissue by using dental pulp stem cells, a collagen scaffold, and dentin matrix protein 1 after subcutaneous transplantation in mice. *Journal of endodontics,* 2008. 34 (4): p. 421-426.

[97] McLaughlin, R.P. and Bennet J.C., *Orthodontic treatment mechanics and the preadjusted appliance.* 1993: Wolfe Publishing.

[98] Darabara, M.S., Bourithis, L.I., Zinelis, S., and Papadimitriou, G.D., Metallurgical characterization, galvanic corrosion, and ionic release of orthodontic brackets coupled with Ni-Ti archwires. *Journal of Biomedical Materials Research Part B: Applied Biomaterials,* 2007. 81 (1): p. 126-134.

[99] Inami, T., Tanimoto, Y., Minami, N., Yamaguchi, M., and Kasai, K., Color stability of laboratory glass-fiber-reinforced plastics for esthetic orthodontic wires. *The Korean Journal of Orthodontics,* 2015. 45 (3): p. 130-135.

[100] Elayyan, F., Silikas, N., and Bearn, D., Ex vivo surface and mechanical properties of coated orthodontic archwires. *European Journal of Orthodontics,* 2008. 30 (6): p. 661-667.

[101] Jancar, J., Dibenedetto, A.T., and Goldberg, A.J., Thermoplastic fiber-reinforced composites for dentistry: Part II Effect of moisture on flexural properties of unidirectional composites. *Journal of Materials Science: Materials in Medicine,* 1993. 4 : p. 562-568.

[102] Imaia, T., Watari, F., Yamagata, S., Kobayashi, M., Nagayama, K., and Nakamura, S., Effects of water immersion on mechanical properties of new esthetic orthodontic wire. *American journal of orthodontics and dentofacial orthopedics, 1999.* 116 (5): p. 533-538.

[103] Zufall, S.W., Kennedy, K.C, and Kusy, R.P., Frictional characteristics of composite orthodontic archwires against stainless steel and ceramic brackets in the passive and active configurations. *Journal of Materials Science: Materials in Medicine,* 1998. 9 : p. 611-620.

[104] Ohtonen, J., Vallittu, P.K., and Lassila, L.V.J., Effect of monomer composition of polymer matrix on flexural properties of glass fiber-reinforced orthodontic archwire. *The European Journal of Orthodontics,* 2013. 35 (1): p. 110-114.

[105] Lim, K.F., Lew, K.K.K. and Toh, S.L., Bending stiffness of two aesthetic orthodontic archwires: an in vitro comparative study. *Clinical materials,* 1994. 16 (2): p. 63-71.

[106] Imai, T., Yamagata, S., Watari, F., Kobayashi, M., Nagayama, K., Toyoizumi, H., Uga, M., and Nakamura, S., Temperature-dependence of the mechanical properties of FRP orthodontic wire. *Dental materials journal,* 1999. 18 (2): p. 167-175.

[107] Neumann, P., Bourauel, C., and Jäger, A., Corrosion and permanent fracture resistance of coated and conventional orthodontic wires. *Journal of Materials Science: Materials in Medicine, 2002.* 13 : p. 141-147.

[108] Silva, D.L.d., Mattos, C.T., Simão, R.A., de Oliveira Ruellas, A.C., Coating stability and surface characteristics of esthetic orthodontic coated archwires. *The Angle Orthodontist,* 2013. 83 (6): p. 994-1001.

[109] Zegan, G., Sodor, A., and Munteanu, C., Surface characteristics of retrieved coated and nickel-titanium orthodontic archwires. *Rom J Morphol Embryol,* 2012. 53 (4): p. 935-939.

[110] Tanimoto, Y., Inami, T., Yamaguchi, M., Nishiyama, N., and Kasai, K., Preparation, mechanical, and in vitro properties of glass fiber-reinforced polycarbonate composites for orthodontic application. *Journal of Biomedical Materials Research Part B: Applied Biomaterials,* 2015. 103 (4): p. 743-750.

[111] Filho, H.L., Maia, L.H., Araujo, M.V., Elias, C.N., Ruellas, A.C.O., Color stability of aesthetic brackets: ceramic and plastic. *Australian orthodontic journal,* 2013. 29 (1): p. 13-20.

[112] Silva, A.V.M.d., Mattos, G.V.d., Kato, C.M., and Normando, D., In vivo color changes of esthetic orthodontic ligatures. *Dental Press Journal of Orthodontics, 2012.* 17 : p. 76-80.

[113] Kim, S.H. and Lee, Y.K., Measurement of discolouration of orthodontic elastomeric modules with a digital camera. *The European Journal of Orthodontics,* 2009. 31 (5): p. 556-562.

[114] Mendonça, M.R.d., Fabre, A.F., Goiatto, M.C., Cuoghi, O.A., Martins, L.P., and Verri, A.C.G., Spectrophotometric evaluation of color changes of esthetic brackets stored in potentially staining solutions. RPG. *Revista de Pós-Graduação,* 2011. 18 (1): p. 20-27.

[115] Lee, Y.K., Changes in the reflected and transmitted color of esthetic brackets after thermal cycling. *American Journal of Orthodontics and Dentofacial Orthopedics,* 2008. 133 (5): p. 641. e1-641. e6.

[116] Jancar, J., Dibenedetto, A.T., Hadziinikolau, Y., Goldberg, A.J., and Dianselmo, A., Measurement of the elastic modulus of fiber-reinforced composites used as orthodontic wires. *Journal of Materials Science: Materials in Medicine,* 1994. 5 : p. 214-218.

[117] Goldberg, A.J., Burstone, C.J., and Koenig, H.A., Material Science Plastic Deformation of Orthodontic Wires. *Journal of Dental Research,* 1983. 62 (9): p. 1016-1020.

[118] Goldberg, A.J. and Burstone, C.J., The use of continuous fiber reinforcement in dentistry. *Dental materials,* 1992. 8 (3): p. 197-202.

[119] Deeksheetha, P. and Kumar, N., Composites in Orthodontics-A Literature Review. *International Journal of Pharmaceutical Research and Applications,* 1980. 6 (1): p. 309-309.
[120] Watari, F., Yamagata, S., Imai, T., Nakamura, S., and Kobayashi, M., The fabrication and properties of aesthetic FRP wires for use in orthodontics. *Journal of materials science,* 1998. 33 : p. 5661-5664.
[121] Suzuki, Y., Maekawa, Z., Hamada, H., Kibune, M., Hojo, M., and Ikuta, N., Influence of adsorption behavior of a silane coupling agent on interlaminar fracture in glass fiber fabric-reinforced unsaturated polyester laminates. *Journal of Materials Science,* 1992. 27 : p. 6782-6790.
[122] Kusy, R.P. and Whitley, J.Q., Influence of archwire and bracket dimensions on sliding mechanics: derivations and determinations of the critical contact angles for binding. *The European Journal of Orthodontics,* 1999. 21 (2): p. 199-208.
[123] Arici, N., Akdeniz, B.S., Oz, A.A., Gencer, Y., Tarakci, M., and Arici, S., Effectiveness of medical coating materials in decreasing friction between orthodontic brackets and archwires. *Korean Journal of Orthodontics,* 2021. 51 (4): p. 270-281.
[124] Feldner, J.C., Sarkar, N.K., Sheridan, J.J., and Lancaster, D.M., In vitro torque-deformation characteristics of orthodontic polycarbonate brackets. *American Journal of Orthodontics and Dentofacial Orthopedics,* 1994. 106 (3): p. 265-272.

Biographical Sketches

Orhan Cicek, PhD

Affiliation: Department of Orthodontics, Zonguldak Bulent Ecevit University, 67600, Zonguldak, Türkiye

Business Address: Department of Orthodontics, Zonguldak Bulent Ecevit University, 67600, Zonguldak, Türkiye

Research and Professional Experience: Orthodontics applications

Professional Appointments: Clinical sciences

Publications from the Last 3 Years:

1. Investigation of the Relationship between Sagittal Skeletal Nasal Profile Morphology and Malocclusions: A Lateral Cephalometric Film Study. Yunus Ocak, Orhan Cicek, Nurhat Ozkalayci, Hande Erener. *Diagnostics,* 2023 https://dx.doi.org/10.3390/diagnostics13030463

2. Effect of Different Types of Adhesive Agents on Orthodontic Bracket Shear Bond Strength: A Cyclic Loading Study. Irfan Eser, Orhan Cicek, Nurhat Ozkalayci, Mehmet Yetmez, Hande Erener. *Materials*, 2023 https://dx.doi.org/10.3390/ma16020724
3. Evaluation of the Corrosion Resistance of Different Types of Orthodontic Fixed Retention Appliances: A Preliminary Laboratory Study. Büşra Kumrular, Orhan Cicek, Ilker Emin Dağ, Bariş Avar, Hande Erener. *Journal of Functional Biomaterials*, 2023 https://dx.doi.org/10.3390/jfb14020081
4. Comparison of the Mesiodistal Angulations of Canine and Molar Teeth in Different Types of Orthodontic Malocclusions: A Retrospective Study. Orhan Cicek, Hakan Yilmaz, Büşra Demir Cicek. *Diagnostics,* 2023 https://dx.doi.org/10.3390/diagnostics13071351
5. Do Different Types of Adhesive Agents Effect Enamel Demineralization for Orthodontic Bonding? An In Vitro Study. Raif Murat Demircioğlu, Orhan Cicek, Füsun Cömert, Hande Erener. *Coatings,* 2023 https://dx.doi.org/10.3390/coatings13020401

Nurhat Ozkalayci, PhD

Affiliation: Department of Healthcare Management, Sinop University, Boyabat, 57200, Sinop, Türkiye

Business Address: Department of Healthcare Management, Sinop University, Boyabat, 57200, Sinop, Türkiye

Research and Professional Experience: Orthodontics applications

Professional Appointments: Clinical sciences

Publications from the Last 3 Years:

1. Investigation of the Relationship between Sagittal Skeletal Nasal Profile Morphology and Malocclusions: A Lateral Cephalometric Film Study. Yunus Ocak, Orhan Cicek, Nurhat Ozkalayci, Hande Erener. *Diagnostics*, 2023 https://dx.doi.org/10.3390/diagnostics13030463
2. Effect of Different Types of Adhesive Agents on Orthodontic Bracket Shear Bond Strength: A Cyclic Loading Study. Irfan Eser, Orhan Cicek, Nurhat Ozkalayci, Mehmet Yetmez, Hande Erener. *Materials*, 2023 https://dx.doi.org/10.3390/ma16020724
3. Comparison the Effects of Lingual and Labial Fixed Orthodontic Appliances: A Retrospective Study. Nurhat Ozkalayci, Orhan Çiçek, Yunus Ocak. *The Journal*

of Turkish Dental Research, 2023 https://dx.doi.org/10.58711/turkishjdentres.vi.1245419
4. Investigation the Retrospective Cephalometric Growth Changes of Sagittal Airway in Skeletal Class 2 Patients with Completed Functional Therapy. Ibrahim Berat Ay, Orhan Cicek, Nurhat Ozkalayci, *Journal of Health Sciences*, 2023 https://dx.doi.org/10.34108/eujhs.1163116
5. Evaluation of Publication Rates of Orthodontics Speciality and Doctorate Theses Written Between 2016-2020. Nurhat Ozkalayci, Irem Yolcu Kir, Melis Töz Ertop, Orhan Cicek. *Journal of Higher Education and Science*, 2022 https://dx.doi.org/10.5961/higheredusci.1066345

Mehmet Yetmez, PhD

Affiliation: Zonguldak Bulent Ecevit University

Business Address: Department of Mechanical Engineering, Zonguldak Bulent Ecevit University, 67100, Zonguldak, Türkiye

Research and Professional Experience: Solid mechanics, Fracture Mechanics, Biomechanics

Professional Appointments:
Design Engineer, TAI-Turkish Aerospace Industries Inc., Ankara, Türkiye

Honors:

- Ihsan Dogramaci Supreme Success Award. Hacettepe University, 1994
- The First Prize Publication Award in the Field of Engineering Sciences. Bulent Ecevit University, 2015.

Publications from the Last 3 Years:

1. Uzun U, Yetmez M, Akinci N. Nano-Mechanical Characterization of Single-Crystal Bi2Se3 Topological Insulator. *Micro&Nano Letters*, 2021; 16(3):203-212. DOI: 10.1049/mna2.12025 (JCR2020: Q4)
2. Kocar O, Yetmez M, Baysal E, Ozyigit HA. Mechanical Behaviour of a Friction Welded AA6013/AA7075 Beam. *Materials Testing*, 2022; 64(2):284-293. DOI: 10.1515/mt-2021-2041 (JCR2021: Q2)
3. Inik T, Karakus S, Yetmez M. Failure Analysis of a Double Edge Cracked Curved Composite Beam. *Materialwissenschaft und Werkstofftechnik*. 2022; 53(4):388-393. DOI: 10.1002/mawe.202100328. (JCR2021: Q4)

4. Uzun U, Lamuta C, Yetmez M. Nanoindentation Creep Behavior of Single-Crystal Bi_2Se_3 Topological Insulator. *Physica Status Solidi B*. 2022; 259(4):1-9. DOI: 10.1002/pssb.202100481 (JCR2021: Q3)
5. Oktar FN, Unal S, Gunduz O, Ben-Nissan B, Macha IJ, Akyol S, Duta L, Ekren N, Altan E, Yetmez M. Marine-Derived Bioceramics for Orthopedic, Reconstructive and Dental Surgery Applications. *Journal of the Australian Ceramic Society*. 2023; 59(1):57-81. DOI: 10.1007/s41779-022-00813-3 (JCR2021: Q3)
6. Eser I, Cicek O, Ozkalayci N, Yetmez M, Erener H. Effect of Different Type of Adhesive Agents on Orthodontic Bracket Shear Bond Strength: A Cycling Loading Study. *Materials*. 2023; 16(2), Article ID: 724, 12 pages. DOI: 10.3390/ma16020724 (JCR2021: Q3)
7. Uzun U, Lamuta C, Yetmez M. Friction and Wear Characteristics of Bismuth Selenide Topological Insulator. *Materials Letters*. 2023; 344: 134402. DOI:10.1016/j.matlet.2023.134402. (JCR2021: Q3)

Chapter 4

Investigating the Effects of Multi-Walled Carbon Nanotubes on Polyvinyl Alcohol with Hydroxyapatite Derived from Calcined Cuttlefish (Sepia Officinalis) Bone

Tsi-Yin Foo[1]
Lee Tin Sin[1,*]
Soo-Tueen Bee[2,†]
Jun-Ven Lim[2,]
Chong-Yu Low[3]
and Chi Ma[4]

[1]Department of Chemical Engineering, Lee Kong Chian Faculty of Engineering and Science, Universiti Tunku Abdul Rahman, Jalan Sungai Long, Bandar Sungai Long, Kajang, Malaysia
[2]Department of Mechanical and Material Engineering, Lee Kong Chian Faculty of Engineering and Science, Universiti Tunku Abdul Rahman, Jalan Sungai Long, Bandar Sungai Long, Kajang, Malaysia
[3]Department of Petrochemical Engineering, Faculty of Engineering and Green Technology, Jalan Universiti, Bandar Barat, Kampar, Perak, Malaysia
[4]School of Materials Science and Engineering, Shenyang University of Chemical Technology, Shenyang Economical and Technological Development Zone, Shenyang, China

[*] Corresponding Author's Email: leets@utar.edu.my.
[†] Corresponding Author's Email: beest@utar.edu.my.

In: Biocomposites
Editor: James R. Bush
ISBN: 979-8-89113-267-2

Abstract

Since polyvinyl alcohol (PVOH/PVA) presents some drawbacks in bone scaffolding applications, the aim of this study is to determine the effects of multi-wall carbon nanotubes (MWCNT) and calcined Sepia Officinalis cuttlefish bone (SOCB) hydroxyapatite as reinforcing material. The advantage of using cuttlebone is its availability and is biodegradable. The maximum tensile strength is at 3 phr MWCNT loading, with 42.65 MPa for 2 phr SOCB hydroxyapatite and 41.43 MPa for 5 phr SOCB hydroxyapatite. Scanning electron microscopy (SEM) images suggested that agglomeration of MWCNT in PVOH matrix tends to occur at high loading (5 phr) of SOCB hydroxyapatite and good distribution was induced with increasing loading of MWCNT. X-ray diffraction (XRD) results showed that PVOH peaks were observed at the range of 19°-21° and crystallinity of PVOH/MWCNT composite is the highest at low (2 phr) amount of SOCB hydroxyapatite. Fourier transform infrared spectroscopy (FTIR) analysis showed that the bonding between molecules increased with increasing loading of MWCNT at low (2 phr) loading of SOCB hydroxyapatite. Meanwhile, differential scanning calorimetry (DSC) results showed that the melting temperature elevated with increasing loading of MWCNT at both 2 phr and 5 phr SOCB hydroxyapatite. In summary, the mechanical and thermal properties increased with higher loading level of MWCNT but the matrix tends to agglomerate at higher SOCB hydroxyapatite content. In conclusion, the addition of 2 phr SOCB hydroxyapatite showed the most promising interaction between molecules.

Keywords: polyvinyl alcohol, carbon nanotube, hydroxyapatite, cuttlefish bone

Abbreviations

MWCNT	multi-wall carbon nanotube
PVOH	polyvinyl alcohol
PVA	polyvinyl alcohol
SWCNT	single-wall carbon nanotubes
SCOB	Sepia Officinalis cuttlefish bone
SEM	Scanning Electron Microscope
XRD	X-Ray Diffraction Analysis
FTIR	Fourier Transform Infrared Spectroscopy
DSC	Differential Scanning Calorimetry
OH	hydroxyl

TGA	thermogravimetric analysis
EDX	electron dispersive X-ray

1. Introduction

In recent studies, natural polymers such as polyvinyl alcohol (PVOH) were limited in the ability to obtain extensive commercial application due to their weakness in mechanical properties [1]. Significant improvement of the properties such as toughness, strength and stiffness can be accomplished through the incorporation of various kinds of organic and inorganic fillers. In time, inorganic fillers were substituted by natural composites due to environmental issues. In order to attain high mechanical characteristics at reduced filler loading levels, natural composites such as cuttlebone and nanofiller such as multi-walled carbon nanotube (MWCNT) were used as reinforcing materials for PVOH in this research study.

Composite material consists of two or more materials that are able to work together to give unique properties. Natural composites can be found in animal bones. Composite synthesized from collagen and hydroxyapatite are known as candidate for tissue and bone scaffolding materials due to the fact that their composition is analogous to the extracellular matrix of natural bone [2]. Cuttlebone is considered as a natural composite material which can be found naturally and it is better than artificial composite. Hydroxyapatite is a bioceramic which is widely used for bone tissue repair due to its biocompatibility, exceptional mechanical strength and osteoconductivity [3]. Hydroxyapatite can be obtained through synthetic and natural ways. The natural method to obtain hydroxyapatite is thermal calcination, which is used to extract and isolate hydroxyapatite from animal bones [4]. In this research, cuttlebone from the common cuttlefish or European common cuttlefish (Sepia Officinalis) was treated under thermal calcination to obtain hydroxyapatite powder.

Relevant studies have been made in the field of reinforcement of polymer compounds with the use of composite material as well as nanofiller [5-7]. One of the methods to improve the compressive strength of hydroxyapatite and bone composite is by blending them with synthetic polymer [2]. Among these synthetic polymers, PVOH has received most of the attention due to its physiochemical properties and biocompatibility, along with better thermal stability, tensile strength as well as flexibility [8]. However, the weak mechanical properties of PVOH often leads to failure in some engineering applications. Hence, the mechanical properties of PVOH could be improved with biocompatible reinforcement in the PVOH matrix [8]. Involvement of carbon-based biomaterials in polymer matrix contributes to high mechanical and biological properties [9].

Additionally, MWCNT has received a lot of the attention from researchers due to their remarkable mechanical and electrical properties and their ability to act as one of the most promising nanofillers for polymer composite [10]. The excellent properties of MWCNT can be exploited if they are homogeneously embedded into other materials in the form of composites or light-weight matrices [11]. MWCNT/polymer composites with high mechanical strength can be achieved only in the case of good interaction of MWCNT within the matrix [12]. There are two basic types of carbon nanotubes which are single-wall carbon nanotubes (SWCNTs) and MWCNTs. MWCNT were used in this research due to better enhancement in mechanical, thermal and electrical properties. MWCNTs consist of multiple concentric layers of graphene composing it, causing it to be stronger than SWCNTs. Also, the open end of MWCNTs has a fast electron rate which contributes to better electrical properties. Besides, MWCNTs has lower production cost as material modification. Over the past few years, attentions have been given to the development of reinforced polymer composite. Among other carbon compounds, MWCNT will be the most suitable choice to blend in with polymer composite. Animal bone which act as a natural composite possess the capability to further improve the mechanical properties of polymer composite. The aims and objectives were to study the properties of PVOH/MWCNT composite after adding SOCB hydroxyapatite. The study included bonding interaction, morphology of composite, mechanical, chemical and thermal properties with the target for biomaterial for artificial implant applications.

2. Materials and Methodology

2.1. Materials

The materials used in this research were Sepia Officinalis cuttlefish bone (SOCB) hydroxyapatites, PVOH and MWCNT. Raw cuttlebones of 1 kg were obtained from a seafood wholesaler located at Kedah. The cuttlebones were readily to be calcined since they were in a clean and dry condition. Hence, no pre-treatment was required in this research study. PVOH was obtained from

Sekisui Specialty Chemical Co. Ltd. The fully hydrolysed PVOH has a viscosity of 4% at 20ºC and hydrolysis of 98.4 mol %. The type of carbon nanotubes used in this research study was multiwall type (MWCNT). The purity of the MWCNT was more than 90% and they were obtained from Nanjing XFNANO Materials Tech Co. Ltd, China.

2.2. Formulation

The amount of SOCB hydroxyapatites used is 2 phr or 5 phr, with the amount of PVOH used being constant at 100 phr, along with different amount of MWCNT being added to the PVOH in this research study. The manipulated variable was the concentration of MWCNT, ranging from 1 phr, 2phr and 3 phr. Table 1 shows the composition of PVOH-MWCNT added with SOCB hydroxyapatite which were used in this research study.

Table 1. Composition of PVOH/MWCNT and Calcined Cuttlebone for Preparation of Sample

PVOH (phr)	Calcined Cuttlebone (phr)	MWCNT (phr)
100	2	1
100	2	2
100	2	3
100	5	1
100	5	2
100	5	3

* phr: Parts per hundred resin.

2.3. Sample Preparation

Raw cuttlebones were purchased online. The bones were crushed into powder form using a crusher in order to ease the sample preparation process. The treated cuttlebones were filled into the silica crucible and placed inside the programmable furnace from room temperature to 900°C for a minimum of 4 hours. After the calcination process, the cuttlebones were cooled to at least 100 °C. The SOCB hydroxyapatites were then ready to use in casting process.

The preparation of polyvinyl alcohol composite was done using solution casting method. For example, one of the sample formulations which involved 2 phr of SOCB hydroxyapatite, 100 phr of PVOH and 1 phr of MWCNT were first mixed together. The mixture was stirred at 800 rpm and was heated to 90 °C - 95 °C using hot water bath for 1 hour until the PVOH was completely dissolved. The mixture was then poured into the telfon mold and was dried in the oven with temperature of 60 °C to form scaffold. Six samples with different formulation were prepared under the same experimental conditions. The samples were sealed for storage and were kept at the temperature of 25 °C for conditioning purpose.

2.4. Characterization Techniques

There were several methods used to test the morphology, mechanical, thermal and chemical properties of carbon nanotubes composite.

2.4.1. Tensile Test

The tensile strength of the MWCNTs/PVOH composite was tested using Instron Universal Testing Machine (Model 4302 Series IX). The test components were Young's modulus, elongation at break and tensile strength. The device used to standardise the sample dimension was ASTM D882. The typical dimension of each samples was 6 cm x 1 cm x 1 cm. The tensile measurement for each different formulation sample were performed for six sets of samples and the average of these six measurements were taken. The tensile tests were conducted with a strain rate of 50 mm/min and cell load of 5 kN.

2.4.2. Scanning Electron Microscopy Analysis (SEM)

SEM operated using an electron beam with the voltage of 20 kV to 30 kV to scan the surface morphology of the sample composites. The type of device used was Hitachi Scanning Electron Microscope (BS 340 TESLA) with a chosen magnification power of 500x to 3000x. A thin layer of gold foil was coated around the sample composite before being scanned by the SEM. The samples were placed facing up in the microscope. Additionally, all of the figures showing SEM images of fractured sample were taken with magnification of x1000.

2.4.3. X-Ray Diffraction (XRD) Test

The Shimadzu XRD 6000 Diffractometer was used to conduct the crystallinity test of the MWCNT composite. The operating condition of XRD was specified to 2θ range of 5° to 70°, current of 30 mA, voltage of 40 kV and scanning length from 0 – 3 cm. The polymer was coated with a thin layer of gold and it was placed on the goniometer which will rotate according to the operating condition. When the goniometer was rotating, the change of XRD intensities was being recorded. Based on the scanning pattern obtained, the crystallinity and crystallite size of the composite was then calculated. The crystallite size L, was calculated using Scherrer's formula as shown in Equation (1). While the d-spacing, d was calculated using the Bragg's equation as shown in Equation (2).

$$L = \frac{K\lambda}{\beta \cos\theta} \tag{1}$$

$$d = \frac{\lambda}{2\sin\theta} \tag{2}$$

Where λ is the x-ray wavelength (1.542 Å), θ is the angle between the beam and the normal to the reflecting plane, β is the peak width, and K is a constant (taken as 1)

2.4.4. Fourier Transform Infrared Spectroscopy (FTIR)

The chemical properties of MWCNT/PVOH with SOCB hydroxyapatite was tested using FTIR. The model of the instrument used was Perkin Elmer Spectrum One. The presence of different functional groups in the MWCNT/PVOH with SOCB hydroxyapatite polymer composite were identified by determining the vibration frequency of the different ratio of SOCB hydroxyapatite added with MWCNT. The samples were scanned under the region of 4000 cm^{-1} to 400 cm^{-1}. The data transmitted from FTIR was then recorded at the end of experiment.

2.4.5. Differential Scanning Calorimetry (DSC)

DSC was conducted by using Perkin Elmer DSC 7 to determine the thermal properties of PVOH/MWCNT/cuttlebone composite. The composite was heated around 25°C to 240°C, with scanning rate of 2°C min^{-1} for the sample. DSC was used to obtain enthalpy of melting and melting temperature of the composite.

3. Results and Discussion

3.1. Mechanical Properties

3.1.1. Tensile Test

Mechanical properties in terms of tensile strength, Young's modulus and elongation at break of the PVOH/MWCNT composites that has been reinforced using SOCB hydroxyapatite has been studied. Figure 1 shows the tensile strength for all the samples. Based on the results from Figure 1, the tensile strength generally increased when the amount of MWCNT increased (however, it should be noted that this is the case only for composites with 5 phr SOCB hydroxyapatite added to it); in which the maximum tensile strength is 42.65 MPa for 2 phr SOCB hydroxyapatite and 41.43 MPa for 5 phr SOCB hydroxyapatite. Well mixing of PVOH solution into porous MWCNTs network has improved the interfacial bonding between PVOH/MWCNT composite and thus increased the mechanical strength.

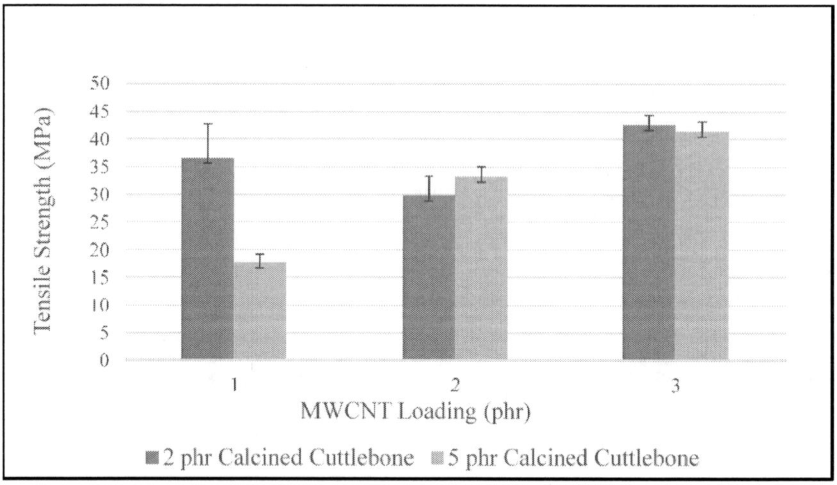

Figure 1. Tensile Strength of PVOH/MWCNT Composite with Different Loading of Calcined Cuttlebone and MWCNT.

The tensile strength has decreased from 36.65 MPa to 29.82 MPa from 1 phr to 2 phr of MWCNT at low loading (2 phr) of SOCB hydroxyapatite. The decreasing mechanical properties can be explained by a few reasons. Firstly, the presence of MWCNTs have restricted the chain mobility of PVOH. Besides that, another reason is the complete infiltration of PVOH into

MWCNTs network [13]. Additionally, poor dispersion of MWCNTs in PVOH, weak interaction between PVOH and MWCNTs and bad orientation of MWCNTs and PVOH molecular chain are other reasons which can lead to decreasing tensile strength [14]. Unlike the previous result, tensile strength of PVOH/MWCNT composite has increased from 29.82 MPa to 42.65 MPa when the loading of MWCNT increased from 2 phr to 3 phr.

The tensile strength of PVOH/MWCNT composite has increased from 17.69 MPa to 33.26 MPa from 1 phr to 2 phr of MWCNT at high loading (5 phr) of SOCB hydroxyapatite. The tensile strength increased from 33.26 MPa to 41.43 MPa from 2 phr to 3 phr of MWCNT. The result indicates that the tensile strength of PVOH/MWCNT composite increased with increasing MWCNT due to high strength and high aspect ratio of carbon nanotubes [14]. The load of external force can be dispersed along the direction of carbon nanotubes when subjected to external force.

Hydroxyapatite in cuttlebone could also enhance the mechanical properties of PVOH/MWCNT composite through interfacial interaction between the polymers. The stronger the interfacial interaction, the better the tensile strength. However, when high loading of SOCB hydroxyapatite is added into the PVOH/MWCNT composite, the tensile strength has decreased as depicted in Figure 1. The tensile strength of PVOH/MWCNT composite is higher on average for 2 phr SOCB hydroxyapatite as compared to 5 phr SOCB hydroxyapatite. The result indicates that hydroxyapatite could served as a reinforced material and could improve the mechanical properties.

A significant drop in tensile strength can be observed when incorporating high loading (5phr) of SOCB hydroxyapatite into PVOH/MWCNT composite compared to low loading (2phr) of SOCB hydroxyapatite. This is because the increasing of particle size and variation of irregular structures of particles causes stress concentration in PVOH/MWCNT composite which lead to low tensile strength.

Additionally, the decreasing mechanical properties could be explained upon the addition of SOCB hydroxyapatite into PVOH/MWCNT composite. The formation of agglomerates was due to the high viscosity of the matrix and Van der Waals forces among MWCNT [15]. The ability of composite to resist applied stress was reduced with the formation of agglomerates. The degree of uniformity in the composite also decreases due to agglomeration of untreated nanoparticles with high surface area. In short, agglomeration weaken the mechanical properties of the composite.

Besides discussing the effect of MWCNT and SOCB hydroxyapatite on the PVOH matrix, it is also worth comparing the difference in tensile strength between the nanocomposites and pure PVOH. As it is seen in the results above, the tensile strength for the PVOH/MWCNT nanocomposites that incorporated with SOCB hydroxyapatite has a maximum of 42.65 MPa and a minimum of 17.69 MPa. This is in comparison with the tensile strength of pure PVOH, which is 7.1 MPa [16]. This immense improvement in tensile strength has proven that both MWCNT and SOCB hydroxyapatite has beneficial interaction effect with the PVOH matrix and is able to enhance the mechanical properties of pure PVOH.

3.1.2. Young's Modulus

The Young's modulus of each sample was shown in Figure 2. At 2 phr SOCB hydroxyapatite, the Young's modulus has decreased when the amount of MWCNT increased from 1 phr to 2 phr. This phenomenon can be explained by the agglomeration of MWCNT in nanoparticles which causes the tensile strength to decrease. Low Young's modulus indicates that the interaction of MWCNT with PVOH is not well as compared to 3 phr of MWCNTs. When the loading of MWCNT increased to 3 phr, the Young's modulus has increased due to well dispersion of MWCNT in PVOH matrix [17].

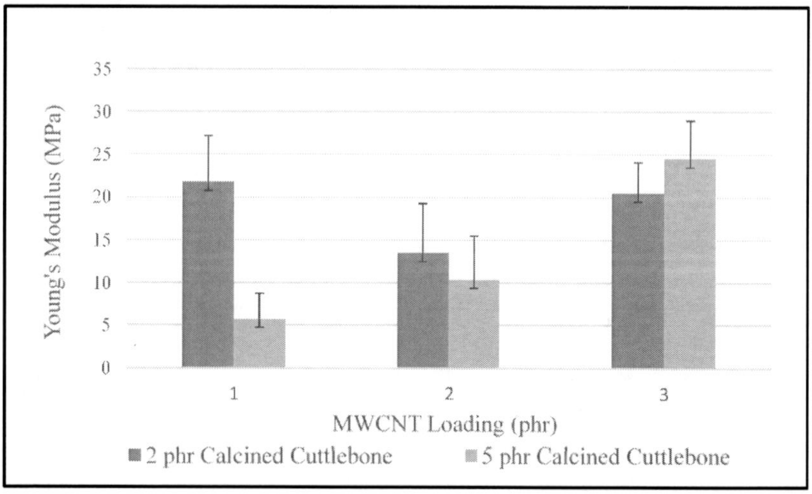

Figure 2. Young's Modulus of PVOH/MWCNT Composite with Different Loading of Calcined Cuttlebone and MWCNT.

At 5 phr SOCB hydroxyapatite, the Young's modulus increased with increasing concentration of MWCNT from 1 phr to 3 phr. Increasing modulus indicates that MWCNT is well dispersed in PVOH matrix. Agglomeration of

1 phr MWCNT in PVOH matrix led to lower Young's modulus when compared to 2 phr and 3 phr of MWCNT. The tear strength increased and less agglomeration occurred in the matrix as the loading of MWCNT increased. The result indicates that appropriate dispersion and interaction of MWCNT in PVOH matrix has improved the tearing strength [17].

A significant drop in the Young's modulus graph can be observed when incorporating high loading (5phr) of SOCB hydroxyapatite into PVOH/MWCNT composite compared to low loading (2phr) of SOCB hydroxyapatite at lower loading levels of MWCNT. This is mainly due to the dispersion of larger particles (SOCB hydroxyapatite) in PVOH matrix which causes interlocking between particles in the composite.

3.1.3. Elongation at Break

The results of elongation at break for each sample were shown in Figure 3. At 2 phr SOCB hydroxyapatite, the elongation at break increased with increasing MWCNT loading. PVOH with SOCB hydroxyapatite composite is stretcher with higher amount of MWCNT. Small particles of SOCB hydroxyapatite were added into the polymer matrix which leads to improvement of chain sliding between the polymers. The increment of elongation at break indicates that SOCB hydroxyapatite is well mixed and compatible in PVOH/MWCNT composite. Small particles of SOCB hydroxyapatite create lubrication effect with the polymer composite when force is being applied. Strong interfacial binding force between the particles has improved the movement of polymer chain thus increase the elongation at break.

However, at 5 phr SOCB hydroxyapatite, the elongation at break increased at the initial stage and decreased for further incorporation of MWCNT from 2 phr to 3 phr. The elongation at break depends upon the molecular chain's flexibility. The changes on the flexibility is restricted by MWCNT which decreased the elongation of break from 170.47% to 148.33%. This is mainly due to the irregular shape of particles which tend to form stress concentration regions easily.

In the case of elongation at break, the value is lower at 5 phr SOCB hydroxyapatite as compared to 2 phr SOCB hydroxyapatite at both 1 phr MWCNT and 3 phr MWCNT. However, elongation at break is the highest at 2 phr MWCNT for 5 phr SOCB hydroxyapatite. Large particle of SOCB hydroxyapatite tends to agglomerate which is easily to form stress concentration regions. Stress concentration is the main reason for low mechanical properties.

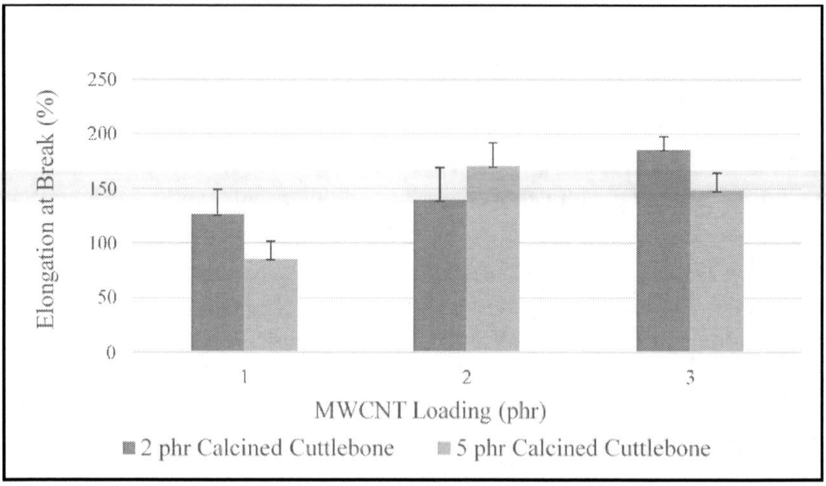

Figure 3. Elongation of Break of PVOH/MWCNT Composite with Different Loading of Cuttlebone and MWCNT.

3.2. Scanning Electron Microscopy

The morphology of PVOH/MWCNT with SOCB hydroxyapatite was studied by using SEM micrographs and the results were shown in Figure 4 (a-f). Elongation ability was significantly observed. In general, before adding in SOCB hydroxyapatite into the polymer matrix, the polymer matrix had a smooth and continuous surface with no agglomeration occurred. When the amount of MWCNT is low in a polymer matrix, the surface of the nanocomposite is smooth and fewer rough particles were formed on the surface of polymer. However, the composite had a rough structure as the amount of MWCNT increased in the polymer composite.

At 2 phr SOCB hydroxyapatite, the surface of PVOH/MWCNT composite with SOCB hydroxyapatite became rough and large particle can be observed as the loading of MWCNT increased. At 1 phr of MWCNT, uneven surface and numerous elongated grooves on the composite fiber surface were observed as shown in Figure 4(a). Agglomeration were observed in PVOH/MWCNT composite with SOCB hydroxyapatite which would impact the mechanical properties of the polymer matrix.

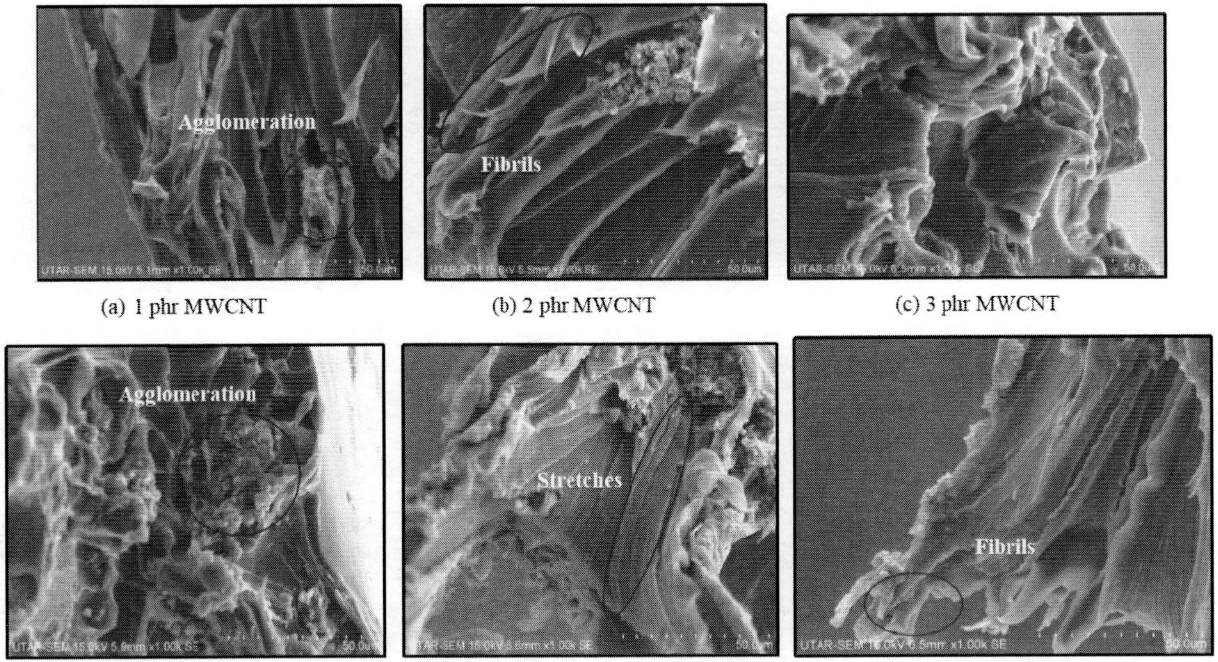

(a) 1 phr MWCNT (b) 2 phr MWCNT (c) 3 phr MWCNT

(d)1 phr MWCNT (e) 2 phr MWCNT (f) 3 phr MWCNT

Figure 4. Surface Morphology of PVOH/MWCNT Composite at 2 phr of Calcined Cuttlebone added with Different Loading of MWCNT (a) 1 phr MWCNT (b) 2 phr MWCNT (c) 3 phr MWCNT and at 5 phr of Calcined Cuttlebone added with Different Loading of MWCNT (d) 1 phr MWCNT (e) 2 phr MWCNT (f) 3 phr MWCNT.

As the loading of MWCNT increased from 1 phr to 2 phr, more voids or defects appeared between agglomerates and polymer matrix [12]. The present of voids in the polymer matrix is the main reason which lead to stress concentration and would cause a sharp declination of tensile strength at 2 phr of MWCNT. However, unlike the previous sample with 1 phr of MWCNT, there were more fibrils with shorter length and the surface was smoother as shown in Figure 4(b). Hence, the elongation at break started to increase at 2 phr of MWCNT.

As the loading of MWCNT increased from 2 phr to 3 phr, less fibrils can be observed as compared to low loading of MWCNT as shown in Figure 4(c). The result indicates that MWCNT was homogeneously dispersed in polymer matrix resulted in high mechanical strength as discussed in tensile properties.

As shown in Figure 4 (a)(b)(c), less agglomeration and a more uniform fiber density distribution were observed from the SEM image when compared to Figure 4(d)(e)(f), the dispersion of MWCNT was poor in the matrix where many clumps and agglomerate are present and where the presence of discontinuities at the boundaries between mixing zone containing different concentrations of dispersed nanofibers were observed.

As the loading of SOCB hydroxyapatite increased from 2 phr to 5 phr, more agglomerates and voids were formed on the surface of polymer matrix. This was mainly due to poor intermolecular chain bond between PVOH and causes the polymer matrix to tear into various fibrils as shown in Figure 4(d). In addition, the formation of agglomeration on the surface of polymer specifies a higher interaction between the SOCB hydroxyapatite and MWCNT.

As the amount of MWCNT increased to 2 phr, stretches and fiber grooves were obvious in SEM micrograph as shown in Figure 4(e). Less agglomerates were observed as compared to 1 phr MWCNT. The result indicates that the polymer matrix was good in terms of its ductility where the film resists to deform and lead to good elongation at break as discussed in the previous section.

As the amount of MWCNT increased to 3 phr, more fibrils were presented in the SEM micrograph as shown in Figure 4(f). The fibrils are short and this may due to the agglomerated particles decreasing the intermolecular bonding. Agglomeration which causes lower intermolecular bonding could lead to low tensile strength and elongation at break when compared to low amount of SOCB hydroxyapatite (2 phr).

Additionally, voids were spotted in the polymer matrix. The voids are the pores in the scaffold that were formed after evaporation of water in solvent-rich region during the drying process. Poor interactions between MWCNT and SOCB hydroxyapatite that were incorporated into PVOH when the amount of SOCB hydroxyapatite increased were observed. This leads to the agglomeration with voids being formed on the composite fracture surface due to poor interactions between MWCNT and the SOCB hydroxyapatite when the amount of SOCB hydroxyapatite increased.

From the results as shown in Figure 4(a-f), MWCNT tends to agglomerate as the amount increased. The occurrence of agglomeration is due to Van Der Waal force taking place in between the particles and the shape of aggregation is usually in bundle style. Agglomeration weakens the mechanical properties of a polymer composite. Bad dispersibility of MWCNT in a polymer composite is one of the significant reasons for low mechanical strength of PVOH/MWCNT composite as discussed in mechanical properties.

When the hydroxyapatite content is low, hydroxyapatite particles disperse homogeneously in PVOH matrix. As the content of hydroxyapatite increased, large number of hydroxyapatite particles can be observed. High specific surface area of nanoparticles increased the mechanical properties of the composites and eventually lead to improvement in interfacial bonding strength. However, the matrix will tend to agglomerate as the hydroxyapatite content increased due to its high surface-active energy when minimum hydroxyapatite content is exceeded.

3.3. X-Ray Diffraction

In X-ray diffraction analysis, peak position and peak width are the main observations to be analyzed. Several parameters such as crystallinity, crystal size and d-spacing were studied. Range of 3° to 40° was used to study the deflection peak for different loading of MWCNT and SOCB hydroxyapatite. Figures 5 and 6 shows the XRD pattern for all the samples at 2 phr SOCB hydroxyapatite and 5 phr SOCB hydroxyapatite respectively.

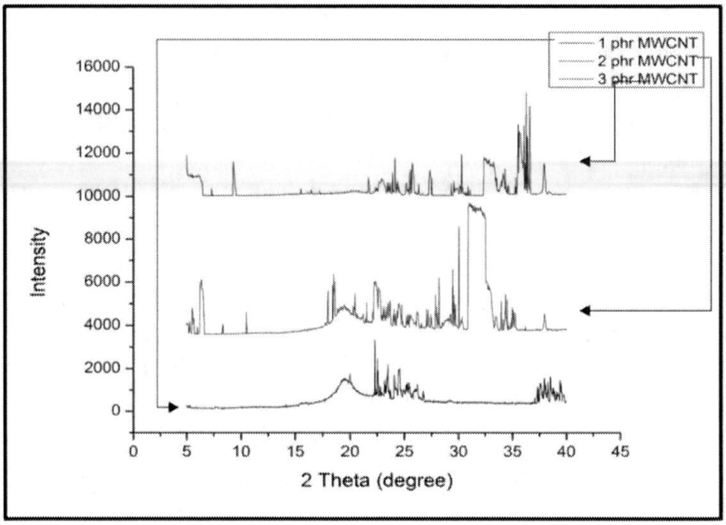

Figure 5. XRD Pattern of PVOH with Different Loading of MWCNT at 2 phr Calcined Cuttlebone.

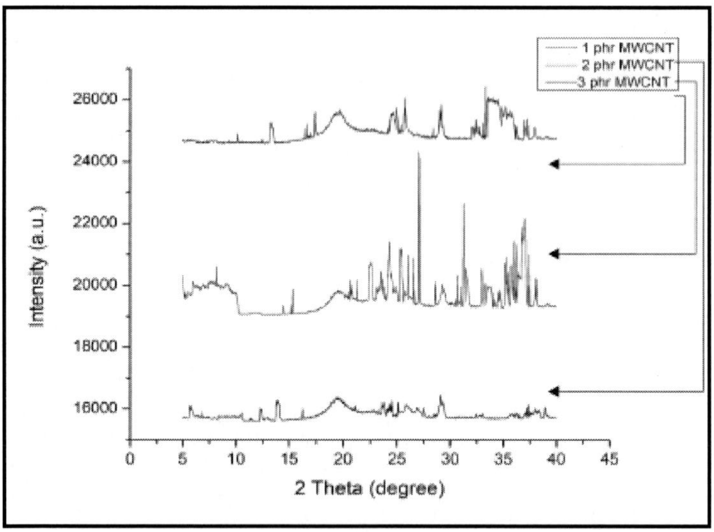

Figure 6. XRD Pattern of PVOH with Different Loading of MWCNT at 5 phr Calcined Cuttlebone.

3.3.1. *Crystallinity of Composite*

At both 2 phr and 5 phr SOCB hydroxyapatite, peaks were observed at the range of 19°-21°. The peak with intensity appears at $2\theta = 19.6°$ which represents (101) crystal plane [18]. Another peak was observed at 22.9° of the "semi-crystalline" PVOH [19]. The diffraction occurred due to "intermolecular interference" between PVOH polymer chains. As the intensity of diffraction increased, the number of PVOH chains packing increased as well. The peak of SOCB hydroxyapatite can be observed around 31.77°.

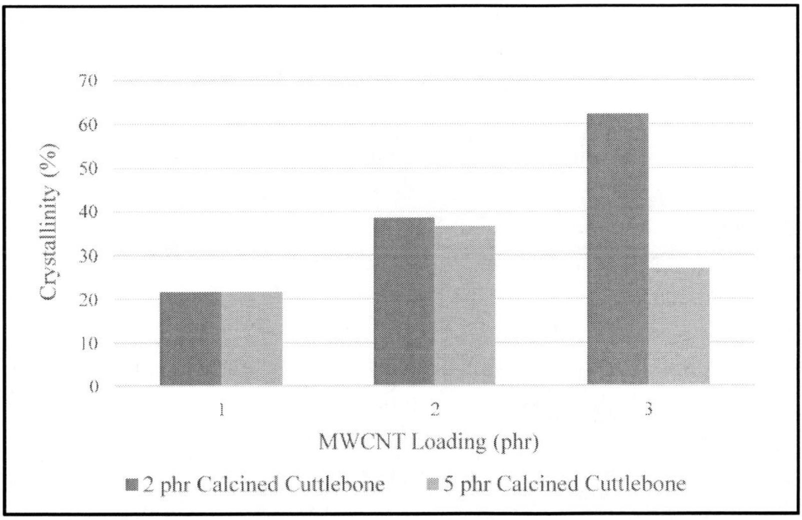

Figure 7. Crystallinity of PVOH with Different Loading of MWCNT and Calcined Cuttlebone.

As the loading of MWCNT increased from 1 phr to 2 phr, more sharp and narrow peaks were observed. Moreover, a broad peak around 30.9° with high intensity was observed in Figure 5. The peak around 19° disappeared from 2 phr to 3 phr of MWCNT. The result was mainly due to the rupture of crystal structure of SOCB hydroxyapatite. Besides, the peak of SOCB hydroxyapatite has shifted from around 30° to 32° with lower intensity. An increase in broadness and decrease in peak intensity have been observed. Therefore, the crystallinity has the highest value as observed in the crystallinity graph as shown in Figure 7. In other words, the broadness of peak decrease with increasing loading of MWCNT. The result indicates that the MWCNT is well dispersed in PVOH matrix and crystalline structure was induced. Crystalline structure is orderly arranged with uniform intermolecular forces in the crystal lattice.

At 5 phr SOCB hydroxyapatite, more sharp and narrow peaks were observed from 1 phr to 2 phr of MWCNT. High amount of crystallite peak indicates that crystal structure was induced in the sample. A good dispersion of MWCNT in PVOH matrix can be one of the reasons which causes strong crystallite peak. Moreover, a sharp peak around 27° was observed at 2 phr MWCNT and the peak has disappeared at 3 phr MWCNT. This is mainly due to the rupture of crystal structure of SOCB hydroxyapatite. The sample's peak broadness increased as the loading of MWCNT increased further from 2 phr to 3 phr, just as it did for 2 phr SOCB hydroxyapatite loading level. Figure 7 shows that broad peaks instead of sharp and narrow peaks were observed. Peak broadening effect in the samples were due to crystal defect caused by non-homogenous interaction between the particles in the polymer matrix. More defects were presented in the crystal structure. One of the reasons which contributes to broadening effect is the disturbance of ordered structured arrangement of polymer chain.

The increase of crystallinity for 2 phr SOCB hydroxyapatite is due the compactness of the matrix as higher loading of MWCNT was added into the polymer composite. At 5 phr SOCB hydroxyapatite, the crystallinity increased from 1 phr to 2 phr of MWCNT then decreased from 2 phr to 3 phr of MWCNT. This is due to crystal defect in polymer composite. As shown in Figure 8, the crystallinity of 5 phr SOCB hydroxyapatite is lower than 2 phr SOCB hydroxyapatite. The crystallinity shows the same trend as the XRD graph obtained from Figure 5 and 6. As the loading of SOCB hydroxyapatite increased from 2 phr to 5 phr, the amount of small crystallite peak decreased. The results indicate that crystallinity decreased as higher amount of nanocomposite is being dispersed in PVOH composite. Dispersion of larger particle in PVOH matrix causes interlocking between particle and reduced the crystallinity of a component. The decrease in crystallinity upon addition of 5 phr of SOCB hydroxyapatite is mainly due to the particle interaction forming aggregates which lower the elongation at break.

3.3.2. Crystallite Size of Composite

The crystallite size of the composite was determined by using Scherrer's equation and the result was tabulated in Table 2. At 2 phr SOCB hydroxyapatite, the crystallite size decreased from 1 phr to 2 phr of MWCNT in the polymer matrix whereas the crystallite size increased further from 2 phr to 3 phr of MWCNT. According to Scherrer's equation, crystallite size is inversely proportional to peak width. In other words, when the peak width increases, the crystallite size decreases. The results showed the same trend as broadness of peak where the crystallite size decreased from 1 phr MWCNT to 2 phr MWCNT and increased from 2 phr MWCNT to 3 phr MWCNT. As shown in Figure 5, the peak broadness increased first then decreased. The results obtained in Table 2 has the same trend as Figure 5.

Table 2. Crystallite Size and d-Spacing at 2 and 5 phr Calcined Cuttlebone

SOCB hydroxyapatite (phr)	MWCNT Loading (phr)	Crystallite Size (Å)	d-spacing (Å)
2	1	761.27	4.44516
2	2	114.39	3.95866
2	3	1055.15	3.92433
5	1	257.28	3.75111
5	2	315.67	3.57054
5	3	126.94	3.87035

At 5 phr SOCB hydroxyapatite, the crystallite size increased when the loading of MWCNT increased from 1 phr to 2 phr in the polymer matrix whereas the crystallite size decreased when the loading of MWCNT increased further from 2 phr to 3 phr. The results showed the same trend as broadness of peak where the peak broadness decreased first then increased as shown in Figure 6. The factor which contributes to the changes in crystallite size was intercalation effect.

Intercalation effect captures the polymer inside the interlayer spacing. The intercalation effect shows the characteristics of dispersion of SOCB hydroxyapatite and MWCNT in PVOH matrix. The d-spacing is the principle for this characteristic as high d-spacing indicates that intercalation was improved and vice versa. Based on the data tabulated in Table 2, d-spacing decreased from 1 phr MWCNT to 3 phr MWCNT whereas there is only a slight change in d-spacing where the value decreased from 1 phr MWCNT to 2 phr MWCNT and increased from 2 phr MWCNT to 3 phr MWCNT.

3.4. Fourier Transform Infrared Spectroscopy

Besides mechanical and physical properties, interaction of bonding between PVOH, MWCNT and SOCB hydroxyapatite was studied. Figure 8 shows the FTIR spectrum at 2 phr SOCB hydroxyapatite whereas Figure 9 shows the

FTIR spectrum at 5 phr SOCB hydroxyapatite. The FTIR spectra of PVOH/MWCNT composite and SOCB hydroxyapatite presented in Figure 8 was the evidences of the formation for nanocomposite. The pattern of FTIR spectrum was consistent and the peak was obvious.

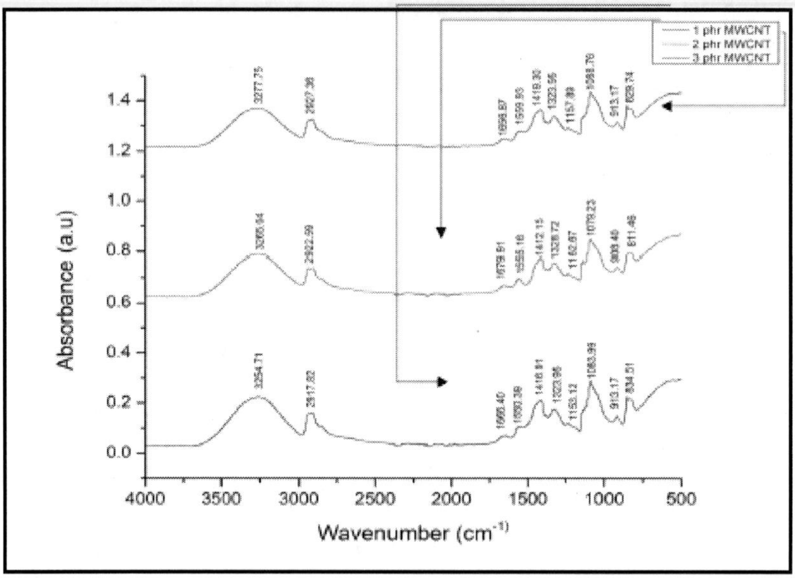

Figure 8. FTIR Spectra of PVOH with Different Loading of MWCNT at 2 phr Calcined Cuttlebone.

As shown in Figure 8, broad peak was observed at the range of 3254 cm^{-1} to 3277 cm-1 indicating the presence of hydrogen-bonded hydroxyl (OH) group of PVOH or O-H in COOH groups and absorbed water [20]. The occurrence of str hing mode in region 2917 cm^{-1} to 2927 cm^{-1} were corresponding to two vibration bands which are asymmetric stretching vibration band of CH_3 as well as asymmetric and symmetric stretching bands of CH_2 [14]. These bonds are assigned to C-H stretching mode of the backbone of PVOH. Moreover, stretching mode in region 1079 cm^{-1} to 1088 cm^{-1} indicating the presence of C-O-C groups in PVOH matrix [21]. In addition, the presence of peak at the range of 1656 cm^{-1} to 1679 cm^{-1} was attributed to the C double bond (C = C) stretching vibration in graphitic layers. However, functional MWCNT has stronger peaks in their spectra. The peak at 1323 cm^{-1} to 1328 cm^{-1} indicates the presence of O-H bending vibration in COOH groups.

Furthermore, peak located at the range of 1153 cm^{-1} to 1162 cm^{-1} was attributed to C-O stretching which contributed the crystallinity phase of PVOH matrix. Another important absorption peak located at the range of 1412 cm^{-1} to 1419 cm^{-1} and 811cm^{-1} to 834 cm^{-1} which can be used to compare the crystallinity of polymer composite at different loading of MWCNT corresponds to CH_2 scissoring and rocking deformation [19]. Besides that, the effect of MWCNT on the crystallinity of PVOH can be related to the relative intensity of bands located at 1153 cm^{-1} to 1162 cm^{-1} and 1412 cm^{-1} to 1419 cm^{-1} [19]. An increase of intensity at these peaks indicate a higher crystallinity in PVOH/MWCNT composite. All of the above results indicate that MWCNT and SOCB hydroxyapatite were introduced into PVOH matrix.

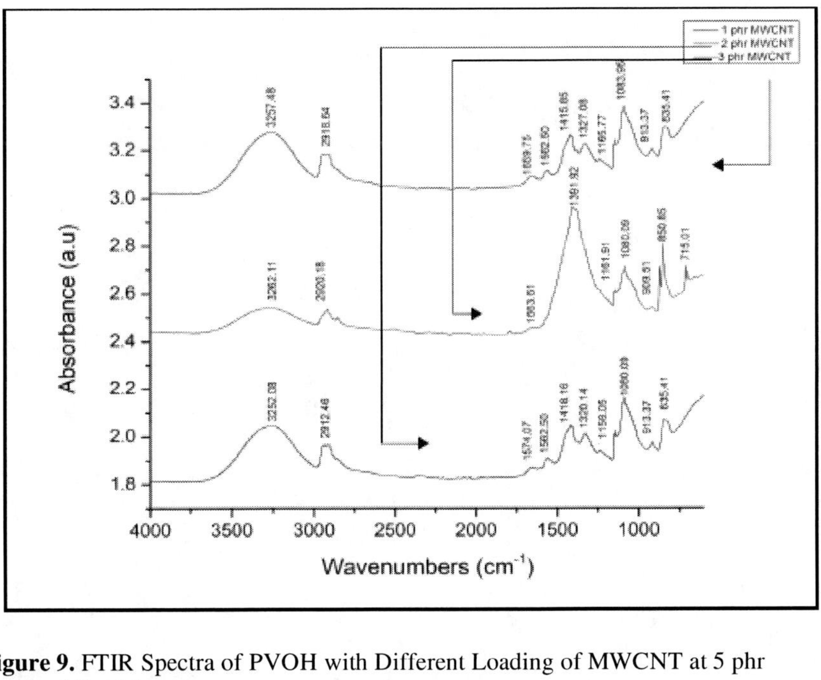

Figure 9. FTIR Spectra of PVOH with Different Loading of MWCNT at 5 phr Calcined Cuttlebone.

In Figure 9, broad peak was observed at the range of 3252 cm^{-1} to 3262 cm^{-1} and the band was attributed to the stretching vibration of -OH group in PVOH. In addition, the peak at the range of 2912 cm^{-1} to 2920 cm^{-1} were due to symmetric and antisymmetric of CH_2 vibration. Besides, the band at the range of 1080 cm^{-1} to 1083 cm^{-1} was attributed to the stretching of C-O group whereas the band at the range of 1562 cm^{-1} was attributed to the C = C stretching in graphitic layer of MWCNT [14]. Stretching of C-O group has proved the esterification process between carboxyl groups of MWCNT and hydroxyl (OH) group of PVOH. Additional bonds located at the range of 835 cm^{-1} to 850 cm^{-1} and 909 cm^{-1} to 913 cm^{-1} proves that meta-disubstitution process had occurred between the benzene rings of MWCNT [20]. Moreover, stretching mode in region 1080 cm^{-1} to 1083 cm^{-1} indicating the presence of C-O-C groups in PVOH matrix. The stretching vibration band at the range of 1391 cm^{-1} to 1418 cm^{-1} was attributed to O-H stretching in COOH group [20]. The pattern of FTIR spectrum at 2 phr MWCNT has a strong and sharp peak at 1391 cm^{-1} whereas the peak was absence in both 1 phr MWCNT and 3 phr MWCNT. The reason may due to well dispersion of MWCNT in PVOH matrix which causes strong wavenumber in the sample.

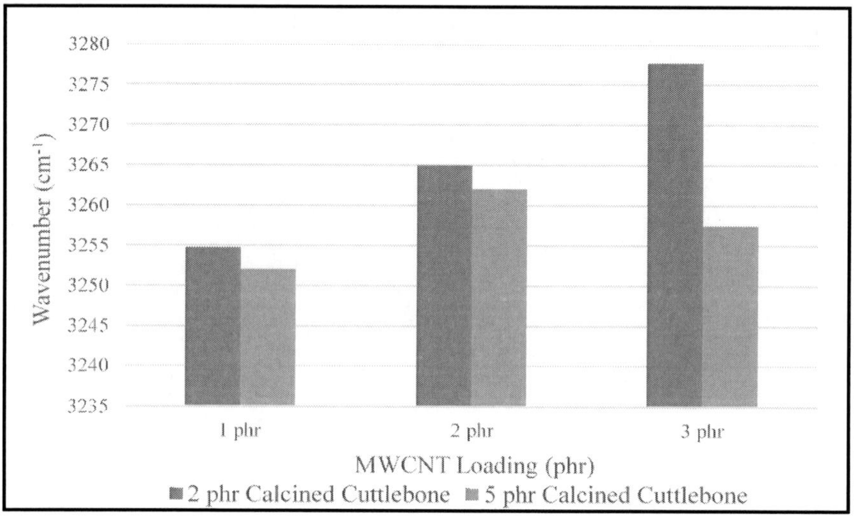

Figure 10. Hydroxyl Group of PVOH with Different Loading of MWCNT and Calcined Cuttlebone.

Broad peak around 3252 cm^{-1} to 3277 cm^{-1} corresponds to OH group. As the loading of SOCB hydroxyapatite increased from 2 phr to 5 phr, the wavenumber for hydroxyl group has a lower value as compared to low loading of SOCB hydroxyapatite (2 phr) based on Figure 10. A decrement in wavenumber indicates that the interaction between OH group is bad. However, bad interaction of OH group denotes that the hydrogen bonding is strong. Therefore, high loading (5 phr) of SOCB hydroxyapatite leads to stronger hydrogen bond as the OH bonding between the PVOH/MWCNT matrix with

SOCB hydroxyapatite was being disrupted. Besides, at 2 phr SOCB hydroxyapatite, the wavenumber increased from 1 phr to 3 phr of MWCNT. On the contrary, at 5 phr SOCB hydroxyapatite, the wavenumber number increased when the loading of MWCNT increased from 1phr to 2 phr and decreased from 2 phr to 3 phr.

Peak at the range of 2912 cm^{-1} to 2927 cm^{-1} corresponds to stretching of C-H bond between the polymer matrix. Based on Figure 11, at 2 phr SOCB hydroxyapatite, the wavenumber increased from 1 phr to 3 phr of MWCNT. According to Kuo [21], physical properties such as amorphous and crystalline structure is one of the factors which affect the strength of molecular bonding within the particles of a sample and it is not only limited to chemical properties such as the type of bonding within particles.

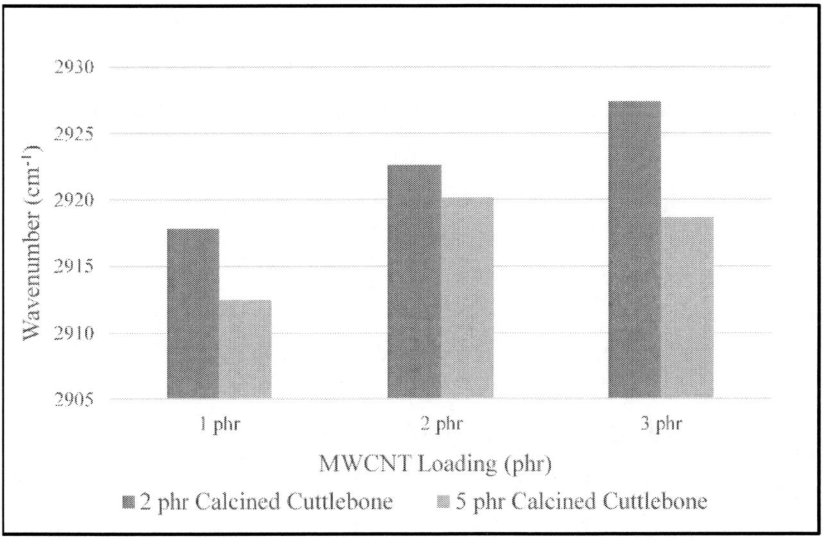

Figure 11. C-H Bonds of PVOH with Different Loading of MWCNT and Calcined Cuttlebone.

Besides that, an increase in the wavenumber indicates that high crystallinity was induced. The result indicates that the crystallinity has increased. The results were proven in XRD crystallinity test as shown in Figure 7. The rise in wavenumber is due to high energy required to overcome the vibration in a confined structure and caused by high shifting of absorption spectra by high crystallinity structure [19]. However, as the loading of SOCB hydroxyapatite increased from 2 phr to 5 phr, the wavenumber is relatively low as compared to low loading of SOCB hydroxyapatite (2 phr). This was due to the dispersion of larger particle in PVOH matrix which causes interlocking between particle and reduced the crystallinity of a component. The result indicates that the crystallinity is lower at high loading (5 phr) of SOCB hydroxyapatite and it was proven in XRD characteristic test.

3.5. Differential Scanning Calorimetry

The DSC test was performed to study the thermal properties of PVOH with different loading of MWCNT and SOCB hydroxyapatite by determining the melting temperature and enthalpy of melting of the composite. Figure 12 and 13 shows the melting temperature curve for all the samples at 2 phr SOCB hydroxyapatite and 5 phr SOCB hydroxyapatite respectively.

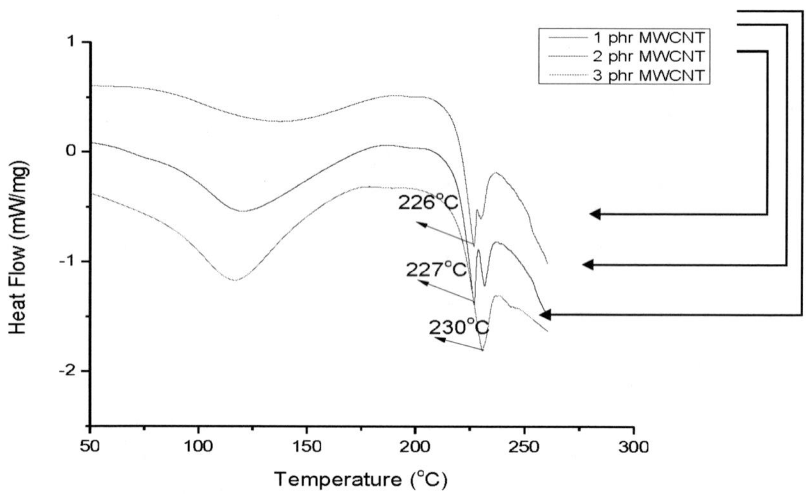

Figure 12. DSC Result for PVOH with Different Loading of MWCNT at 2 phr Calcined Cuttlebone.

3.5.1. Melting Temperature

As shown in Figure 12, as the loading of MWCNT increased from 1 phr to 2 phr, the melting temperature increased from 226°C to 227°C. The melting temperature increased further from 227°C to 230°C when the loading of MWCNT increased from 2 phr to 3 phr. In other words, the curve shifted to right from 1 phr to 3 phr of MWCNT. Peak shifting to right indicates that high temperature is needed to melt the composite. The result indicates that while adding small amount of MWCNT into the PVOH matrix may not have a significant effect on the melting temperature, the addition of large amount of MWCNT can result in an increase in the melting temperature which causes the PVOH matrix to be more rigid due to strong bonding between SOCB hydroxyapatite and PVOH/MWCNT composite.

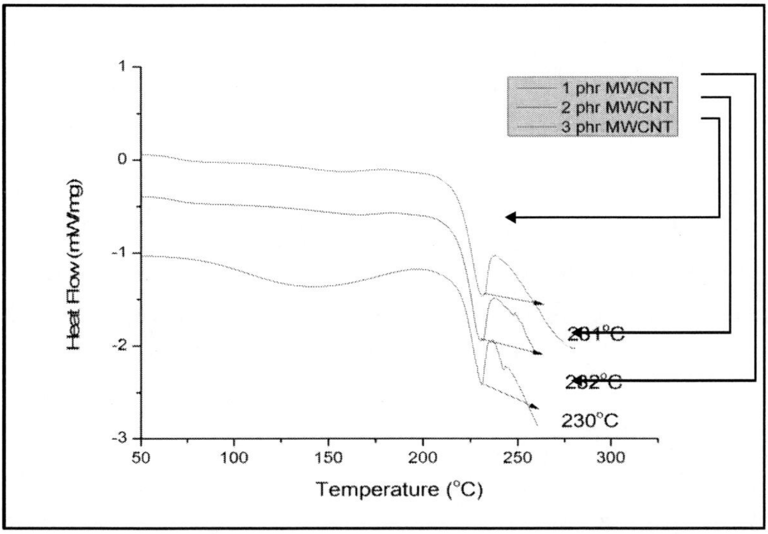

Figure 13. DSC Result for PVOH with Different Loading of MWCNT at 5 phr Calcined Cuttlebone.

As shown in Figure 13, as the loading of MWCNT increased from 1 phr to 2 phr, the melting temperature increased from 231 °C to 232 °C. However, the melting temperature decreased from 232 °C to 230 °C when the loading of MWCNT increased from 2 phr to 3 phr. The curve shifted to right from 1 phr MWCNT to 2 phr MWCNT and shifted to left again from 2 phr MWCNT to 3 phr MWCNT. The decrease in melting temperature is due to agglomeration of MWCNT in PVOH matrix. In addition, another reason which causes decrement in melting temperature is due to the restriction of "chain mobility" in the polymer matrix. The mobility is being restricted by MWCNT in PVOH matrix due to strong hydrogen bonding between the oxygen-containing groups of PVOH/MWCNT composite [14]. Besides, strong interfacial interaction between SOCB hydroxyapatite and PVOH/MWCNT composite in terms of efficient stress transfer has strengthened the mechanical properties as shown in the previous tensile test which is seen in Figure 1.

In general, at high loading (5 phr) of SOCB hydroxyapatite, the melting temperature increased as well. As seen in Figure 14, the melting temperature is higher at 5 phr SOCB hydroxyapatite as compared to 2 phr SOCB hydroxyapatite. High melting temperature indicates that the bonding between SOCB hydroxyapatite and PVOH/MWCNT composite is strong.

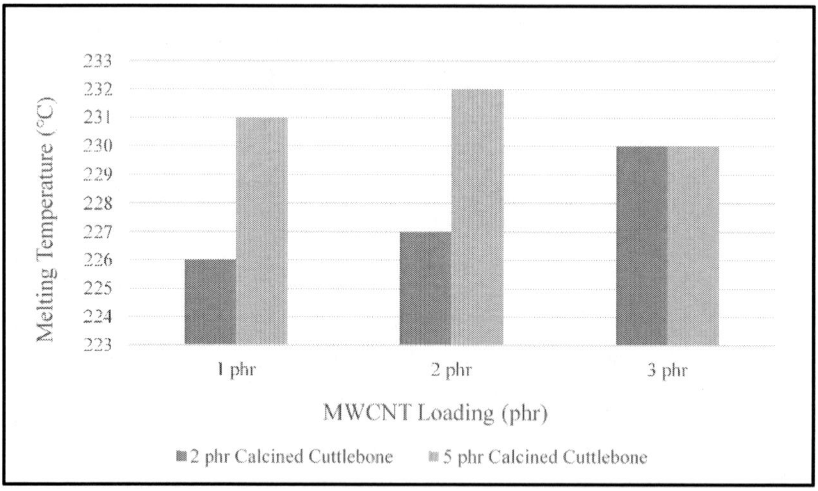

Figure 14. Melting Temperature for PVOH with Different Loading of MWCNT and Calcined Cuttlebone.

3.5.2. Enthalpy of Melting

The data of enthalpy of melting was recorded and graph was plotted as shown in Figure 15. Figure 15 shows the comparison of enthalpy of melting for PVOH matrix with different loading of MWCNT at 2 phr and 5 phr SOCB hydroxyapatite.

At 2 phr SOCB hydroxyapatite, the enthalpy of melting decreased from 1 phr MWCNT to 3 phr MWCNT. The decline in enthalpy of melting indicates the composite is amorphous, thus the elongation at break increased and the results were proven in tensile test which is shown in Figure 3. This is in line with the observation in XRD of 1 phr where amorphous structure tends to increase the degree of freedom of the polymer matrix. Hence, a less cohesive and more disorderly composite due to addition of higher loading of MWCNT tends to increase the elongation at break.

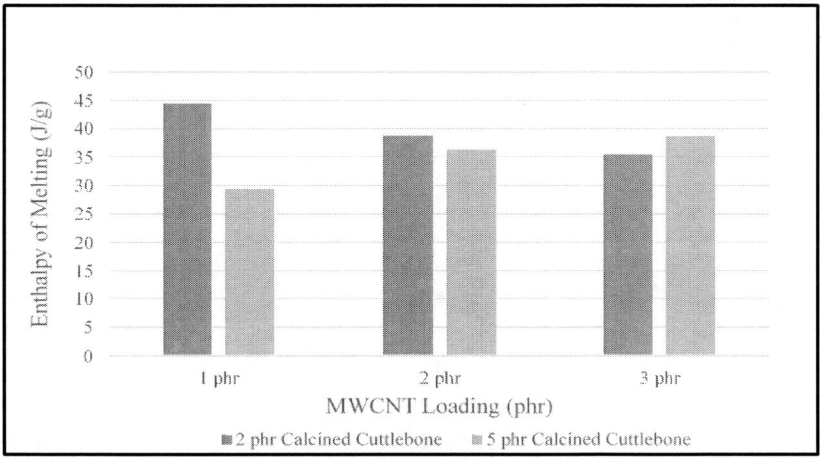

Figure 15. Enthalpy of Melting for PVOH with Different Loading of MWCNT and Calcined Cuttlebone.

At 5 phr SOCB hydroxyapatite, the enthalpy of melting increased from 1 phr MWCNT to 3 phr MWCNT. An increment of enthalpy of melting indicates that the intermolecular interactions between SOCB hydroxyapatite and MWCNT is strong. Hence, higher energy is needed to break the bonding between the composites.

Conclusion and Future Perspectives

It was discovered that the incorporation of SOCB hydroxyapatite at both 2 phr and 5 phr has significantly increased the tensile strength of PVOH/MWCNT at 3 phr MWCNT. The result also indicates that the incorporation of SOCB hydroxyapatite served as reinforcement phase and improved the mechanical properties of a component. At high loading of SOCB hydroxyapatite (5 phr), the Young's modulus increased with increasing MWCNT. Increasing modulus indicates that MWCNT is well dispersed in PVOH matrix. The mechanical properties of these nanocomposites also exceeded the properties of pure PVOH, for example the minimum tensile strength of the nanocomposites (17.69 MPa) is larger than that of pure PVOH (7.1 MPa). This observation was in line with SEM results denoted that the surface of PVOH/MWCNT composite with SOCB hydroxyapatite became rough and large particle can be observed as the loading of MWCNT increased. Importantly, XRD tests depicted that SOCB hydroxyapatite and MWCNT are well dispersed into PVOH matrix. At 2 phr SOCB hydroxyapatite, the broadness of peak decreased with increasing loading of MWCNT. The result indicates that MWCNT is well dispersed in PVOH matrix and a crystalline structure was induced. At 5 phr SOCB hydroxyapatite, more sharp and narrow peaks were observed at the initial stage but the sample's peak broadness increased as the loading of MWCNT increased further. Broadening effect was due to the disturbance of ordered structured arrangement of polymer chain [22]. For infrared spectroscopy (FTIR) analysis showed that the wavenumber of O-H group the wavenumber number increased first then decreased. A decrease in wavenumber indicates that the hydrogen bonding is strong. At 5 phr SOCB hydroxyapatite, the enthalpy of melting increased with increasing MWCNT loading. In summary, the mechanical and thermal properties of the polymer nanocomposite increased with higher loading level of MWCNT but the agglomerations tend to form within the matrix at higher SOCB hydroxyapatite content. Thus, it is concluded that the addition of 2 phr SOCB hydroxyapatite with 3 phr MWCNT can induced the strongest interaction of bonding between molecules. As a result, this means that this nanocomposite has the potential to be a biocompatible bone scaffolding material due to its inertness and mechanical strength, which can be applied in medical fields [23]. However, this polymer composite still possesses some disadvantages. One of them being the formation of agglomerates on the polymer matrix as MWCNT increases (as seen in the SEM images), which leads to brittleness of the composites [24]. For future perspectives, Studies on the characterization of PVOH added with MWCNT and SOCB hydroxyapatite should be carried out by using thermogravimetric analysis (TGA) in order to study the thermal stabilities in details. Moreover, electron dispersive X-ray (EDX) should be carried out on SOCB hydroxyapatite to analyze the chemical composition and the Ca/P weight ratio of the derived hydroxyapatite [23].

References

[1] Jiang, X.C., Xia, C., Ye, D., Liu, L. Properties of poly(vinyl alcohol) plasticized by magnesium chloride. *Chem. J. Chin. Univ.* **2012**, *33*, 1872–1876.

[2] Tontowi, A.E., Perkasa, D.P., Siswomihardjo, W., Darwis, D. Effect of Polyvinyl Alcohol (PVOH) Blending and Gamma Irradiation on Compressive Strength of FHAp/FGel Composite as Candidate of Scaffold. *Int. J. Eng. Technol.* **2018**, *8*, 108-116.

[3] Venkatesan, J., Kim, S.-K. 2013. Hydroxyapatite from Marine Fish Bone. In *Marine Biomaterials*, Kim S.-K, Eds., CRC Press: Boca Raton, U.S.A, 2013; pp.17-28.

[4] Venkatesan, J., Rekha, Anil, S., Bhatnagar, I., Sudha, P.N., Dechsakulwatana, C., Kim, S.K., Shim, M.S. Hydroxyapatite from Cuttlefish Bone: Isolation, Characterizations, and Applications. *Biotechnol. Bioprocess Eng.* **2018**, *23*, 383-393.

[5] Bee, S., Sin, L.T., Ooi Ker Qi, N., Ratnam, C.T., Rahmat, A.R. Interactive effects of carbon nanotube and montmorrilonite reinforcement polyvinyl alcohol composite system. *J. Vinyl Addit. Technol.* **2018**, *26*, 77–89.

[6] Islam, M.R., Pias, S.M., Alam, R.B., Khondaker, S.I. Enhanced electrochemical performance of solution-processed single-wall carbon nanotube reinforced polyvinyl alcohol nanocomposite synthesized via solution-cast method. *Nano Express* **2020**, *1*, 030013.

[7] Zhang, R.L., Chen, C., Li, J., Wang, X. Preparation of montmorillonite @ carbon composite and its application for U(VI) removalfrom aqueous solution. *Appl. Surf. Sci.* **2015**, *349*, 129–137.

[8] Kaur, T., Thirugnanam. A., Kar, S. Microwave-assisted synthesis of porous chitosan–modified montmorillonite–hydroxyapatite composite scaffolds. *Int. J. Biol. Macromol.* **2016**, *82*, 628-636.

[9] Kaur, T., Thirugnanam, A. Tailoring in vitro biological and mechanical properties of polyvinyl alcohol reinforced with threshold carbon nanotube concentration for improved cellular response. *RSC Adv.* **2016**, *6*, 39982–39992.

[10] Choudhary, V., Gupta, A. Polymer/Carbon Nanotube Nanocomposites. In *Carbon Nanotubes - Polymer Nanocomposites*, Yellampalli S., Eds., IntechOpen: London, U.K, 2011; pp.65-90.

[11] Spitalsky, Z., Tasis, D., Papagelis, K., Galiotis, C. Carbon nanotube–polymer composites: chemistry, processing, mechanical and electrical properties. *Prog. Polym. Sci.* **2010**, *35*, 357-401.

[12] Li, J. Multiwalled Carbon Nanotubes Reinforced Polypropylene Composite Material. *J. Nanomater.* 2017, 2017, 1-5.

[13] Yee, M.J., Mubarak, N.M., Khalid, M., Abdullah, E.C., Jagadish, P. Synthesis of polyvinyl alcohol (PVOH) infiltrated MWCNTs buckypaper for strain sensing application. *Sci Rep.* **2018**, *8*, 17295 .

[14] Lai, D., Wei, Y., Zou, L., Xu, Y., Lu, H. Wet spinning of PVOH composite fibers with a large fraction of multi-walled carbon nanotubes. *Prog. Nat. Sci.* **2015**, *25*, 445-452.

[15] Ferreira, F.V., Franceschi, W., Menezes, B.R.C., Brito, F.S., Lozano, K., Coutinho, A.R., Cividanes,L.S; Thim, G.P. Dodecylamine functionalization of carbon nanotubes to improve dispersion, thermal and mechanical properties of polyethylene based nanocomposites. *Appl. Surf. Sci.* **2017**, *410*, 267-277.

[16] Mohamed, A., Kahder, M., AlSaad, K., AlMeer, S. Properties of nanoclay PVA composites materials. *QScience Connect* **2013**, *2013*, 1–9.

[17] Yaser, S., Toraj, M. 2013. Effects of CNTs Content on Physicochemical and Pervaporation Separation Properties of PVOH Membranes. *Sep Sci Technol.* **2013**, *48*, 716-727.

[18] Hajeeassa, S.K., Hussein, M.A., Anwar, Y., Tashkandi, N.Y., Al-amshany, Z.M. Nanocomposites containing polyvinyl alcohol and reinforced carbon-based nanofiller: A super effective biologically active material. *Nanobiomedicine.* **2018**, *5*, 1-12.

[19] Diouri, N., Baitoul, M. Effect of carbon nanotubes dispersion on morphology, internal structure and thermal stability of electrospun poly(vinyl alcohol)/carbon nanotubes nanofibers. *Opt Quantum Electron.* **2013**, *46*, 259-269.

[20] Malikov, E.Y., Muradov, M.B., Akperov, O.H., Eyvazova,G.M., Puskás, R., Madarász, D., Nagy, L., Kukovecz, Á., Kónya, Z. Synthesis and characterization of polyvinyl alcohol based multiwalled carbon nanotube nanocomposites. *Physica E.* **2014**, *61*, 129-134.

[21] Kuo, S.W. Hydrogen-bonding in polymer blends. *J. Polym. Res.* **2008**, *15*, 459-486.

[22] Lawton, K., Le, H., Tredwin, C., Hand, R.D. Carbon nanotube reinforced Hydroxyapatite Nanocomposites As Bone Implants: Nanostructure, Mechanical Strength And Biocompatibility. *Int. J. Nanomedicine.* **2019**, *2019*, 7947-7952.

[23] Zhou, H., Lee, J. Nanoscale hydroxyapatite particles for bone tissue engineering. *Acta Biomaterialia.* **2011**, *7*, 2769-2781.

[24] Khan, A.S., Hussain, A.N., Sidra, L., Sarfarz, Z., Khalid, H., Khan, M., Manzoor, F., Shahzadi, L., Yar, M., Rehman, I.U. Fabrication and in vivo evaluation of hydroxyapatite/carbon nanotube electrospun fibers for biomedical/dental application. *Mat. Sci. Eng.* C. **2017**, *80*, 387-396.

Chapter 5

The Production of Bioceramics from Natural Sources and Controlled Drug Release Applications

Dilek Bilgic Alkaya[1,2,*], **PhD**
and Serap Ayaz Seyhan[1,2], **PhD**
[1]Department of Analytical Chemistry, Marmara University Faculy of Pharmacy, Basıbuyuk, Istanbul, Turkey
[2]Center for Nanotechnology & Biomaterials Application and Research (NBUAM), Marmara University, Turkey

Abstract

Traditional drug applications are notable for their limited efficiency, low biodistribution, and limited selectivity. Currently, in the treatment of many diseases, delivering the therapeutic compound to the targeted tissue is an important problem. A controlled drug release system is a system in which the release rate of the drug can be determined in a certain region or systematically for a certain period of time. These systems protect the drug against external effects such as photodegradation and pH and also reduce the frequency of drug administration. Additionally, by transporting the drug to the targeted area, the impact of the drug on surrounding tissues and organs is reduced. Thus, while increasing patient satisfaction, toxicity and possible side effects are also reduced. In recent years, various drug release profiles can be obtained by adding biodegradable polymers with different degradation rates and mechanisms to ceramic materials. Hydroxyapatite (HA, $Ca_{10}(PO_4)_6(OH)_2$) is a bioceramic material with an apatite-like bone structure and is found in

[*] Corresponding Author's Email: dbilgic@marmara.edu.tr.

In: Biocomposites
Editor: James R. Bush
ISBN: 979-8-89113-267-2

bone and tooth composition. In addition to naturally-synthesized HA, which mainly uses animal bones or coral reefs containing various calcium phosphate groups, and due to its biocompatibility, bioactivity, non-toxicity, and ability to mimic the mineral composition of bone, it presents a promising platform for drug delivery. Ongoing research aims to further optimize the drug loading and release properties of naturally-synthesized HA and explore its potential in targeted and personalized medicine. Therefore, this section provides information on various literature studies reported on drug release studies of naturally-synthesized HA composites.

Keywords: biocomposite, hydroxyapatite, drug release

Introduction

For many years, the development of new molecules to treat various diseases has been the focus of medical research. However, these efforts often face challenges such as time constraints, economic burdens, and unexpected results, the main problems facing the pharmaceutical industry. To address these challenges, new approaches have emerged that aim not only to treat diseases but also to improve patients' quality of life.

Nanotechnology, after more than two decades of scientific innovation through research and development, has revolutionized medicine, particularly in the areas of diagnostic methods, imaging, and drug delivery. Nanotechnology has significant potential in the field of medicine, including drug delivery systems, tissue engineering structures, implants, and pharmaceutical therapeutics (Sim and Wong, 2021).

The Food and Drug Administration (FDA) has approved certain nanotechnology-based products. In the field of medical nanotechnology, there are currently 51 products based on this technology that are used in clinical practice. These nano-drugs are primarily developed for drugs with low water solubility and high toxicity. These nanoformulations often improve the pharmacokinetic properties of the drug while reducing toxicity. Of these ongoing nanomaterials, 18 relate to chemotherapy, 15 to antimicrobial agents, and 28 to various medical applications (Patra et al., 2018).

When there is no internal route for drug delivery, external therapeutic methods such as radiation therapy and surgical procedures are used. These methods are often used interchangeably or in combat disease against diseases. Nanotechnologies are making a compelling contribution in this area by developing new modes of drug delivery, and some of these methods have been shown to be effective in the clinical setting and are already being used clinically (Sim and Wong, 2021).

Controlled-release systems have been shown to improve treatment outcomes in many ways, including increasing therapeutic efficacy, reducing toxicity, improving patient compliance, and enabling entirely new medical therapies (Gao et al., 2023). As the therapeutic landscape evolves toward next-generation therapeutics such as proteins, peptides, nucleic acids, and living cells, drug delivery technologies have also evolved to meet unique delivery requirements (Gao et al., 2023). This technology allows for the long-term delivery of small molecule drugs, reduces the frequency of drug administration, and improves compliance and patient adherence. This practice has revolutionized the treatment and management of many diseases. Bioceramics are known for their high biocompatibility and high chemical stability. Natural ceramic biomaterials have the ability to bond with the implant-tissue interface. In the early 1920s, it was discovered that the inorganic phase of bones and teeth is mainly composed of calcium hydroxyapatite, which led to the widespread use of hydroxyapatite (HA) as a bioactive ceramic in the biomedical field.

Hydroxyapatite (HA) is a naturally occurring mineral form of calcium apatite. HA is a ceramic material that is biocompatible, bioactive, osteoconductive, and can form direct bonds with living tissue. HA exhibits significantly better biocompatibility and biodegradability than other nanoparticles. It shows pH-dependent dissolution and higher biocompatibility. It is also very stable and less toxic. Due to its importance in clinical applications and tissue engineering, HA has been the subject of biomaterials research for many years. HA is considered a suitable carrier for controlled release of compounds as it is the most stable CaP under physiological conditions (Lara-Ochoa et al., 2021).

HA can be obtained from natural or synthetic sources. Synthetic HA is obtained by mixing reactive precursors of calcium (Ca^{2+}) and phosphate (PO_4^{3-}). Natural HA, on the other hand, includes all biomaterials such as animal bones, fish scales and bones, shells, minerals, and plants (Obada et al., 2022). Both natural and synthetic hydroxyapatite are among the most widely studied materials due to their biocompatibility, bioactivity, low cost, and compatibility with existing applications.

Natural HA, unlike synthetic HA, contains characteristic trace elements such as Na^+, Zn^{2+}, Mg^{2+}, K^+, Si^{2+}, Ba^{2+}, F^-, and CO_3^{2-}. In addition, there are differences between chemically synthesized hydroxyapatite and hydroxyapatite obtained from natural sources in terms of cost, Ca/P ratio, source, trace elements, and processing time. For these reasons, the natural source of hydroxyapatite has become more attractive for biomedical applications (Obada et al., 2022).

This section provides a general overview of the properties, synthesis, and applications of hydroxyapatite from natural sources, as well as the advantages and application prospects of drug delivery systems based on HA.

Natural-Based Hydroxyapatite

Hydroxyapatite $Ca_{10}(PO_4)_6(OH)_2$, can be synthesized from inorganic compounds or natural organic-based materials. Both types of biomaterials are biocompatible, osteoconductive, bioactive, and important materials for biomedical applications. However, the main problem with hydroxyapatite synthesized from inorganic calcium and phosphorus sources is the high cost, processing time, and lack of trace elements (Akram et al., 2014; Kalpana and Nagalakshmi, 2022).

Natural-based hydroxyapatite (HA) refers to hydroxyapatite from biological and recycled sources that have the potential to be converted into calcium phosphate materials. Natural hydroxyapatite is free of impurities, has high crystallinity, and is environmentally friendly (Hussin et al., 2022). Due to its biocompatibility, bioactivity, rapid bone regeneration processes, osseointegration, and similarity to the composition of natural bone, natural hydroxyapatite is generally preferred over synthetic hydroxyapatite (Arokiasamy et al., 2022).

Natural-based HA can be obtained through several sources:

8. Biological Sources*:* Examples of biological sources commonly used to obtain HA include cattle, pigs, chickens, goats, camel bones, and teeth (Pu'ad et al., 2019; Bulus et al., 2018; Ramesh et al., 2018). Bones usually undergo a demineralization process to remove organic components, leaving behind mineralized hydroxyapatite. This natural HA retains the inherent properties and structure of the original bone, making it an ideal biomimetic material.

   Marine organisms (such as corals) (Ben-Nissan, 2003), seashells (such as sea urchins) (Agaogullaria et al., 2012), Mediterranean mussels (Macha et al., 2013; Kel et al., 2012), nacre (Oktar et al., 2015), cuttlefish, fish scales (Pu'ad et al., 2019) also produce hydroxyapatite structures. Natural HA, derived from marine sources often has unique properties and may have a porous structure that can be beneficial for certain applications.

9. Recycled Sources: The recycling of agricultural waste has contributed to many areas of life, from its use as a feedstock for the production of biobased products to its subsequent use as a source of renewable energy. Another approach is to obtain natural-based HA from recycled sources, such as eggshells. These materials can be processed to obtain hydroxyapatite, thereby reducing waste and utilizing available resources (Zhang et al., 2012; Macha et al., 2015; Abdulrahman et al., 2014; Santos et al., 2009; Siva Rama Krishna et al., 2007).

Natural-based hydroxyapatite has several advantages for biomedical applications:

1. Biocompatibility: HA, which is derived from natural sources, is very similar to the mineral composition of human bone and is therefore highly biocompatible. It integrates well with surrounding tissues and promotes cell attachment and proliferation.
2. Bioactivity: Natural-based HA has the ability to bond directly to bone through chemical interactions rather than physical or mechanical attachment, which means it can stimulate the body's natural healing processes. It leads to osseointegration by forming a bond with living bone tissue and improving the overall biocompatibility of implants.
3. Structural Mimicry: Due to its similarity to the structure and properties of bone, including the presence of micro and nano-scale features, natural-based HA enables better integration with the surrounding tissue, and facilitates efficient nutrient exchange and tissue regeneration.

HA Synthesis Methods

Commercial synthetic hydroxyapatite powders can be obtained by chemical precipitation, solid-state reactions, emulsions, sol-gel processes, microemulsions, and hydrothermal processes because they are inexpensive and simple. However, the resulting material may lack important ions. Natural sources, such as mammalian bones or fish bones and scales, provide readily available presence of important ions in xenografts and offer low processing costs, making them a preferred option for applications research. In addition, shells are considered valuable calcium sources for hydroxyapatite production due to their abundance and economic feasibility (Hussin et al., 2022).

Hydroxyapatite (HA) has a composition range characterized by the calcium-phosphorus (Ca/P) ratio. This ratio is important for the properties and behavior of HA obtained from natural sources, especially bone. Because, this efficiency depends on factors such as phase purity, size distribution, extraction technique, calcination temperature and type of bone.

HA can be synthesized by various techniques such as mechanochemical, chemical precipitation, biomimetic mineralization, hydrothermal and sol-gel methods using natural resources such as mammals, aquatic organisms, shell and minerals. The choice of method depends on both the intended application and the desired properties of the final product. Chemical precipitation is a widely used method in which calcium and phosphate ions are mixed in an aqueous solution at a certain pH and temperature in a controlled manner. Sol-gel synthesis uses the precursor solution, which forms a gel by hydrolysis and condensation and is then calcined to obtain hydroxyapatite. In biomimetic mineralization, organic templates or biological molecules are used to promote the formation of hydroxyapatite. Hydrothermal synthesis uses high pressure and temperature to produce a crystalline material (Arokiasamy et al., 2022).

The formation of natural calcium phosphate minerals is highly influenced by variable environmental conditions, leading to their formation in different compositions and qualities (Doğdu, 2021).

Examples of HA Applications from Natural Sources

Biological Sources

Natural hydroxyapatite is typically derived from biological sources or wastes such as mammalian bone, marine, shell sources (e.g., cockle, clam, eggshell, and seashell), plants and algae.

Mammalian Bones

When they compared the extracted HA with synthetic HA, they reported that natural HA has better potential in bone tissue engineering studies.

Bulus et al. (2018) synthesized a bioceramic grade from hydroxyapatite and tricalcium phosphate (TCP), which are minerals found in human tooth enamel. XRD analysis of the bioceramics synthesized from enamel, was used to determine the composition of the the crystal structures (HA, FA (Florapatit), TCP). When evaluating the FTIR results, it was found that the structure of bioceramics is consistent with the presence of functional groups in their composition. The analysis EDS confirmed the presence and amounts of the elements calcium (Ca) and phosphorus (P) in the synthesized bioceramics and found that the crystal structures of HA, FA, and TCP corresponded to the stoichiometric molar ratio (Bulus et al., 2018; Oktar et al. 2015) also reported that the hydroxyapatite material used was derived from calcined human teeth. The sodium-free bioglass material is obtained from reagent grade chemicals. A high degree of porosity was obtained in blanks made from HA, bioglass and HA-bioglass composites.

In his study, Ramesh prepared hydroxyapatite by sintering from waste bones of cattle, goats, and chickens, and investigated its structure. The bones, which had been purified of organic matter, were exposed to various temperatures ranging from 600°C to 1000°C. As a result, HA produced from cattle bone exhibited good thermal stability, while goat and chicken bones exhibited phase instability, with the presence of TCP observed after 700°C (Ramesh et al., 2018).

In a comparative study by Bahrololoom et al. (2009), HA was extracted from bovine femur bone, sheep femur bone, and sheep skull flat bone. The Ca/P ratios obtained ranged from 1.46 to 2.01. They concluded that HA with these Ca/P ratios could be used in orthopedic and dental applications.

Marine Biological Waste

In addition, marine biowaste such as mollusc shells, shellfish, cuttlefish and fishbone containing $CaCO_3$ can be converted into bio-HA, CaP ceramic, which is widely used in orthopedics and dental surgery biomedical devices (Borciani et al., 2023).

In a study by Lee, Lee and Yoon (2007), HA was extracted from *inkfish* bones, and the Ca/P ratios observed ranged from 1.38 to 1.82. Based on their findings, they suggested that HA materials within this Ca/P range could be suitable for biomedical applications. Ozawa and Suzuki (2002) isolated HA from waste bones of *Pagrus* major but could not determine the Ca/P ratio due to calcium deficiency. However, they reported that the obtained HA is an economical and biologically/environmentally compatible ceramic material, suggesting the potential of fish-derived HA as a ceramic source. Boutinguiza et al. (2012) extracted HA from *Xiphias gladius* and *Thunnus thynnus*, resulting in Ca/P ratios of 1.87 and 1.86, respectively. They highlighted that HA with a high Ca/P ratio, as obtained in their study, could be utilized in bone healing applications. Latif et al. (2020) extracted HA from *Thunnus thynnus* fish and emphasized its high biocompatibility and suitability for biomedical applications, noting a Ca/P ratio of 1.60 and the presence of elements such as Ca, Mg, Sr, Na, K, and Zn. Venkatesan et al. (2011) conducted the alkaline extraction of HA from *Thunnus obesus* bones and discussed its applicability not only in biomedical fields but also as an environmental indicator, highlighting HA with a Ca/P ratio of 1.76.

Pal et al. (2017) extracted HA along with TCP from the bones of *Lates calcarifer* fish at high calcination temperatures (1200 °C) and reported a Ca/P ratio of 1.62 for the calcined sample. They emphasized its potential use in orthopedic implants and tissue engineering applications. Biazar et al. (2020) extracted HA from caspian kutum bones and obtained a Ca/P ratio of 1.82. They proposed that HA extracted, including bone grafts, could serve as an economical alternative to synthetic HA. Hasan, Ghazali, and Mohtar (2021) found a Ca/P ratio of 1.67 in their HA extraction from Amblygaster sirm fish bones, highlighting the potential commercialization of fish bones with high efficiency in HA extraction for various applications as a biomaterial. Athinarayanan, Periasamy, and Alshatwi (2020) extracted HA from Lethrinus lentjan fish bones and reported a Ca/P ratio of 2.33. They suggested that the obtained HA, with its biocompatibility, could be utilized in bioimaging and bone tissue engineering applications. Surya et al. (2021) obtained a Ca/P ratio of 1.67 with their HA extraction from fish bones of Sardinella longiceps, emphasizing. The nontoxic, effective, and biocompatible nature of the extracted HA as bone graft for biomedical applications. Wijedasa et al. (2020) performed the extraction of HA from fish bones of Salmo salar and Lutjanus campechanus, obtaining Ca/P ratios ranging from 1.56 to 1.62. They concluded that the extracted HA showed promising properties for periodontal tissue regeneration.

These studies demonstrate the potential of extracting HA from various fish and animal bones for biomedical and environmental applications, as well as the suitability of the reported Ca/P ratios for specific purposes, including bone healing and medical applications.

Kongsri et al. (2013) produced high purity nano-sized HA with a Ca/P ratio of 1.66. They used alkaline heat treatment for HA extraction from Tilapia nilotica fish scales and found that HA extracted from fish was more cost-effective and beneficial than synthetic HA. Likewise, PonOn et al. (2016) used alkaline hydrolysis to extract HA from Probarbus jullieni flakes. They showed pure HA with a Ca/P of 2.01. They also investigated the physicochemical, bioactive and biological properties of the extracted HA. Alparslan, Baygar and Baygar (2017) examined HA isolated from the mixture of Dicentrarchus labrax and Sparus aurata flakes and found the Ca/P ratio to be 1.70. They suggested that low-crystalline HA has a higher biodegradation rate and metabolic activity than high-crystalline HA. Deb and Deoghare (2019) extracted HA from Catla catla fish scales and reported Ca/P ratios of 1.68, 1.76 and 1.79 for acid, alkali and alkali acid treatments, respectively. They emphasized that the resulting HA showed no cytotoxic effects on mammalian cell lines and suggested its potential as an alternative source of biomaterials for tissue engineering applications.

These studies demonstrate the potential of extracting HA from fish sources for various biomedical applications. The physical physiology of fish scales, with more collagen content compared to bone, provides greater flexibility and softness, making extraction easier compared to other methods. The reported Ca/P ratios highlight the suitability of the obtained HA materials for specific uses in bioceramics, biomaterials, tissue engineering, and bone reconstruction. Additionally, in terms of hydroxyapatite extraction, it holds significant potential for transforming unwanted waste into a useful and functional material.

Besides that, the production of HA such as sea urchins, were reported by authors (Kel et al., 2012; Gómez et al., 2020; Komalakrishna et al., 2017; Rajabiyan et al. 2021; Li et al., 2021) In these studies, it has been reported that raw materials obtained from the processing of sea urchins can be benefited by providing added value for the biomedical industry.

Agricultural Recycled Source

Agricultural wastes such as eggshell, fruit are considered to be a rich source for calcium as carbonates and oxides that qualify them as excellent sources of hydroxyapatite Furthermore, extensive scientific studies have also been conducted on eggshells. HA is the most extensively studied bioceramic and has been used in bone and dental implantation since the 1960s. There are over 1000 articles published in the literature each year. Although eggshells are not the first natural source for the synthesis of hydroxyapatite, it provides a cost effective and a renewable as well as a sustainable source of biological HA (Wu et al., 2013.; Rivera et al., 1999; Siva Rama Krishna et al., 2007; Dasgupta et al., 2004; Hussain and Sabiruddin, 2021; Ibraheem et al., 2019)

Drug Release

Conventional release systems rely on passive delivery methods, where drugs are administered to the body, but they may be subjected to pH changes or enzymatic degradation. This can result in fluctuating drug concentrations in the blood plasma, leading to ineffective levels, toxic levels, or a failure to maintain effectiveness in the targeted region over an extended period. In contrast, controlled release systems enhance therapeutic activity while reducing side effects, allowing for a reduction in the required dosage amount during treatment. Additionally, they can extend the dosing interval, improving the patient's quality of life.

Controlled release systems provide a method for releasing active substances from a system at a predetermined rate, specific timing, and desired quantity. These systems involve modifying the routes of drug administration or targeting specific systems, thereby altering various parameters of the drug, such as release profile, passage capacity through biological carriers (dependent on particle size), biodistribution, and metabolism. Through the use of controlled release systems, the pharmacokinetics and pharmacodynamics of drugs can be altered to enhance the efficacy of drug therapy and provide several advantages over traditional delivery systems (Adepu and Ramakrishna, 2021).

The use of ceramic materials as drug delivery vehicles has been extensively studied in various biomedical applications. Hydroxyapatite (HA)-based porous drug delivery systems have the advantage of sustained and continuous drug delivery with minimal side effects. HA enables localized drug delivery and is suitable for drugs with short half-lives. Additionally, HA is favored as a drug carrier in many biological applications due to its non-inflammatory and non-toxic nature (Patra et al., 2018).

Table 1. Drug delivery systems based on hydroxyapatite

Drugs/Peptides/Proteins/DNA	Physical form of hydroxyapatite	Method of drug immobilization	Application	Reference
Testosterone enanthate, cyclosporin A, human serum albumin	Microparticles, granules, and nanoparticles	Soaking, adsorption	Delivery	Mizushima et al.2006
Bovine serum albumin	Microparticles, granules, and nanoparticles	Soaking, adsorption	Delivery	Wu et al. 2012
Cytochrome, bovine serum albumin, lysozyme HCl	Microparticles, granules, and nanoparticles	Adsorption	Delivery	Ishihara et al. 2010; Tomoda et al. 2010
Fibrinogen, Insulin, type I collagen, bovine serum albumin	Porous ceramics	Adsorption	Diabetes	Zhu et al. 2010
Fish sperm DNA	Microparticles, granules, and nanoparticles	Encapsulation	Gene therapy	Bertran et al. 2014
APWHLSSQYSRT, STLPIPHEFSRE, VTKHLNQISQSY	Compressed disks	Adsorption	functional peptides	Segvich et al. 2009
ciprofloxacin	Nonaparticles	Adsorption	Bacterial Infection, Delivery	Munir et al. 2018
Vancomycin	Microparticles, granules, and nanoparticles	Soaking Encapsulation	Delivery	Lin et al. 2013 Ye et al. 2010
Carvedilol	Microparticles, granules, and nanoparticles	Immersion/solvent evaporation	Delivery	Zhao et al. 2011
5-Fluorouracil	Microparticles, granules, and nanoparticles	Spray drying	Delivery	Santos et al. 2009
Doxycycline HCl	Microparticles, granules, and nanoparticles	Soaking, adsorption	Delivery	Wang et al. 2010
Ibuprofen	Microparticles, granules, and nanoparticles Porous ceramics	Soaking, adsorption	Delivey Arthritis	Baradari et al. 2012 Ryabenkova et al. 2017

Table 1. (Continued)

Drugs/Peptides/Proteins/DNA	Physical form of hydroxyapatite	Method of drug immobilization	Application	Reference
Dexamethasone	Microparticles, granules, and nanoparticles	Vacuum impregnation	Delivery	Hong et al. 2011
Rifampicin	Porous ceramics	Encapsulation	Delivery	Seyhan et al. 2022
Amoxicillin	Porous ceramics	Encapsulation	Delivery	Alkaya et al. 2022
Gentamicin	Porous ceramics	Incorporation Vacuum impregnation	Delivery	Tut et al. 2022 Loca et al. 2011
Doxorubicin	Porous ceramics Nanoparticles	Vacuum impregnation Adsorption	Delivery Breast cancer, Delivery	Pelss et al. 2011 Yang et al. 2013
Cisplatin	Porous ceramics	Adsorption	Cancer	Barroug and Glimcher 2002
Ampicilin	Porous ceramics	Adsorption	Bacterial infection	Queiroz et al. 2001
Riboflavine sodium phosphate, metoprolol tartrate	Compressed tablets	Soaking, adsorption	Delivery	Cosijns et al. (2007)

Various techniques are reported in the fabrication of porous HA bioceramics, including burn-out of pore formers, polymer template production, gel casting, sponge replication technique, ice templating, application of natural resources, 3D printing, selective sintering, and additive technologies (Loca et al., 2015). Also, the use of bioceramics as drug delivery systems has gained attention. Numerous strategies have been developed, particularly for the creation of drug delivery systems based on porous HA. HA is an attractive biomaterial for drug release due to its chemical affinity to the inorganic phase of bone and its ability to adsorb and release therapeutic agents. Comprehensive research has been conducted in recent years to investigate the incorporation of various drugs, proteins, and peptides into HA (Table 1).

Drug release systems utilizing HA can be prepared through various strategies, including physical/mechanical interactions, chemical linking, or sorption. The most straightforward method of incorporating drugs into the scaffold involves sorption mechanisms. Different techniques can be used to increase the adsorbed drug, such as nailing onto porous scaffolds or vacuum impregnation of aqueous solutions. In addition, biomolecules can be immobilized on biomaterials using biodegradable polymers, hydrogels or xerogels, or coatings on bioceramic scaffolds as sources of HA and drugs (Loca et al., 2015).

Currently, there is significant research focused on developing ceramic/polymer scaffolds. Given the diversity of polymers with different degradation and mechanisms, the addition of biodegradable polymers may affect the drug release profile. Commonly used polymers for drug delivery include polylactic/polyglycolic acid, carbohydrates (starch, chitin, alginate, chitosan), proteins (such as gelatin, collagen, fibrin), and other polymers (nylon, poly (propylene fumarate), poly(ethylene) glycol, polycarbonate, polypropylene). By combining tailored polymer degradability and ceramic scaffold porosity, it becomes possible to design scaffolds with desired degradation and drug release properties. Cross-linking agents like calcium chloride (for alginate) and genipin (for chitosan) are utilized, along with various solvents to ensure suitable pH and polymer solubility. The biocompatibility of the polymer and subsequently the drug delivery system relies on all components involved in its development (Habraken et al., 2007; Wei et al., 2011; Li et al., 2012; Laird et al., 2012; Loca et al., 2015).

An alternative approach to prepare polymer/HA drug delivery systems involves the use of HA porous microspheres. The drug release mechanism from polymer-coated HA microspheres is typically controlled by membrane diffusion. Various methods can be employed to coat the microspheres, such as emulsion-solvent evaporation (Wang et al., 2010), spray-drying (Santos et al., 2009), or immersion-drying (Venkatasubbu et al., 2013). These coating techniques enable sustained drug release from the drug delivery system, addressing the challenge of initial burst release.

In recent years, the immobilization of biomolecules on biomaterials has been studied extensively. Biodegradable polymers, hydrogels or xerogels are often used as a link between drugs and HA (Zhang et al., 2010).

Another idea is to use chitosan as a binder. Chitosan provides amino and hydroxyl groups that can be easily impregnated and functionalized into porous HA. Compounds derived from chitosan can be used as effective drug carriers due to their good adsorption properties.

These studies aim to understand the release kinetics, release mechanisms and factors affecting drug release from hydroxyapatite. Different drugs (hydrophobic and hydrophilic) including antibiotics, anti-inflammatory, anti-proliferative and anti-inflammatory drugs were loaded on hydroxyapatite and their release profiles were evaluated (Kim et al., 2010; Munır et al., 2021).

Chemicals released from hydroxyapatite can occur through various processes such as diffusion, ion exchange, separation, or a combination of these processes. Mathematical models such as zero order, first order, Higuchi and Korsmeyer-Peppas models were used to analyze and explain the drug release kinetics of hydroxyapatite.

Kinetic Models Used in Controlled Drug Release Systems

Drug release is dependent on various factors such as shape, carrier pore size, drug/carrier interactions, connectivity, polymer dissolution rate, drug/polymer interactions, and others. Therefore, determining the drug release mechanism is necessary. Different mathematical models are used to determine this mechanism (Loca et al., 2015).

These models provide information about the underlying mechanisms governing drug release and assist in the design and optimization of drug delivery systems. Various kinetic models are commonly used in drug release studies, including:

1. Zero-order kinetics: This model represents drug dissolution from non-disintegrating dosage forms that release the drug slowly, assuming a constant release rate over time independent of drug concentration. It

can be expressed by the equation $Q = K_0t$; where Q is the oscillation rate, K_0 is the oscillation rate, and t is the time.

2. First-order kinetics: This model is used to explain the absorption and/or elimination of various therapeutic agents. The amount of released drug is proportional to the remaining drug in the matrix. Therefore, the amount of released drug tends to decrease as a function of time. It can be explained by the equation $Ln\ (C/C_0) = -K_1t$, where C is the chemical concentration at time t, C_0 is the initial chemical concentration, K_1 is the constant release rate, and t is time.
3. Higuchi model: This model describes the release of a drug from a system that diffuses throughout the matrix. According to the equation $Q = K_H.t^{1/2}$, it assumes that the oscillation rate is proportional to the square root of time, where Q is the oscillation velocity, K_H is the Higuchi constant, and t is the time.
4. Hixson-Crowell model: This model is valid for dosage forms where dissolution occurs on parallel planes of the dosage form surface; this surface shrinks proportionally over time while maintaining a constant geometric shape. It can be expressed by the equation $W^{1/3} - W_0^{1/3} = K_{HC}.t$, where W is the value of the additive, W_0 is the original value of the drug, K_{HC} is the Hixson-Crowell rate constant and t is the time (Bruschi, 2015).

These are just a few examples of kinetic models used in drug release control. Selection of the appropriate model depends on the specific characteristics of drug delivery and the respective release process.

In summary, hydroxyapatite has been widely studied for its drug delivery potential. Its unique properties, biocompatibility and similarity to natural bone make it a promising product for controlled drug release.

Conclusion and Discussion

Consequently, the use of hydroxyapatite-based nanocomposites in biomedical applications has attracted great interest due to their ability to interact with body components. Scientists continue to research and develop the best materials that can be used for bone scaffolds while delivering drugs to specific cells. The mechanical properties of porous materials are influenced by both the material's strength and the porous structure. HA has low solubility under physiological conditions and can be used as a carrier for local drug distribution through both surgical implantation and injection. HA structures, when used as a matrix for delivering either a single therapeutic agent or multiple drugs, can provide different release rates and quantities. Drug carriers composed of multiple binding agents will enable drug release for various purposes. Full investigation of the behavior of these materials is important and further research is required to understand the relevant mechanisms, including drug release, of nanocomposites that exhibit biocompatibility, biodegradation, osteocytes integration, and scaffold behavior, particularly in vitro and in vivo. In addition, the evaluation of the desired properties of the scaffold structure is important for the development of hydroxyapatite-based scaffolds in terms of drug release. These advances help develop biomedical applications and treatments.

Disclaimer

The information provided within this book chapter is for general informational purposes only. The authors of this chapter have made reasonable efforts to ensure the accuracy and reliability of the content. While we try to keep the information up-to-date and correct, there are no representations or warranties, express or implied, about the completeness, accuracy, reliability, suitability or availability with respect to the information, products, services, or related graphics contained in this book chapter for any purpose. The authors disclaim any liability, loss, or risk incurred as a direct or indirect consequence of the use and application of the information contained in this book chapter.

References

Abdulrahman I, Tijani HI, Mohammed BA, Saidu H, Yusuf H, Jibrin MN. & Mohammed S. From garbage to biomaterials: an overview on egg shell based hydroxyapatite. *J. Mater*, (2014) 802467.

Adepu S, Ramakrishna S. Controlled Drug Delivery Systems: Current Status and Future Directions. *Molecules*. (2021) (26) (19):5905.

Agaogullari D, Kel D, Gokce H, Duman I, Ovecoglu ML, Akarsubasi AT, et al. Bioceramic Production from Sea Urchins. *Acta Phys Pol A*. (2012) 121 (1): 23-26.

Akram M, Ahmed R, Shakir I, Ibrahim WAW, & Hussain R. Extracting hydroxyapatite and its precursors from natural resources. *Journal of Materials Science* (2014) 49: 1461-1475

Alkaya DB, Cesur S, Seyhan SA. Characterization and in vitro release kinetics of chitosan based biocomposites from Scotch Bonnets. *Journal of the Indian Chemical Society* (2022) (99) (7), 100525.

Alparslan Y, Baygar T, Baygar T. Extraction, characterization and antimicrobial activity of hydroxyapatite from seabass and seabream scale. *Food and Health* (2017) 3(3): 90-96.

Arokiasamy P, Abdullah MMAB, Abd Rahim SZ, Luhar S, Sandu AV, Jamil NH, Nabiałek M. Synthesis methods of hydroxyapatite from natural sources: A review. *Ceramics International* (2022) (48) (11): 14959-14979.

Athinarayanan J, Periasamy VS, Alshatwi AA. Simultaneous fabrication of carbon nanodots and hydroxyapatite nanoparticles from fish scale for biomedical applications. *Materials Science and Engineering: C* (2020) 117: 111313.

Bahrololoom ME, Javidia M, Javadpoura S, Ma J. Characterisation of natural hydroxyapatite extracted from bovine cortical bone ash. *Journal of Ceramic Processing Research* (2009) 10: 129–138.

Baradari H, Damia C, Dutreih-Colas M, Laborde E, Pecout N, Champion E, Chulia D, Viana M. Calcium phosphate porous pellets as drug delivery systems: effect of drug carrier composition on drug loading and in vitro release. *J. Eur. Ceram. Soc.* (2012) 32: 2679–2690.

Barroug A, Glimcher MJ. Hydroxyapatite crystals as a local delivery system for cisplatin: Adsorption and release of cisplatin in vitro. *J. Orthop. Res.* (2002) 20: 274–280.

Ben-Nissan B. Natural bioceramics: from coral to bone and beyond. *Curr Opin Solid St M.* (2003) (7) (4-5): 283-288.

Bertran O, Del Valle LJ, Revilla-López G, Chaves G, Cardús L, Casas MT, Casanovas J, Turon P, Puiggalí J, Alemá C. Mineralization of DNA into nanoparticles of hydroxyapatite. *Dalt. Trans.* (2014) 43: 317–327.

Biazar E, Daliri JM, Heidari KS, Navayee AD, Kamalvand M, Sahebalzamani M, Farajpour LF. Characterization and biocompatibility of hydroxyapatite nanoparticles extracted from fish bone. *Journal of Bioengineering Research* (2020) 2(2):10-19.

Boutinguiza M, Pou J, Comesaña R., Lusquiños F, De Carlos A, León B. Biological hydroxyapatite obtained from fish bones. *Materials Science and Engineering C* (2012) 32(3): 478-486

Bulus E, Mansuroglu D, Ismık D, Sahin YM, Oktar FN, Gunduz O, and Gokce H. Bioceramic synthesis and characterization to be used in major tissue engineering applications. Electric Electronics, Computer Science, Biomedical Engineerings' Meeting (EBBT) IEEE, (2018) :1-4

Borciani G, Fischetti T, Ciapetti G, Montesissa M, Baldini N, Graziani G. Marine biological waste as a source of hydroxyapatite for bone tissue engineering applications. *Ceramics International* (2023) (49) (2): 1572-1584.

Bruschi ML. (2015). *Strategies to modify the drug release from pharmaceutical systems.* Woodhead Publishing.: 63-86

Cosijns A, Vervaet C, Luyten J, Mullens S, Siepmann F, Van Hoorebeke L, Masschaele B, Cnudde V, Remon JP. Porous hydroxyapatite tablets as carriers for low-dosed drugs. *Eur. J. Pharm. Biopharm.* (2007) 67: 498–506.

Dasgupta PA, Singh S, Adak S, Purohit KM. Synthesis and characterization of hydroxyapatite produced from eggshell. *Proceedings of the International Symposium of Research Students on Materials Science and Engineering* (2004): 1–6.

Deb P, Deoghare AB. Effect of acid, alkali and alkali–acid treatment on physicochemical and bioactive properties of hydroxyapatite derived from catla catla fish scales. *Arabian Journal for Science and Engineering* (2019) 44(9): 7479-7490.

Doğdu S A. İstilacı deniz türlerinden biyomateryal eldesi ve biyomedikal endüstrisindeki potansiyel kullanımı (Obtaining biomaterial from invasive marine species and its potential usage in biomedical industry). İskenderun Technical University Institute of Graduate Studies, İskenderun, (2021)115,

Gao J, Karp JM, Langer R, Joshi N. The future of drug delivery. *Chemistry of Materials* (2023) 35(2): 359-363.

Gómez Vázquez NS, Luque Morales PA, Gomez Gutierrez CM, Nava Olivas OdJ, Villarreal Sánchez RC, Vilchis Nestor AR, Chinchillas Chinchillas MdJ. Hydroxyapatite Biosynthesis Obtained from Sea Urchin Spines (Strongylocentrotus purpuratus): Effect of Synthesis Temperature. *Processes* (2020) 8 (4) 486.

Hasan MR, Ghazali MSM, Mohtar NF. Correlation of heating profile with calcination temperature for the extraction of nano hydroxyapatite (Nano-HAp) derived from bone. *Journal of Mechanical Engineering and Science* (2021) 15(1): 7807- 7823.

Habraken WJEM, Wolke JGC, Jansen JA. Ceramic composites as matrices and scaffolds for drug delivery in tissue engineering. *Adv. Drug Deliver. Rev.* (2007) 59: 234–248.

Hussain S, Sabiruddin K. Synthesis of eggshell based hydroxyapatite using hydrothermal method. *In IOP Conference Series: Materials Science and Engineering* (2021) 1189 (1): 012024. IOP Publishing.

Hussin MSF, Abdulah HZ, Idris MI, Wahap MAA. Extraction of natural hydroxyapatite for biomedical applications—A review. *Heliyon* (2022) e10356.

Hong MH, Son JS, Han KM, Oh DS, Lee YK. Drug-loaded porous spherical hydroxyapatite granules for bone regeneration. *Mater. Sci. Mater. Med.* (2011) 22: 349–355.

Ibraheem, S. A., Audu, E. A., Adudu, J. A., Barminas, J. T., Ochigbo, V., Igunnu, A., & Malomo, S. O. (2019). Novel pectin from Parkia biglobosa pulp mediated green route synthesis of hydroxyapatite nanoparticles. *Surfaces and Interfaces*, 17, 100360.

Ishihara S, Matsumoto T, Onoki T, Uddin MH, Sohmura T, Nakahira A. Regulation of the protein-loading capacity of hydroxyapatite by mercaptosuccinic acid modification. Acta Biomater. (2010) 6: 830–835.

Kalpana M, Nagalakshmi R. Nano hydroxyapatite for biomedical applications derived from chemical and natural sources by simple precipitation method. *Applied Biochemistry and Biotechnology* (2022) 1-17.

Kel D, Gokce H, Bilgic D, Agaogulları D, Duman I, Ovecoglu ML, et al. Production of Natural HA from Land Snails. *Key Eng Materials* (2012) 493:287-292.

Kim CW, Yun YP, Lee HJ, et al. In situ fabrication of alendronate-loaded calcium phosphate microspheres: controlled release for inhibition of osteoclastogenesis. *J Control Release* (2010) 147(1): 45–53.

Komalakrishna H, Jyoth TS, Kundu B, Mandal S. Low Temperature Development of Nano-Hydroxyapatite from Austromegabalanus psittacus, Star fish and Sea urchin, *Materials Today: Proceedings* (2017) 4 (11): 11933-11938.

Kongsri S, Janpradit K, Buapa K, Techawongstien S, Chanthai S. Nanocrystalline hydroxyapatite from fish scale waste: Preparation, characterization and application for selenium adsorption in aqueous solution. *Chemical Engineering Journal* (2013) 215: 522-532.

Laird DF, Mucalo MR, Dias GJ. Vacuum-assisted infiltration of chitosan or polycaprolactone as a structural reinforcement for sintered cancellous bovine bone graft. *J. Biomed. Mater. Res. A* (2012) 100: 2581–2592.

Latif AFA, Mohd Pu'ad NAS, Ramli NAA, Muhamad MS, Abdullah HZ, Idris MI, Lee TC. Extraction of Biological Hydroxyapatite from Tuna Fish Bone for Biomedical Applications. *Materials Science Forum* (2020)1010: 584–589.

Lara-Ochoa S, Ortega-Lara W, Guerrero-Beltrán CE. Hydroxyapatite nanoparticles in drug delivery: physicochemistry and applications. *Pharmaceutics* (2021) 13(10): 1642.

Lee S, Lee Y, Yoon Y. Characteristics of calcium phosphate powders synthesized from cuttlefish bone and phosphoric acid. *Journal of Ceramic Processing Research* (2007) 8(6): 427.

Li XY, Nan KH, Shi S, Chen H. Preparation and characterization of nanohydroxyapaitite/ chitosan cross-linking composite membrane intended for tissue engineering. *Int. J. Biol. Macromol.* (2012) 50: 43–49.

Li J, Liu M, Qiu Y, Gan Y, Jiang H, Liu B, Wei H, Ma N. Urchin-like Hydroxyapatite/Graphene Hollow Microspheres as pH-Responsive Bone Drug Carriers. *Langmuir* (2021) 37 (14): 4137-4146.

Lin K, Liu P, Wei L, Zou Z, Zhang W, Qian Y, Shen Y, Chang J. Strontium substituted hydroxyapatite porous microspheres: surfactant-free hydrothermal synthesis, enhanced biological response and sustained drug release. *Chem. Eng. J.* (2013) 222: 49–59.

Loca D, Locs J, Dubnika A, Zalite V, Berzina-Cimdina L. *Porous hydroxyapatite for drug delivery. In Hydroxyapatite (Hap) for biomedical applications*. Woodhead Publishing (2015) (pp. 189-209).

Loca D, Locs J, Salma K, Gulbis J, Salma I, Berzina-Cimdina L. Porous hydroxyapatite bioceramic scaffolds for drug delivery and bone regeneration. *IOP Conf. Ser. Mater. Sci. Eng*. (2011) 18: 192019.

Macha IJ, Ozyegin LS, Chou J, Samur R, Oktar FN, Ben-Nissan B. An Alternative Synthesis Method for Di Calcium Phosphate (Monetite) Powders from Mediterranean Mussel (Mytilus galloprovincialis) Shells. *J Aust Ceram Soc.* (2013) (49) (2): 122-128.

Macha IJ, Ozyegin LS, Oktar FN, Ben-Nissan B. Conversion of Ostrich Eggshells (Struthio camelus) to Calcium Phosphates. *Journal of the Australian Ceramic Society* (2015) 51(1).

Munir MU, Salman S, Javed I, Bukhari SNA, Ahmad N, Shad NA, Aziz F. Nano-hydroxyapatite as a delivery system: overview and advancements. *Artificial Cells, Nanomedicine, and Biotechnology* (2021) 49(1): 717-727.

Munir MU, Ihsan A, Sarwar Y, Bajwa SZ, Bano K, Tehseen B, ... & Khan WS. Hollow mesoporous hydroxyapatite nanostructures; smart nanocarriers with high drug loading and controlled releasing features. *International journal of pharmaceutics* (2018) 544(1): 112-120.

Mizushima Y, Ikoma T, Tanaka J, Hoshi K, Ishihara T, Ogawa Y, Ueno A. Injectable porous hydroxyapatite microparticles as a new carrier for protein and lipophilic drugs. *J. Control. Release* (2006) 110 (2): 260–265.

Obada DO, Alıyu A, Kuta UM, Lısıya, A, Ogenyı VF, Aquatar MO, ... & Adam I. Trends in the development of hydroxyapatite from natural sources for biomedical applications. *WIT Transactions on Ecology and the Environment* (2022) 257: 199-210.

Oktar F, Tuyel U, Demirkol N, Gunduz O, Samur R, Kannan S, et al. A new safe method to produce bioceramic nano-powders from nacre venus verrucosa. *Artificial Organs Conference, Journal of The Australian Ceramic Society* (2015) 51(1): 125 – 133 133 FYROM, 2010: International Journal of Artificial Organs, (Special Issue) 2010:467-467.

Ozawa M, Suzuki S. Microstructural development of natural hydroxyapatite originated from fish-bone waste through heat treatment. *Journal of the American Ceramic Society* (2002) 85(5):1315-1317.

Pal A, Paul S, Choudhury A R, Balla V K, Das M ve Sinha A. Synthesis of hydroxyapatite from Lates calcarifer fish bone for biomedical applications. *Materials Letters* (2017) 203:89-92.

Patra JK, Das G, Fraceto LF, Campos EVR, Rodriguez-Torres MDP, Acosta-Torres LS, ... & Shin HS. Nano based drug delivery systems: recent developments and future prospects. *Journal of nanobiotechnology* (2018) 16(1): 1-33.

Pelss J, Loca D, Berzina-Cimdina L, Locs J, Lakevics V. Release of anticancer drug doxorubicin from biodegradable polymer coated porous hydroxyapatite scaffolds. *Adv. Mat. Res.* (2011) 284–286: 1770–1773.

Pu'ad NM, Koshy P, Abdullah HZ, Idris MI., Lee TC. Syntheses of hydroxyapatite from natural sources. *Heliyon* (2019) 5(5): e01588.

Pon-On W, Suntornsaratoon P, Charoenphandhu N, Thongbunchoo J, Krishnamr N, Tang I M. Hydroxyapatite from fish scale for potential use as bone scaffold or regenerative material. *Materials Science and Engineering C* (2016) 62:183- 189.

Queiroz AC, Santos JD, Monteiro FJ, Gibson IR, Knowles JC. Adsorption and release studies of sodium ampicillin from hydroxyapatite and glass-reinforced hydroxy-apatite composites. *Biomaterials* (2001) 22: 1393–1400.

Rajabiyan A, Shakiba Maram N, Ghatrami ER, Zarei Ahmady A. Preparation of magnetic methotrexate nanocarrier coated with extracted hydroxyapatite of sea urchin (Echinometra mathaei). *Main Group Chemistry* (2021) 20(4): 447-461.

Ramesh S, Loo ZZ, Tan CY, Chew WK, Ching YC, Tarlochan F and Sarhan AA. Characterization of biogenic hydroxyapatite derived from animal bones for biomedical applications. *Ceramics International* (2018) 44 (9): 10525-10530.

Rivera EM, Araiza M, Brostow W, Castano VM, Dıaz-Estrada JR., Hernández R, Rodrıguez, JR. Synthesis of hydroxyapatite from eggshells. *Materials Letters* (1999) 41(3): 128-134.

Ryabenkova Y, Jadav N, Conte M, Hippler MFA, Reeves-Mclaren N, Coates PD, Twigg P, Paradkar A. Mechanism of Hydrogen-Bonded Complex Formation between Ibuprofen and Nanocrystalline Hydroxyapatite. *Langmuir* (2017) 33: 2965–2976.

Santos C, Rovath CF, Franke RP, Almeida MM, Costa MEV. Spray-dried hydroxya-patiet-5-fluorouracil granules as a chemotherapeutic delivery system. *Ceram. Int.* (2009) 35: 509–513.

Segvich SJ, Smith HC, Kohn DH. The adsorption of preferential binding peptides to apatite-based materials. *Biomaterials* (2009) 30: 1287–1298.

Seyhan SA, Alkaya DB, Cesur S, Sahin A, Gunduz O. Preparation and characterization of pure natural hydroxyapatite derived from seashells for controlled drug delivery. *J Aust Ceram Soc* (2022) 58: 1231–1240.

Sim S, Wong NK. Nanotechnology and its use in imaging and drug delivery. *Biomedical reports* (2021) 14(5): 1-9.

Siva Rama Krishna D, Siddharthan A, Seshadri SK, Sampath Kumar TS. A novel route for synthesis of nanocrystalline hydroxyapatite from eggshell waste. *Journal of Materials Science: Materials in Medicine* (2007)18(9): 1735–1743.

Surya P, Nithin A, Sundaramanickam A, Sathish M. Synthesis and characterization of nano-hydroxyapatite from Sardinella longiceps fish bone and its effects on human osteoblast bone cells. *Journal of the Mechanical Behavior of Biomedical Materials* (2021) 119: 104501.

Tomoda, K., Ariizumi, H., Nakaji, T., Makino, K., 2010. Hydroxyapatite particles as drug carriers for proteins. Colloids and Surfaces B: Biointerfaces 76, 226–235.

Tut TA, Cesur S, Ilhan E, Sahin A, Yildirim OS, Gunduz O. Gentamicin-loaded polyvinyl alcohol/whey protein isolate/hydroxyapatite 3D composite scaffolds with drug delivery capability for bone tissue engineering applications. *European Polymer Journal* (2022) 179: 111580.

Venkatasubbu GD, Ramasamy S, Avadhani GS, Ramakrishnan V, Kumar J. Surface modification and paclitaxel drug delivery of folic acid modified polyethylene glycol functionalized hydroxyapatite nanoparticles. *Powder Technol.* (2013) 235: 437–442.

Venkatesan J, Qian ZJ, Ryu B, Thomas N, Kim SKA. comparative study of thermal calcination and an alkaline hydrolysis method in the isolation of hydroxyapatite from Thunnus obesus bone. *Biomedical Material*, (2011) 6(3): 035003.

Wang S, Wang X, Xu H, Abe H, Tan Z, Zhao Y, Guo J, Naito M, Ichikawa H, Fukumori Y. Towards sustained delivery of small molecular drugs using hydroxyapatite microspheres as the vehicle. *Adv. Powder Technol.* (2010) 21: 268–272.

Wijedasa NP, Broas SM, Daso RE, Banerjee IA. Varying fish scale derived hydroxyapatite bound hybrid peptide nanofiber scaffolds for potential applications in periodontal tissue regeneration. *Materials Science and Engineering: C* (2020) 109: 110540.

Wei L, Cai C, Lin J, Wang L, Zhang X. Degradation controllable biomaterials constructed from lysozyme-loaded Ca-alginate microparticle/chitosan composites. *Polymer* (2011) 52: 5139–5148.

Wu X, Song X, Li D, Liu J, Zhang P, Chen X. Preparation of mesoporous nanohydroxyapatite using a surfactant template method for protein delivery. *J. Bionic Eng.* (2012) 9 (2): 224–233.

Yang YH, Liu CH, Liang YH, Lin FH, Wu KCW. Hollow mesoporous hydroxyapatite nanoparticles (hmHANPs) with enhanced drug loading and pH-responsive release properties for intracellular drug delivery. *J. Mater. Chem. B* (2013) 1: 2447–2450.

Ye F, Guo H, Zhang H, He X. Polymeric micelle-templated synthesis of hydroxyapatite hollow nanoparticles for a drug delivery system. *Acta Biomater.* (2010) 6 (6): 2212–2218.

Zhang C, Li C. Huang S, Hou Z, Cheng Z, Yang P, Peng C, Lin J. Self-activated luminescent and mesoporous strontium hydroxyapatite nanorods for drug delivery. *Biomaterials* (2010) (31): 3374–3383.

Zhang Y, Liu Y, Ji X, Banks CE, Zhang W. Conversion of natural egg-shell to 3D flowerlike hydroxyapatite agglomerates for highly sensitive detection of As3 + ions. Materials Letters (2012) [78] (0): 120- 123.

Zhao Q, Wang T, Wang J, Zheng L, Jiang T, Cheng G, Wang S. Templatedirected hydrothermal synthesis of hydroxyapatite as a drug delivery system for the poorly water-soluble drug carvedilol. *Appl. Surf. Sci.* (2011) 257 (23): 10126–10133.

Zhu XD, Zhang HJ, Fan HS, Li W, Zhang XD. Effect of phase composition and microstructure of calcium phosphate ceramic particles on protein adsorption. *Acta Biomater.* (2010) 6: 1536–1541.

Chapter 6

Dental Bioceramics: From Lab to Clinical Practice

Kerim Emre Öksüz*, PhD

[1]Sivas Cumhuriyet University, Department of Metallurgical and Materials Engineering, Sivas, Türkey
[2]Sivas Cumhuriyet University, Institute of Science and Technology, Department of Bioengineering, Sivas, Türkey

Abstract

Functional ceramics known as bioceramics have gained significant interest in the field of regenerative medicine. They have been widely used as bone reconstructive materials in orthopedics, oral and maxillofacial surgery. One of the most commonly used materials in clinical applications are those based on calcium phosphate ($Ca_3(PO_4)_2$) and calcium silicate (Ca_2SiO_4), which have been available on the market for several decades. Currently, there are ongoing works to develop bioceramics for various clinical applications in regenerative medicine, including dentistry. Bioceramics have shown promise in periodontal treatment, prevention of relapse, nerve regeneration, drug delivery technology, and even in esthetic medicine and cosmetics. For restorations in the anterior of the mouth, glass ceramics are an alternative approach to provide both esthetics and strength, while (Al_2O_3) alumina and (ZrO_2) zirconia-based structural ceramics are utilized as crown cores and bridge frameworks. However, they require the application of an esthetic veneering ceramic for clinical use in tooth structure replacement. This chapter discusses the advantages of dental bioceramics, their compositions, biological evolutions, and how their physical properties relate to clinical findings. The future perspective and challenges in using bioceramics for next-generation regenerative therapy are also discussed.

---

* Corresponding Author's Email: emre.oksuz@cumhuriyet.edu.tr, kerimemreoksuz@gmail.com.

In: Biocomposites
Editor: James R. Bush
ISBN: 979-8-89113-267-2

Keywords: bioceramics, clinical applications, dental ceramics, maxillofacial surgery, orthopedic, regenerative therapy

Introduction

The category of ceramics dedicated to the restoration and renewal of afflicted segments within the musculoskeletal framework is categorized as bioceramics. Over time, bioceramics have transformed into a multifaceted group of biomaterials, currently comprising three fundamental classifications: bioinert high-strength ceramics, (Aluminium oxide (Al_2O_3), Zirconium oxide (ZrO_2)) bioactive ceramics (Hydroxyapatite (HA), Tricalcium phosphate (TCP), specific formulations of silicate and phosphate glasses known as bioactive glasses, and glass-ceramics like apatite-wollastonite constitute the array of materials) fostering direct chemical bonds not only with bones but also with the supple tissues of living organisms, and an array of bioresorbable ceramics (Calcium carbonate ($CaCO_3$), Carbonate apatite (CO_3Ap), and β-tricalcium phosphate (β-TCP)) that actively engage in the metabolic processes within an organism, yielding anticipated outcomes. Within the domain of dental care, ceramics have evolved to encompass a spectrum of variations, including glasses, glass-ceramics, and structural ceramics. These diverse ceramic variations find utility across a wide spectrum of dental applications, spanning from inlays and onlays to crowns and bridges. Additionally, they have extended their utility to encompass metal implant fixture abutments and, more recently, to the innovation of one-piece all-ceramic implants. This remarkable array of ceramic applications highlights the profound impact of bioceramics in the realms of both orthopedic and dental disciplines, poised to advance medical practice and improve patient outcomes (Shenoy and Shenoy, 2010; Mc Entire, 2017; Surya Raghavendra et al., 2017; Öksüz, 2019; Öksüz et al., 2023). Dental ceramics, particularly porcelain, are primarily valued for their aesthetic attributes. However, while highly aesthetic porcelains possess exquisite visual appeal, they generally lack robust mechanical strength, necessitating their amalgamation with structural reinforcements either in the form of metals or high-strength ceramics when employed for posterior crowns or bridges. Another avenue explored to marry aesthetics with durability involves glass-ceramics, offering a potential solution for anterior (front) dental restorations. While investigations into their use for posterior teeth are ongoing, glass-ceramics are showcasing promise in this domain. On the other hand, structural ceramics, while exhibiting durability, are challenged by their inherent lack of translucency and the requirement for a white coloration to be deemed suitable. Ceramic materials grounded in Al_2O_3 and ZrO_2 form the basis of structural ceramics, serving as crown cores and bridge frameworks; however, their effective use mandates the application of an esthetically pleasing veneering ceramic to mimic natural tooth structure (Thompson and Rekow, 2008; Dong and Xu, 2023). The upcoming chapters of this book will delve into the realm of innovative dental ceramics, exploring their diverse compositions. The ensuing exploration will encompass the intricate interplay between the physical characteristics of these ceramics and their clinical implications. Moreover, as the horizon of dental ceramics continues to evolve, the text will proffer insights into the prospective directions and advancements anticipated in this dynamic field.

Innovative Bioceramics for Endodontics

Calcium Silicate (Ca_2SiO_4) Based Bioceramics

Portland cement (PC) is derived through the calcination process of a blend of limestone originating from the region of Portland in England and silicon-rich clay-like materials (Naiana Viana Viola et al., 2012). Renowned for its affordability, PC shares a comparable primary composition with Mineral Trioxide Aggregate (MTA), with the exception of the absence of bismuth oxide and the presence of elevated levels of calcium aluminate and calcium sulfate. Much like MTA, Portland cement is accessible in both grey and white variants (Torabinejad et al., 2017). Both PC and MTA, besides their compositional resemblance, are available in diverse color variations ranging from grey to white providing practitioners with flexibility when selecting materials for various applications.

Ordinary grey PC exhibits less discoloration than grey MTA, while white MTA and white PC show comparable discoloration levels (Lenherr et al., 2012). In comparison to white PC, MTA displays greater solubility and improved resistance to washout in various solutions (Vivan et al., 2010; Formosa et al., 2012). PC, like MTA, possesses antibacterial and antifungal properties against a range of microorganisms (*E. faecalis, M. luteus, S. aureus, S. epidermidis, P. aeruginosa* and *C. Albicans*) (Parirokh and Torabinejad, 2010). White and grey MTA demonstrate similar sealing efficacy as root end filling materials, comparable to white and grey PC, when assessed through dye penetration (Shahi et al., 2009). Notably, as a perforation repair material, white

PC outperforms white and grey MTA in terms of sealing ability as determined by protein leakage. *In vitro* cell culture studies, outcomes vary depending on the cell type. PC, similar to MTA, demonstrates no genotoxicity or cytotoxicity concerns with fibroblasts (Fayazi et al., 2011). However, MTA displays superior proliferation and migration effects on human bone marrow-derived mesenchymal stem cells (MSCs) compared to PC (D'Antò et al., 2010). Biomineralization appears more pronounced in MTA than in PC at 30 and 60 days (Dreger et al., 2012).

Clinical and radiographic evaluations of pulpotomy procedures using PC and MTA reveal successful outcomes; yet, PC leads to greater root canal obliteration (Sakai et al., 2009). Biomineralization achieved with PC falls short in both effectiveness and duration when compared to MTA, a crucial factor for a bioactive material (Dreger et al., 2012).

Mineral Trioxide Aggregate (MTA-Bioactive Calcium Silicate Cement)

The initial bioceramic material known as MTA cement possesses osseoconductive, inductive, and biocompatible properties. Originally formulated as a root-end filling substance, it has since found application in pulp capping, pulpotomy, apexogenesis, creating apical barriers in teeth with open apexes, mending root perforations, and as a root canal filling agent. To address tooth discoloration associated with grey MTA, white MTA was introduced under the name ProRoot MTA by Dentsply Endodontics in Tulsa, OK, USA (Tuba and Emine, 2011). The grey color of the first version was attributed to Fe ions, which were subsequently eliminated to achieve the white variant. The setting reaction occurs through hydration, resulting in hydrated Ca_2SiO_4 and $Ca(OH)_2$ release over time. Its integration into the biological system is facilitated by Ca ions, which, upon contact with phosphate ions in the body, lead to the formation of hydroxyapatite. Comparatively, White MTA was discovered to contain 54.9% less Al_2O_3, 56.5% less MgO, and 90.8% less FeO than grey MTA. This points to the reduction in FeO as the likely cause for the change in color (Jitaru et al., 2016; Roberts et al., 2008).

MTA exhibits a compressive strength of 40 MPa at 24 hours and 67 MPa at 3 weeks. Its setting mechanism involves an exothermic reaction, with hydration transforming its powder into mature cement paste over time. Key reactions involve tricalcium silicate (Ca_3SiO_5) and dicalcium silicate

(Ca_2SiO_4) reacting with water to form Ca_2SiO_4 hydrates (C-S-H) and $Ca(OH)_2$. MTA's bioactivity is attributed to powder hydration causing Ca^{+2} dissolution, resulting in reactions that form C-S-H, $Ca(OH)_2$, and ultimately apatite. $CaCl_2$ accelerates MTA's setting, while NaCIO impedes $Ca(OH)_2$ formation. The recommended powder-to-liquid ratio is 3:1. MTA's flexural strength is lower than glass ionomer or zinc phosphate cement, rendering it unsuitable as a luting agent. Studies reveal that a 4-mm MTA thickness (apical barrier) resists displacement better than 1-mm thickness (Zapf et al., 2015; Islam et al., 2006; Ravindran and Jeevanandan, 2023). Moist cotton placement over setting MTA for 24 hours significantly boosts flexural strength (~14.27 ± 1.96 MPa) (Walker et al., 2006). MTA generally demonstrates less micro leakage than traditional materials in apical restorations and is equivalent to Zinc oxide eugenol for furcation perforation repair. Grey and white MTA show similar results to gutta-percha in root canal obturation micro leakage studies. At least 3 mm of MTA after root-end resection shows minimal leakage, while 2 mm or less thickness leads to increased leakage (Roberts et al., 2008). MTA's physical properties can be affected by crystal size; smaller particles enhance contact with liquid, improving early strength and handling. MTA is non-mutagenic, non-neurotoxic, and supports the production of signaling molecules, as confirmed by animal and human studies. It exhibits anti-inflammatory effects on pulp tissue and confirmed osteoinductive and osteoconductive effects (Parirokh and Torabinejad, 2010).

Biodentine (Bioactive Dental Material-Calcium Silicate Cement)

Biodentine, produced by Septodont in Saint Maur des Fosses, (France) is a Ca_2SiO_4-based material crafted using MTA-based cement technology. This enhances various properties like physical attributes and ease of use. The setting process resembles MTA, involving the creation of Ca_2SiO_4 hydrate gel (C–S–H) and $Ca(OH)_2$. Notably, $CaCO_3$ functions as a nucleation point for Ca_2SiO_4 hydrate gel, curtailing the induction phase, resulting in quicker setting and improved microstructure formation. Additionally, a water-soluble polymer reduces cement viscosity, enhancing its manageability (Malkondu et al., 2014).

Biodentine offers a working time of up to 5-6 min., with initial setting taking 9–12 min. and final setting concluding at 45 minutes. This shortened setting period is an improvement over other Ca_2SiO_4 materials and is attributed to the incorporation of $CaCl_2$ into the mixing liquid, a factor also proven to accelerate MTA setting time (Mandeep Kaur, 2014). Notably, Biodentine exhibits rapid growth in compressive strength, exceeding 100 MPa within the first hour and surpassing 200 MPa at 24 hours surpassing most Glass Ionomer Cements (GIC). Over days, it further strengthens to reach 300 MPa after a month, comparable to natural dentin (Arora, 2013). Biodentine's elastic modulus of 22 GPa closely resembles dentin's 18.5 GPa (Priyalakshmi and Ranjan, 2014). Microhardness measures 50 VHN after 2 hours, increasing to 70 VHN within a month, similar to natural dentin (Arora, 2013). Its micromechanical adhesion arises from an alkaline reaction that dissolves organic tissues from dentin tubules. This alkaline environment facilitates the entry of Biodentine into exposed dentin canaliculi, anchoring it through microscopic cones, yielding a stable, bacteria-resistant seal. In comparison to MTA, Biodentine exhibits higher push-out bond strength at 24 hours. Interestingly, blood contamination adversely impacts MTA Plus's bond strength, irrespective of setting time, whereas Biodentine remains unaffected by blood contamination, regardless of the setting duration (Priyalakshmi and Ranjan, 2014). Biodentine exhibits a bending strength of 34 MPa after 2 hours, surpassing conventional GIC at 5-25 MPa, resin-modified GIC at 17-54 MPa, and falling below composite resin at 61-182 MPa (Parirokh and Torabinejad, 2010). This indicates that while Biodentine bending resistance exceeds that of traditional GIC, it remains below that of composite resin (Mandeep Kaur, 2014). The setting phase of Biodentine releases $Ca(OH)_2$ ions, elevating the pH to 12.5, effectively impeding microorganism growth and dentin disinfection (Arora, 2013).

Biodentine is safe, not influencing cell differentiation or specific cell functions. It stimulates TGF-β1 (a growth factor) release from pulp cells, fostering processes like angiogenesis, progenitor cell recruitment, differentiation, and mineralization (Mandeep Kaur, 2014).

Generex A and Generex B (Tricalcium Silicate Cement)

Generex A and Generex B (Dentsply Tulsa Dental Specialties, Tulsa, OK, USA) are both Ca_2SiO_4-based substances that share certain parallels with ProRoot MTA. However, they diverge in terms of mixing methodology, utilizing distinct gels instead of the water employed in MTA. Notably, Generex A displays contrasting handling characteristics compared to MTA. When mixed, Generex A takes on a dough-like consistency, simplifying the process of forming a rope-like mass, akin to intermediate restorative dental material (Porter et al., 2010; Washington et al., 2011). On the other hand, Generex B, an experimental root canal sealer, exhibits intriguing properties. It has been observed to generate HA within SBF, and its tissue irritability is minimal.

These qualities are particularly advantageous for a material that could unintentionally extend beyond the root apex and into the surrounding bone during root canal obturation procedures. In contrast, commonly employed zinc oxide-eugenol sealers are cytotoxic and are less favorable for periapical tissues if they extend beyond the apical foramen (Zmener et al., 2010; Camps and About, 2003).

EndoBinder (Calcium Aluminate Cement)

An innovative endodontic cement known as EndoBinder (Binderware, São Carlos, SP, Brazil) has emerged, utilizing a calcium aluminate base. This cement has been meticulously crafted to not only retain the favorable attributes and clinical applications associated with MTA but also to overcome its inherent drawbacks. The driving force behind EndoBinder's development is the elimination of MTA's undesirable characteristics. The formulation of EndoBinder emphasizes elevated levels of purity, effectively eradicating any traces of free MgO and CaO. These elements are often responsible for the material's unwanted expansion, a challenge that has been effectively addressed in the design of EndoBinder.

Additionally, the inclusion of (Fe_2O_3), which tends to cause tooth darkening, has been eliminated from the composition. In recent comparisons with other materials, EndoBinder has notably demonstrated a satisfactory tissue reaction. Its biocompatibility has been assessed through testing in the subcutaneous tissue of rats, yielding positive outcomes. This signifies that EndoBinder not only aligns with MTA in terms of desired biological responses but also triumphs over certain shortcomings associated with traditional MTA formulations (Aguilar et al., 2012).

Capasio (Calcium-Phospho-Aluminosilicate-Based Cement)

Capasio (Primus Consulting, Bradenton, FL, USA) is formulated from bismuth oxide, dental glass, and calcium alumino-silicate, complemented by a gel incorporating silica and polyvinyl acetate. In a recent research, it was determined that both Capasio and MTA exhibit the ability to encourage apatite deposition upon exposure to synthetic tissue fluid, indicative of their mineralization potential. Additionally, the same group of researchers concluded that Capasio, when utilized as a root-end filling substance, shows a higher likelihood of penetrating dentinal tubules. Another experimental study conducted a comparison among Generex A, Generex B, and Capasio employing primary osteoblasts as test subjects.

Among these materials, Generex A was identified as the sole new-generation endodontic material that facilitated the growth of primary osteoblasts. Interestingly, MTA was the exclusive agent to foster nodule formation. Remarkably, only Generex A and MTA supported consistent cell growth and proliferation throughout the entire experimental duration (Eid et al., 2013; Wei et al., 2012).

Calcium Phosphate ($Ca_3(PO_4)_2$) Based Dental Bioceramics

Compounds or biocomposites containing $Ca_3(PO_4)_2$, commonly used in dental applications for treating bony defects, have demonstrated the ability to promote osteogenesis and facilitate new bone formation. Biocomposites derived from Hydroxyapatite, β-Tricalcium Phosphate, and Bioglass, synthesized using various production methods (Öksüz et al., 2021; Öksüz et al., 2019; Öksüz, 2018), have exhibited the capacity to chemically bond with the host bone, facilitated by the formation of a $Ca_3(PO_4)_2$-rich layer at the interface (Bansal et al., 2014). Notably, the incorporation of $Ca_3(PO_4)_2$ as filler within an active restorative material embedded in a polymer binder has been developed. This innovative approach stimulates the repair of tooth structure by gradually releasing significant amounts of Ca and P ions over time. However, it's important to highlight that a key limitation of dental ceramics containing $Ca_3(PO_4)_2$ is their relatively low strength, rendering them susceptible to fatigue fractures and failure in loadbearing scenarios (Skrtic et al., 2000; Skrtic et al., 2004).

A schematic drawing of tooth anatomy is presented in Figure 1, accompanied by schematic representations of both innovative and traditional applications of bioceramic materials used in endodontics.

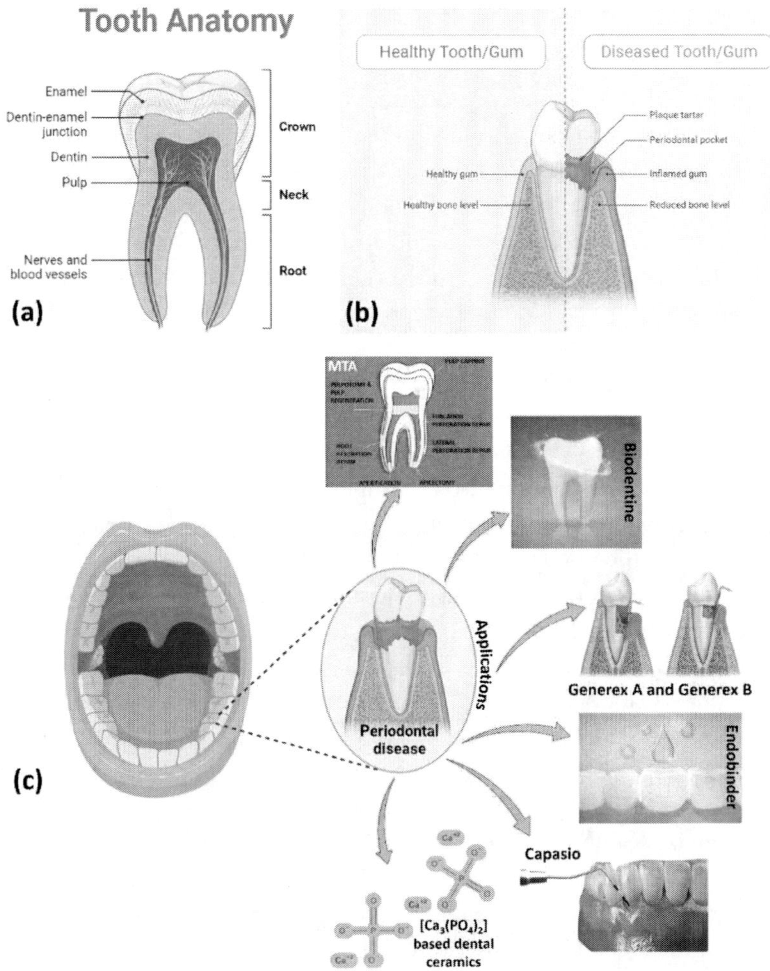

Figure 1. (a) A schematic drawing of a tooth anatomy, (b) Healthy tooth/gum-diseased tooth/gum and (c) Schematic illustration of bioceramic technology and bioceramic materials application in endodontics. The concept of bioceramic technology and its application in endodontics involves the utilization of advanced ceramic materials that possess biocompatible properties for various dental procedures. Bioceramic materials, specifically engineered to interact favorably with biological systems, are integrated into endodontic practices to address a range of challenges, including root canal treatment, pulp capping, apexogenesis, and perforation repair. These specialized ceramics are designed to not only offer functional benefits but also to enhance the overall biocompatibility, bioactivity, and clinical performance of endodontic treatments. As bioceramic technology continues to evolve, its application in endodontics holds the promise of improving treatment outcomes and contributing to the long-term success and preservation of dental health.

Aluminum Oxide (Al_2O_3)

α-alumina (Al_2O_3), commonly referred to as corundum, stands out as an excellent biomaterial due to its unparalleled stability. This exceptional material represents the highest oxidative state of Al metal, rendering it impervious to further oxidative reactions and attaining the utmost levels of chemical and physical stability. The formidable ionic and covalent bonds formed between Al^{3+} and O^{2-} ions contribute to its impressive attributes, including a high melting point, exceptional hardness, and resistance to potent inorganic acids such as orthophosphoric and hydrofluoric acids (Piconi, 2015; Öksüz et al., 2015). With a crystal structure consisting of hexagonal close-packed layers of oxygen atoms, α-Al_2O_3 accommodates two-thirds of octahedral sites for Al atoms. The surface layer enriched with O^{2-} anions facilitates the chemisorption of OH^+ groups, thereby fostering the bonding of H_2O molecules or proteins. This unique surface wettability, surpassing that of numerous metallic alloys, leads to the formation of a fluid film at interfaces within joint connections of prostheses featuring bioceramic components.

This combination of high hardness (2000-2200 GPa) and exceptional wettability positions α-Al_2O_3 as the optimal choice for prostheses and dental components (Öksüz and Şahin, 2016; Piconi, 2015; Şahin and Öksüz, 2013; Öksüz, 2015; Şahin and Öksüz, 2016; Öksüz, 2016). Incorporating trace amounts of MgO (less than 0.1 mol%) during sintering assists in controlling grain growth. This deliberate addition of MgO prevents the formation of foreign phases within Al_2O_3, while meticulous oversight of MgO levels prevents the emergence of leachable glassy phases at grain boundaries by monitoring impurity concentrations like SiO_2 and NaO. Additionally, the interaction between CaO impurities and water can trigger fatigue failure in Al_2O_3 components (Özer et al., 2016; Öksüz and Özer, 2017; Walter, 1992). When Al_2O_3 bioceramics are implanted in soft tissues, they prompt the development of a connective tissue layer encircling the ceramic, gradually maturing into a more robust layer over time. In instances where Al_2O_3 implants are placed within non-loaded bone conditions, a swift generation of bone occurs in direct proximity to the ceramic implant, facilitated by its surface texture and interactions with cells. Despite the absence of chemical bonding, the existence of a proteoglycan layer at the bone-biomaterial interface fosters bone apposition. While micro motions of the implant can lead to an increased thickness of the fibrous capsule, the deposition of bone is additionally stimulated by the mechanical stresses generated through loading.

Upon injection into diverse animal sites in hard tissue, Al_2O_3 particles trigger a tissue reaction featuring fewer granulocytes and lymphocytes compared to metals. Al_2O_3 elicits a gentler tissue response than metals or polymers, with indications that it does not incite the fibrosis often observed after the injection of polymers or metals. Long-term studies spanning over 3.5 years reveal a nonspecific carcinogenic reaction following the implantation of Al_2O_3 ceramics, primarily attributed to surface geometry rather than other factors. Notably, subcutaneous injection of Al_2O_3 powder into rats does not initiate any carcinogenic response (Christel, 1992).

Zirconium Oxide (ZrO_2)

The introduction of Zirconium dioxide (ZrO_2) in hard tissue aimed to harness its exceptional toughness, earning it the moniker "ceramic steel" ZrO_2, a metal oxide bioceramic, serves as a viable substitute for Al_2O_3 due to its substantially higher fracture toughness. Notably, ZrO_2 possesses the highest fracture toughness among monolithic ceramics. It exhibits polymorphism, undergoing distinct transformations upon cooling from molten to room temperature, with three polymorphs: monoclinic, tetragonal, and cubic phases, along with a high-pressure orthorhombic form (The monoclinic phase transforms into the tetragonal phase above 1170°C. It is stable up to this temperature, then above 2370°C, the cubic phase exists up to the melting point at 2680°C). These transformations entail a significant volume change of 3-5% upon cooling from the tetragonal to the monoclinic phase (Li and Hastings, 2016). ZrO_2 ceramics have high density, strength, and fracture toughness, which correspondingly result in lower microhardness and elastic modulus compared to other ceramics. While its application as a biomaterial has been primarily in hip joint replacements, ZrO_2 implants demonstrate outcomes similar to Ti implants in terms of osteointegration and biocompatibility.

Furthermore, ZrO_2 possesses distinct advantages such as a strong affinity to bone tissue, noncarcinogenic properties, and absence of oncogenic effects, rendering it an excellent choice for various implant applications. Of note, ZrO_2 grains can serve as nucleation sites for calcium-based mineral development, a key consideration for designing implant materials like hydroxyapatite (HA), a crucial bone component. Unlike ZrO_2, Al_2O_3 lacks affinity toward calcium (Ca) or hydrogen phosphate $[HPO_4]^{2-}$ ions. This is evidenced by the cauliflower-like growth of calcium phosphate ($Ca_3(PO_4)_2$) minerals extending across the entire surface of ZrO_2 (Soon et al., 2016; Li and Hastings, 2016).

ZrO_2 as a bioceramic faces potential limitations in terms of degradation and radiation. Accelerated tetragonal-to-monoclinic (T→M) phase transformation has been observed in aqueous environments, though its behavior in biological settings, particularly under dynamic loads, remains unclear (Matsui et al., 2014). This degradation could lead to surface deterioration of ZrO_2 bioceramics, potentially causing cracks due to significant volume changes that exceed material limits. To counteract this, researchers have proposed incorporating stabilizing additives like CaO, MgO, and Y_2O_3 to maintain ZrO_2 in either the tetragonal or cubic phase. Partially stabilized zirconia (PSZ), a monoclinic phase and/or a mix of tetragonal and cubic phases, and 100% tetragonal tetragonal zirconia polycrystal (TZP) are suggested for dental/medical implant applications (Li and Hastings, 2016).

Recent research has explored ZrO_2 in various medical applications, such as bone cement, bone graft substitutes, dental implants, and prostheses. In contrast, Al_2O_3 had been used for dental purposes but faced removal from the market due to mechanical failures (Soon et al., 2016). ZrO_2 has garnered attention as a coating for metallic replacements, leveraging its physical properties and implant success. A common method for this involves utilizing filtered cathodic arc systems with oxygen plasma to deposit thin ZrO_2 films on metal surfaces. These films of stoichiometric ZrO_2 nanoparticles have exhibited bone-like apatite formation when exposed to SBF, possibly due to their nanoscale surface structure. These ZrO_2 thin films have demonstrated both bioactivity and cytocompatibility, fostering the growth and proliferation of bone marrow mesenchymal stem cells (MSCs) (Afzal, 2014). Coating metallic bone implants with ZrO_2 thin films is a feasible strategy to enhance their biocompatibility and osseointegration. Furthermore, ZrO_2 coating has been linked to improved osseointegration capabilities of implants. UV light has been identified as a means to enhance ZrO_2 properties; monoclinic ZrO_2 thin films were created through microarc oxidation and subsequent UV light exposure, leading to hydrophilic group formation and apatite generation on ZrO_2 surfaces, indicating enhanced bioactivity (Att et al., 2009; Han et al., 2009).

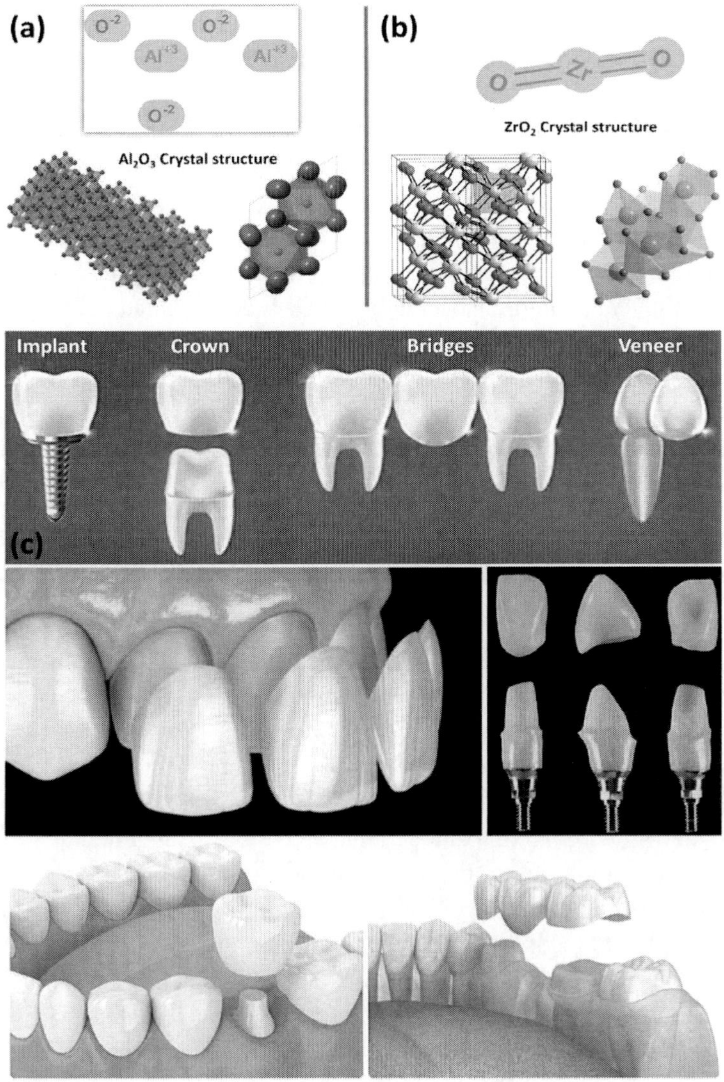

Figure 2. (a) Crystal structure of Aluminum oxide (α-Al_2O_3) and (b) Crystal structure of Zirconium oxide (ZrO_2). (c) Real photographs of ZrO_2 and Al_2O_3 prostheses and crowns for dental medical applications. Actual visual representations in the form of photographs depict prostheses and crowns fabricated from ZrO_2 and Al_2O_3, which are intended for utilization in dental medical contexts. These bioceramic-based dental solutions are designed with precision and expertise to meet the functional, aesthetic, and biocompatibility requirements of modern dental applications, ensuring the restoration of oral health and enhancing the overall well-being of patients.

A ZrO_2-HA composite, utilized as a coating for dental implants, demonstrated superior biocompatibility and mechanical properties compared to ZrO_2 or HA individually (Huang et al., 2013). Nanostructured coatings of fluorinated HA-ZrO_2 and CaO-ZrO_2-SiO_2 biocomposites were electrochemically deposited on Ti, showcasing strong bonding, low dissolution rates, and *in vitro* bioactivity (Yang et al., 2010; Huang et al., 2013). ZrO_2 serves as a promising material for porous bone scaffolds, aiding in bone tissue regeneration in tissue engineering (Chen et al., 2006). Incorporating ZrO_2 into nanocrystalline calcium phosphate (CaP) (e.g., HA, CeO or tricalcium phosphate) alters material strengths, stiffness, fracture toughness, and microhardness (Özer et al., 2017; Rapacz-Kmita et al., 2006). The addition of HA further enhances ZrO_2 bioactivity and osseointegration potential, making the combination ideal for hard tissue engineering. Notably, a microporous ZrO_2-HA composite scaffold, stabilized with Y_2O_3 and HA, exhibited strength akin to cortical bone, along with high protein adsorption and compatibility with cells and tissues (Özer and Öksüz, 2019; Goksen Tosun et al., 2023).

ZrO_2 is utilized as an additive in bone cement due to its toughness, radiopacity, and biocompatibility. It can be blended with methyl methacrylate monomers, poly(methyl methacrylate), and initiators to create a polymerizable bone cement. Incorporating ZrO_2 fibers into acrylic bone cement enhances tensile strength, compressive strength, and fracture toughness. Nanoscale ZrO_2 has shown potential in improving cytocompatibility and osteoblastic activity. Al_2O_3-ZrO_2 composite ceramics have been suggested as a solution for the challenges posed by long-term aging effects and as a practical alternative to monolithic ZrO_2. Sintered nano sized Al_2O_3-ZrO_2 composite powders exhibit biocompatibility without inflammation or wear debris during histological examination, making them suitable for clinical dental applications. Additionally, HA coating on this ceramic composite improves binding ability (Thomas Webster, 2009; Chevalier, 2006; Att et al., 2009). The crystal structures of α-Al_2O_3 and ZrO_2, along with actual photographs of ZrO_2 and Al_2O_3 prostheses and dental crowns designed for medical applications, are provided in Figure 2.

Conclusion

Bioceramics play an increasingly vital role in restoring both the form and function within the oral cavity. The restoration of lost tissue has posed a critical challenge in the realm of regeneration, raising questions about the most effective bioceramics of tissue replacement. The concept of employing conductive strategies introduces a complex dynamic into the regenerative process, carefully balancing the need to facilitate regeneration while also allowing host cells to populate the regrowth site optimally. A remarkable feature of bioceramics is their ability to create a localized environment conducive to cell growth, proliferation, and differentiation. This unique attribute transforms them into instructive extracellular microenvironments that actively contribute to morphogenesis. This inherent bioactivity positions ceramics as a tremendously promising strategy in the field of rehabilitation.

The realm of dental medicine, in particular, benefits significantly from the controlled structures achievable through various fabrication techniques for metal oxide bioceramics. These versatile materials find application across a wide spectrum of uses, with a special emphasis on dental medicine. Their biocompatibility, which categorizes them as both inert and active biomaterials, underlies their integration into composite structures that exhibit markedly enhanced biological performance.

Reviewing into specific metal oxides highlighted within this book chapter, calcium silicate (Ca_2SiO_4) and calcium phosphate ($Ca_3(PO_4)_2$) emerge as standout dental bioceramics. These materials, recognized for their biocompatibility, find substantial utilization in dental applications. Additionally, the clinical landscape has witnessed pioneering applications of bioceramics, notably Al_2O_3 and ZrO_2, which have found their place in orthopedics and dentistry, marking significant milestones in the advancement of bioceramic use within medical practice.

Disclaimer

None.

References

Afzal A. Implantable Zirconia Bioceramics For Bone Repair And Replacement: A Chronological Review. *Mater Express* (2014) 4 (1): 1–12. https://doi.org/10.1166/mex.2014.1148.

Aguilar F. G., Roberti Garcia, L. F., Panzeri Pires-de-Souza, F. C. Biocompatibility of New Calcium Aluminate Cement (EndoBinder). *J Endod* (2012) 38(3): 367–371. https://doi.org/10.1016/j.joen.2011.11.002.

Arora D. V. Bioactive Dentin Replacement. *IOSR JDMS* (2013) 12(4), 51–57. https://doi.org/10.9790/0853-1245157.

Att W., Takeuchi M., Suzuki T., Kubo K., Anpo M., Ogawa T. Enhanced Osteoblast Function On Ultraviolet Light-Treated Zirconia. *Biomaterials* (2019) 30(7): 1273–1280. https://doi.org/10.1016/j.biomaterials.2008.11.024.

Bansal R., Patil S., Chaubey K., Thakur R., Goyel P. Clinical Evaluation of Hydroxyapatite and β-Tricalcium Phosphate Composite Graft In The Treatment of Intrabony Periodontal Defect: A Clinico-Radiographic Study. *J Indian Soc of Periodontol* (2014) 18 (5): 610-620. https://doi.org/10.4103/0972-124x.142455.

Bryan T. E., Khechen K., Brackett M. G., Messer R. L., El-Awady A., Primus C. M., Gutmann J. L., Tay F. R. *In Vitro* Osteogenic Potential of an Experimental Calcium Silicate–based Root Canal Sealer. *J Endod* (2010) 36 (7): 1163–1169. https://doi.org/10.1016/j.joen.2010.03.034.

Camps J., About I. Cytotoxicity Testing of Endodontic Sealers: A New Method. *J Endod* (2003) 29 (9): 583–586. https://doi.org/10.1097/00004770-200309000-00010.

Chen Q. Z., Boccaccini A. R., Zhang H. B., Wang D. Z., Edirisinghe M. J. Improved Mechanical Reliability of Bone Tissue Engineering (Zirconia) Scaffolds by Electrospraying. *J Am Ceram Soc* (2006) 89(5): 1534–1539. https://doi.org/10.1111/j.1551-2916.2006.00935.x.

Chevalier J. What Future for Zirconia As a Biomaterial? *Biomaterials* (2006) 27(4): 535–543. https://doi.org/10.1016/j.biomaterials.2005.07.034.

Christel P. S. Biocompatibility of Surgical-Grade Dense Polycrystalline Alumina. *Clin Orthop Relat Res* (1992) 282: 10-18. https://doi.org/10.1097/00003086-199209000-00003.

D'Antò V., Di Caprio M. P., Ametrano G., Simeone M., Rengo S., Spagnuolo G. Effect of Mineral Trioxide Aggregate on Mesenchymal Stem Cells. *J Endod* (2010) 36(11): 1839–1843. https://doi.org/10.1016/j.joen.2010.08.010.

Dong X., Xu X. Bioceramics in Endodontics: Updates and Future Perspectives. *Bioeng* (2023) 10(3): 354. https://doi.org/10.3390/bioengineering10030354.

Dreger L. A. S., Felippe W. T., Reyes-Carmona J. F., Felippe G. S., Bortoluzzi E. A., Felippe M. C. S. Mineral Trioxide Aggregate and Portland Cement Promote Biomineralization *In Vivo*. *J Endod* (2012) 38(3): 324–329. https://doi.org/10.1016/j.joen.2011.11.006.

Eid A. A., Niu L. N., Primus C. M., Opperman L. A., Pashley D. H., Watanabe I., Tay F. R. *In Vitro* Osteogenic/Dentinogenic Potential of an Experimental Calcium Aluminosilicate Cement. *J Endod* (2013) 39(9): 1161–1166. https://doi.org/10.1016/j.joen.2013.04.005.

Fayazi S., Razmi H., Ostad S.. Effect of ProRoot MTA, Portland Cement, and Amalgam on The Expression of Fibronectin, Collagen I, and TGF-β by Human Periodontal Ligament Fibroblasts *In Vitro*. *IJDR* (2011) 22 (2): 190. https://doi.org/10.4103/0970-9290.84278.

Formosa L. M., Mallia B., Camilleri J. A Quantitative Method for Determining The Antiwashout Characteristics of Cement-Based Dental Materials Including Mineral Trioxide Aggregate. *Int Endod J* (2012) 46(2): 179–186. https://doi.org/10.1111/j.1365-2591.2012.02108.x.

Han Y., Yan Y., Lu C. Ultraviolet-Enhanced Bioactivity of ZrO_2 Films Prepared by Micro-Arc Oxidation. *Thin Solid Films* (2009) 517(5): 1577–1581. https://doi.org/10.1016/j.tsf.2008.09.064.

Huang Y., Yan Y., Pang X. Electrolytic Deposition of Fluorine-Doped Hydroxyapatite/ZrO_2 Films on Titanium for Biomedical Applications. *Ceram Int* (2013) 39(1): 245–253. https://doi.org/10.1016/j.ceramint.2012.06.017.

Islam I., Kheng Chng H., Jin Yap, A. U. Comparison of the Physical and Mechanical Properties of MTA and Portland Cement. *J Endod* (2006) 32(3): 193–197. https://doi.org/10.1016/j.joen.2005.10.043.

Jitaru S., Hodisan I., Timis L., Lucian A., Bud M. The Use of Bioceramics In Endodontics - Literature Review. *Med Pharm Rep* (2016) 89 (4): 470–473. https://doi.org/10.15386/cjmed-612.

Lenherr P., Allgayer N., Weiger R., Filippi A., Attin T., Krastl G. Tooth Discoloration Induced by Endodontic Materials: A Laboratory Study. *Int Endod J* (2012) 45(10): 942–949. https://doi.org/10.1111/j.1365-2591.2012.02053.x.

Li J., Hastings G. W. Chapter 5 Oxide Bioceramics: Inert Ceramic Materials in Medicine and Dentistry. *Handbook of Biomaterial Properties*, (2016) 339–352. https://doi.org/10.1007/978-1-4939-3305-1_21.

Mandeep Kaur, H. S. Biodentine: A Promising Dentin substitute. *JBR J Interdiscip Med Dent Sci* (2014) 2(05): https://doi.org/10.4172/2376-032x.1000140.

Matsui K., Yoshida H., Ikuhara Y. Nanocrystalline, Ultra-Degradation-Resistant Zirconia: Its Grain Boundary Nanostructure and Nanochemistry. *Sci Rep* (2014) 4(1). https://doi.org/10.1038/srep04758.

Mc Entire B. J. Bioceramics: From Clinic to Concept. *Bioceram Dev Appl* (2017) 07(01): 1000e108 https://doi.org/10.4172/2090-5025.1000e108.

Naiana Viana V., Mário Tanomaru F., Paulo Sérgio C. MTA Versus Portland Cement: Review of Literature. *RSBO*, (2012) 8(4): 446–452. https://doi.org/10.21726/rsbo.v8i4.1096.

Öksüz K. E. A Study on Al_2O_3/SiC/B_4C Reinforced Cu-Sn Matrix Composite by Warm Compaction Powder Metallurgy. *Adv Mater Res* (2015) 1128, 123–126. https://doi.org/10.4028/www.scientific.net/amr.1128.123.

Öksüz K. E., Apaydın F., Bozdağ A. E., Çevik M., Özer A. Phase and Morphological Evaluation of Mechanically Activated Sintered YAG Powders. *Procedia Materials Science*, (2015) 11, 44–48. https://doi.org/10.1016/j.mspro.2015.11.067.

Öksüz K., Şahin Y. Microstructure and Hardness Characteristics of Al_2O_3-B_4C Particle-Reinforced Cu Matrix Composites. *Acta Phys Pol A* (2016) 129(4): 650–652. https://doi.org/10.12693/aphyspola.129.650.

Öksüz K. E. Effect of Composition and Sintering Temperature on the Structure and Properties of Porous Bioactive Glass Scaffolds. *Chiang Mai J Sci* (2019) 46(3): 568–578. https://epg.science.cmu.ac.th/ejournal/journal-detail.php?id=10146.

Öksüz K. E., Kilinç S., Özer A. Effect of Calcination on Microstructure Development and Properties of Hydroxyapatite Powders Extracted from Human and Bovine Bones. *Trans Indian Ceram Soc* (2019) *78*(1): 41–45. https://doi.org/10.1080/0371750x.2019.1588170.

Öksüz K. E., Kurt B., Şahin İnan Z. D., Hepokur C. Novel Bioactive Glass/Graphene Oxide-Coated Surgical Sutures for Soft Tissue Regeneration. *ACS Omega*, (2023) 8(24): 21628–21641. https://doi.org/10.1021/acsomega.3c00978.

Öksüz K. E., Özer A. Microstructural and Phase Study of Y_2O_3 Doped Hydroxyapatite/Al_2O_3 Biocomposites, *Dig J Nanomater Biostructures*, (2016) 11: 167-172. https://www.chalcogen.ro/167_EmreO.pdf.

Öksüz K. E., Özkaya N. K., İnan Z. D. A., Özer A. Novel Natural Spider Silk Embedded Electrospun Nanofiber Mats for Wound Healing. *Mater Today Commun*, (2021) *26*: 101942. https://doi.org/10.1016/j.mtcomm.2020.101942.

Öksüz K. E., Properties of Nano Hydroxyapatite Powder Derived From Human Teeth. *Chem Res J* (2018) 3(6): 76-81. https://chemrj.org/download/vol-3-iss-6-2018/chemrj-2018-03-06-76-81.pdf.

Öksüz K., Özer A. Effect of Yttria on the Phase Formation and Sintering of HA-Al_2O_3 Biocomposites. *Acta Phys Pol A* (2017) 131(3): 576–580. https://doi.org/10.12693/aphyspola.131.576.

Özer A., Kumruoğlu L. C., Çelik M., Çağlar E., Öksüz K. E. Sintering Behavior and Microstructural Evaluation of MgO-doped α-alumina (Al_2O_3). *Journal of International Scientific Publications: Materials, Methods & Technologies* (2016) 10: 367-373. https://www.scientific-publications.net/en/article/1001160/.

Özer A., Öksüz K. The Effect of Yttrium Oxide in Hydroxyapatite/Aluminum Oxide Hybrid Biocomposite Materials: Phase, Mechanical and Morphological Evaluation. *Mater Werkst* (2019) 50(11): 1382–1390. https://doi.org/10.1002/mawe.201800141.

Özer U. H., Öksüz K. E., Özer A. Synthesis and Characterization of CeO_2 Doped ZrO_2 Ceramics by a Simple Sol-Gel Route. *Mater Sci Forum* (2017) 907: 56–60. https://doi.org/10.4028/www.scientific.net/msf.907.56.

Parirokh M., Torabinejad M. Mineral Trioxide Aggregate: A Comprehensive Literature Review- Part I: Chemical, Physical, and Antibacterial Properties. *J Endod* (2010) 36(1): 16–27. https://doi.org/10.1016/j.joen.2009.09.006.

Piconi C., Porporati A. A. Bioinert Ceramics: Zirconia and Alumina. *Handbook of Bioceramics and Biocomposites*, (2015) 1–25. https://doi.org/10.1007/978-3-319-09230-0_4-1.

Porter M. L., Bertó A., Primus C. M., Watanabe I. Physical and Chemical Properties of New-generation Endodontic Materials. *J of Endod* (2010) 36(3): 524–528. https://doi.org/10.1016/j.joen.2009.11.012.

Priyalakshmi S., Ranjan M. Review on Biodentine-A Bioactive Dentin Substitute. *IOSR JDMS*, (2014) 13(1): 13–17. https://doi.org/10.9790/0853-13131317.

Rapacz-Kmita A., Ślósarczyk A., Paszkiewicz Z. Mechanical properties of HAp–ZrO_2 composites. *J Eur Ceram Soc*, (2006) 26(8): 1481–1488. https://doi.org/10.1016/j.jeurceramsoc.2005.01.059.

Ravindran V., Jeevanandan G. Comparative Evaluation of the Physical and Antimicrobial Properties of Mineral Trioxide Aggregate, Biodentine, and a Modified Fast-Setting

Mineral Trioxide Aggregate Without Tricalcium Aluminate: An *In Vitro* Study. *Cureus* (2023) 15(8): e42856 https://doi.org/10.7759/cureus.42856.

Roberts H., Toth J., Berzins D., Charlton D. Mineral trioxide aggregate material use in endodontic treatment: A review of the literature. *Dent Mater*, (2008) 24(2): 149–164. https://doi.org/10.1016/j.dental.2007.04.007.

Şahin Y., Öksüz K. E. Effects of Al_2O_3 Nanopowders on the Wear Behavior of NiTi Shape Memory Alloys. *JOM* (2013) 66(1): 61–65. https://doi.org/10.1007/s11837-013-0823-9.

Şahin Y., Öksüz K. E. Tribological Behavior of Al_2O_3 and B_4C Particle-Reinforced Copper Matrix Investigated by The Taguchi Method. *Mater Test* (2016) 58(5): 453–461. https://doi.org/10.3139/120.110868.

Sakai V. T., Moretti A. B. S., Oliveira T. M., Fornetti A. P. C., Santos C. F., Machado M. A. A. M., Abdo R. C. C. Pulpotomy of Human Primary Molars with MTA and Portland Cement: A Randomised Controlled Trial. *Br Dent J*, (2009) 207(3): E5. https://doi.org/10.1038/sj.bdj.2009.665.

Shahi S., Rahimi S., Hasan M., Shiezadeh V., Abdolrahimi M. Sealing Ability of Mineral Trioxide Aggregate and Portland Cement for Furcal Perforation Repair: A Protein Leakage Study. *J Oral Sci* (2009) 51(4): 601–606. https://doi.org/10.2334/josnusd.51.601.

Shenoy A., Shenoy N. Dental ceramics: An update. *J Conserv Dent*, (2010) 13(4): 195. https://doi.org/10.4103/0972-0707.73379.

Skrtic D., Antonucci J. M., Eanes E. D., Eichmiller F. C., Schumacher G. E. Physicochemical Evaluation of Bioactive Polymeric Composites Based on Hybrid Amorphous Calcium Phosphates. *J Biomed Mater Res* (2000) *53*(4): 381–391. http://dx.doi.org/10.1002/1097-4636(2000)53:4<381::aid-jbm12>3.0.co;2-h.

Skrtic D., Antonucci J., Eanes E., Eidelman N. Dental Composites Based on Hybrid and Surface-Modified Amorphous Calcium Phosphates. *Biomaterials*, (2004) *25*(7–8): 1141–1150. https://doi.org/10.1016/j.biomaterials.2003.08.001.

Soon G., Pingguan-Murphy B., Lai K. W., Akbar S. A. Review of Zirconia-Based Bioceramic: Surface Modification and Cellular Response. *Ceram Int* (2016) 42(11): 12543–12555. https://doi.org/10.1016/j.ceramint.2016.05.077.

Surya Raghavendra S., Jadhav G. R., Gathani K. M., Kotadia P. Bioceramics In Endodontics – A Review. *JIUFD* (2017) 51(0): 128-137 https://doi.org/10.17096/jiufd.63659.

Thomas Webster, T. J. Nanofunctionalized Zirconia and Barium Sulfate Particles as Bone Cement Additives. *Int J Nanomedicine* (2009) 5: 1-11. https://doi.org/10.2147/ijn.s7603.

Thompson V. P., Rekow E. D. Dental Ceramics. *Bioceramics and Their Clinical Applications*, (2008) 518–547. https://doi.org/10.1533/9781845694227.2.518.

Torabinejad M., Parirokh M., Dummer P. M. H. Mineral Trioxide Aggregate and Other Bioactive Endodontic Cements: An Updated Overview - Part II: Other Clinical Applications and Complications. *Int Endod J* (2017) 51(3): 284–317. https://doi.org/10.1111/iej.12843.

Tosun N. G., Özer A., Bektaş T., Öksüz K. E., Erden Tayhan S., Özdemir T. Silk sericin-hydroxyapatite nanoribbons toward structurally stable osteogenic scaffolds. *J Aust Ceram Soc* (2023). https://doi.org/10.1007/s41779-023-00909-4.

Tuba U., Emine S. White Mineral Trioxide Aggregate Pulpotomies: Two Case Reports with Long-Term Follow-Up. *Contemp Clin Dent* (2011. 2(4): 381. https://doi.org/10.4103/0976-237x.91809.

Vivan R. R., Zapata R. O., Zeferino M. A., Bramante C. M., Bernardineli N., Garcia R. B., Hungaro Duarte M. A., Tanomaru Filho M., Gomes de Moraes I. Evaluation of the Physical and Chemical Properties of Two Commercial and Three Experimental Root-End Filling Materials. *Oral Surg, Oral Med, Oral Pathol, Oral Radio, Endod* (2010) 110 (2): 250–256. https://doi.org/10.1016/j.tripleo.2010.04.021.

Walker M. P., Diliberto A., Lee C. Effect of Setting Conditions on Mineral Trioxide Aggregate Flexural Strength. *J Endod* (2006) 32(4): 334–336. https://doi.org/10.1016/j.joen.2005.09.012.

Walter I. A. On the Material and the Tribology of Alumina-Alumina Couplings for Hip Joint Prostheses. *Clin Orthop Relat Res*, (1992) 282: 31-46. https://doi.org/10.1097/00003086-199209000-00005.

Washington J. T., Schneiderman E., Spears R., Fernandez C. R., He J., Opperman L. A. Biocompatibility and Osteogenic Potential of New Generation Endodontic Materials Established by Using Primary Osteoblasts. *J Endod* (2011) 37(8): 1166–1170. https://doi.org/10.1016/j.joen.2011.05.011.

Wei W., Qi Y. P., Nikonov S. Y., Niu L. N., Messer R. L., Mao J., Primus C. M., Pashley D. H., Tay F. R. Effects of an Experimental Calcium Aluminosilicate Cement on the Viability of Murine Odontoblast-like Cells. *J Endod* (2012) 38(7): 936–942. https://doi.org/10.1016/j.joen.2012.03.020.

Yang F., Xie Y., Li H., Tang T., Zhang X., Gan Y., Zheng X., Dai K. Human Bone Marrow-Derived Stromal Cells Cultured with A Plasma Sprayed $CaO\text{-}ZrO_2\text{-}SiO_2$ Coating. *J Biomed Mater Res B Appl Biomater* (2010) 95: B(1), 192–201. https://doi.org/10.1002/jbm.b.31702.

Zapf A. M., Chedella S. C., Berzins D. W. Effect of Additives on Mineral Trioxide Aggregate Setting Reaction Product Formation. *J Endod* (2015) 41(1): 88–91. https://doi.org/10.1016/j.joen.2014.07.020.

Zmener O., Pameijer C. H., Kokubu G. A., Grana D. R. Subcutaneous Connective Tissue Reaction to Methacrylate Resin-based and Zinc Oxide and Eugenol Sealers. *J Endod* (2010) 36(9): 1574–1579. https://doi.org/10.1016/j.joen.2010.06.019.

Biographical Sketch

Asst. Prof. Kerim Emre Öksüz

Email: emre.oksuz@cumhuriyet.edu.tr.
Web: https://avesis.cumhuriyet.edu.tr/emre.oksuz.
Address: Sivas Cumhuriyet Üniversitesi, Mühendislik Fakültesi A Binası, Metalürji ve Malzeme Mühendisliği Bölümü, 58140 Kampüs-Sivas, Türkiye

International Researcher IDs:
ScholarID: SZvXz-oAAAAJ&hl=tr.
ORCID: 0000-0001-7424-5930.
ScopusID: 50662051300.
Yoksis Researcher ID: 59245.

Education Information:

Doctorate, Sakarya University, Faculty of Engineering, Department of Metalurgical and Materials Engineering, Türkiye 2011-2018

Postgraduate, Sivas Cumhuriyet University, Mühendislik Fakültesi, Metalürji Mühendisliği, Türkiye 2008-2011

Undergraduate, Sivas Cumhuriyet University, Mühendislik Fakültesi, Metalürji Mühendisliği, Türkiye 2003-2007.

Biography:

K. Emre Öksüz, born in 1984, is currently working as a researcher at Sivas Cumhuriyet University (2023), Sivas, Türkiye. His primary research interest is a materials science, focusing on composite materials, nanomaterials and biomaterials, wear mechanism and characterization techniques. Öksüz finished his Ph.D. studies at Sakarya University, Türkiye, in 2018.

Foreign Languages:

German, B1 Intermediate
English, C1 Advanced.

Certificates, Courses and Trainings:

Vocational Course, Taramalı Elektron Mikroskobu (SEM) Sertifikası, ATÜSEM, Atılım Üniversitesi, Ankara, Atılım Üniversitesi, 2008

Vocational Course, Malzeme Karakterizasyon Teknikleri Sertifikası, ATÜSEM, Atılım Üniversitesi, Ankara, Atılım Üniversitesi, 2008

Foreign Language, Avrupa'da Uluslararası Gençlik Kampları, İngilizce-Fransızca destekli gençlik festivali, programi, (UNESCO/CCIVS–Le Comité de coordination pour le Service Volontaire International), Tunis, TUNUS, by United Nations Educational, Scientific, and Cultural Organization (UNESCO), 2007

Foreign Language, Avrupa'da Uluslararası Gençlik Kampları, Almanca destekli gençlik festivali programi, (UNESCO/CCIVS – Coordinating Commitee for International Voluntary Service), Baden-Würtemberg, Stut gart, Almanya, by United Nations Educational, Scientific, and Cultural Organization (UNESCO), 2006 Foreign Language, Gerçek Dil Merkezi, Almanca Dil Kursu, (3 Ay) M.E.B onaylı dil sertifikası. Sivas, Türkiye, Milli Eğitim Bakanlığı, 2004.

Dissertations:

Doctorate, The Efect of Various Oxides on the Electrical Properties of $BaTiO_3$ Based Ceramics, Sakarya University, Faculty of Engineering, Department of Metallurgical and Materials Engineering, 2018

Postgraduate, Toz Metalurjisi Yöntemiyle Kesici Uç (Soket) Üretilmesi ve Karakterizasyonu, Sivas Cumhuriyet Üniversitesi, Mühendislik Fakültesi, Metalürji ve Malzeme Mühendisliği Bölümü, 2011

Research Areas:

Biomedical Engineering, Biomechanics, Metallurgical and Materials Engineering, Material Science and Engineering, Ceramic Materials, Semiconductor and Superconductor Materials, Mechanical Properties, Electrical and Magnetic Properties, Composites, Polymeric Materials, Biomaterials, Thermal Treatment, Material Characterization, Mechanical Metallurgy, Nanomaterials, Thermal Properties, Engineering and Technology.

Academic Titles/Tasks:

Assistant Professor, Sivas Cumhuriyet Üniversitesi, Mühendislik Fakültesi, Metalürji Mühendisliği, 2010 – Continues.

Academic and Administrative Experience:

Deputy Head of Department, Sivas Cumhuriyet University, Engineering Faculty, Metallurgical and Materials Engineering, 2020-Continues.

Courses:

Electrochemistry and Corrosion, Postgraduate, 2020-2023
Corrosion, Undergraduate, 2020-2023
Casting Technology, Undergraduate, 2020- 2023
Advanced Molding and Casting Techniques, Postgraduate, 2020-2023
Heat Treatments of Metals, Undergraduate, 2020-2023
Mechanical Metalurgy Laboratory, Undergraduate, 2019-2023
Metalography Laboratory, Undergraduate, 2018-2023.

Published journal articles indexed by SCI, SSCI, and AHCI:

[1] Novel bioactive glass/graphene oxide-coated surgical sutures for soft tissue regeneration. Öksüz K. E., Kurt B., Şahin İnan Z. D., Hepokur C. *ACS Omega,* vol. 8, no.24, pp. 21628-21641, 2023 (SCI-Expanded).

[2] Plant-inspired adhesive and injectable natural hydro gels: in vitro and in vivo studies. Bohórquez-Moreno C. D., Öksüz K. E., Dinçer E., Hepokur C., Şen İ. *Biotechnology Letters*, 2023 (SCI-Expanded).

[3] Porous polymer scaffolds derived from bioresources for biomedical applications. Bohorquez-Moreno C. D., Öksüz K. E., Dinçer E. *Celulose Chemistry and Technology*, vol. 57, no.1-2, pp. 107-116, 2023 (SCI-Expanded).

[4] Investigation of physico chemical and biological properties of boron-doped biochar. Ateş A., Aydemir B., Öksüz K. E. *Biomass Conversion and Biorefinery,* 2023 (SCI-Expanded).

[5] Novel natural spider silk embedded electro spun nano fiber mats for wound healing. Öksüz K. E., Özkaya N. K., İnan Z. D. Ş., Özer A. *Materials Today Communications,* vol. 26, 2021 (SCI-Expanded).

[6] The effect of yttrium oxide in hydro xyapatite/aluminum oxide hybrid biocomposite materials: Phase, mechanical and morphological evaluation Der Einfluss von Yttriumoxid in Hydro xylapatit/Aluminiumoxid hybriden Bioverbundwerkstoffen: Gefuge, mechanische und morphologische evaluation. Özer A., Öksüz K. E. *Materialwissenschaft und Werkstofftechnik*, vol. 50, pp. 1382-1390, 2019 (SCI-Expanded).

[7] Effect of composition and sintering temperature on the structure and properties of porous bioactive glass scaffolds. Öksüz K. E. *Chiang Mai Journal of Science*, vol. 46, pp. 568-578, 2019 (SCI-Expanded).

[8] Effect of calcination microstructure development and properties of hydroxyapatite powders extracted from human and bovine bones. Öksüz K. E., Kilinç S., Özer A. *Transactions of the Indian Ceramic Society*, vol. 78, pp. 41-45, 2019 (SCI-Expanded).

[9] Processing and characterization of porous SiC/NiTi aloys for biomedical applications. Şimşir M., Akkan H., Öksüz K. E. *Kovove Materialy-Metallic Materials,* vol. 57, pp. 363-369, 2019 (SCI-Expanded).

[10] Effect of yttria on the phase formation and sintering of HA-Al_2O_3 biocomposites. Öksüz K. E., Özer A. *Acta Physica Polonica A,* vol. 131, pp. 576-579, 2017 (SCI-Expanded).
[11] Influence of ZrO_2 addition on the structure and dielectric properties of $BaTiO_3$ ceramics. Öksüz K. E., Şen Ş., Şen U. *Acta Physica Polonica A*, vol. 131, pp. 197-199, 2017 (SCI-Expanded).
[12] Tribological behavior of Al_2O_3 and B_4C particle-reinforced copper matrix investigated by the Taguchi method. Şahin Y., Öksüz K. E. *Materials Testing*, vol. 58, pp. 453-461, 2016 (SCI-Expanded).
[13] Microstructure and hardness characteristics of Al_2O_3-B_4C particle-reinforced Cu matrix composites. Öksüz K. E., Şahin Y. *Acta Physica Polonica A*, vol. 129, pp. 650-652, 2016 (SCI-Expanded).
[14] Microstructural and phase study of Y_2O_3 doped Hydroxyapatite/Al_2O_3 biocomposites. Öksüz K. E., Özer A. Digest *Journal of Nanomaterials and Biostructures*, vol. 11, pp. 167-172, 2016 (SCI-Expanded).
[15] The effects of aging on the hardness and wear behaviour of 2124 Al alloy/B_4C composites. Öksüz K. E., Oskay K. O. *Acta Physica Polonica A,* vol. 127, pp. 1367-1369, 2015 (SCI-Expanded).
[16] Effects of B_2O_3 addition on the sintering behavior of 0-5-10% ZrO_2 doped $BaTiO_3$ based ceramics. Öksüz K. E., Şen Ş., Şen U. *Acta Physica Polonica A*, vol. 127, pp. 1086-1089, 2015 (SCI-Expanded).
[17] Processing and mechanical properties of B_4C/Diamond impregnated Fe/Co MMC. Öksüz K. E., Şimşir M. *Acta Physica Polonica A*, vol. 127, no.4, pp. 1139-1141, 2015 (SCI-Expanded).
[18] Micro structural evaluation and characterization of solid-state reaction of B_2O_3 doped $BaZr_xTi_{1-x}O_3$ ceramics. Öksüz K. E., Şen Ş., Şen U. *Journal of The Australian Ceramic Society,* vol. 51, pp. 137-143, 2015 (SCI-Expanded).
[19] Effects of Al_2O_3 nanopowders on the wear behavior of NiTi shape memory alloys. Şahin Y., Öksüz K. E. *JOM,* vol. 66, pp. 61-65, 2014 (SCI-Expanded).
[20] Effects of sintering temperature and addition of Fe and B_4C on hardness and wear resistance of diamond reinforced metal matrix composites. Şimşir M., Öksüz K. E. *Journal of Superhard Materials,* vol. 35, no.6, pp. 369-382, 2013 (SCI-Expanded).
[21] Abrasive wear behaviour of SiCp-reinforced Al2124 Alloy composite based on the Taguchi Method. Şahin Y., Öksüz K. E. *Journal of The Balkan Tribological Association*, vol. 19, pp. 520-538, 2013 (SCI-Expanded).
[22] Tribological behaviour of Al2014-Al_2O_3 particle-reinforced composites produced by powder metallurgy method. Şahin Y., Öksüz K. E. *Journal of The Balkan Tribological Association*, vol. 19, pp. 190-201, 2013 (SCI-Expanded).

Articles Published in Other Journals:

[1] Investigation of the structure and hardness properties of *Anodonta anatina* mussel shells. Öksüz K. E., Şereflişan H. *Su Ürünleri Dergisi*, vol. 40, no.2, pp. 132-139, 2023 (Peer-Reviewed Journal).

[2] Microstructure of *Eobania vermiculata* (Müller, 1774): SEM, F-TIR and XRD methods. Öksüz K. E., Şereflişan H. *Journal of Agricultural Production (Online)*, vol. 3, no.2, pp. 42-47, 2022 (Peer-Reviewed Journal).
[3] The effect of ball miling speed on the penetration of Cr and Zr into Cu for production of Cu1CrZr alloy powder. Kul M., Özer A., Öksüz K. E., Kurt N. *International Scientific Journal Machines. Technologies. Materials.*, vol. 13, no.8, pp. 362-366, 2019 (Peer-Reviewed Journal).
[4] Investigations on microstructural and Raman scattering properties of B_2O_3 doped $Ba(Ti_{1-x}Zr_x)O_3$ ceramics. Öksüz K. E., Şen Ş., Şen U. *Chemistry Research Journal,* vol. 3, no.6, pp. 71-75, 2018 (Peer-Reviewed Journal).
[5] Properties of nano hydroxyapatite powder derived from human teeth. Öksüz K. E. *Chemistry Research Journal,* vol. 3, no.6, pp. 76-81, 2018 (Peer-Reviewed Journal).

Books & Book Chapters:

[1] 3D-Printable Materials for Tissue Engineering. Öksüz K.E. in: *Bioprinting the Human*, NOVA Science Publishers Inc., New York, pp. 17-36, 2022.
[2] Section : 11 Surface Modification. Öksüz K. E., Gün T., Şimşir M. in: *High Performance and Optimum Design of Structures and Materials Book,* Wessex Institute of Technology, UK, Editor, WIT Press, London, pp. 545-552, 2014.
[3] Composites Section 3: Hard and Super Hard Materials, Chapter 25 Investigation of Mechanical Behavior of B_4C/Diamond Reinforced Fe/Co Metal Matrix Composites Produced by PM. Şimşir M., Öksüz K. E., Şahin Y. in: *Powder Metalurgy for Automotive Aluminium Aloys Hard and Super Hard Materials Book,* P. Ramakrishan, Editor, Medtec Publisher, Scientific Int. Pvt. Ltd., Pune, pp. 203-209, 2014.

Refereed Congress/Symposium Publications in Proceedings:

[1] Erken evre meme kanseri tedavisinde pertuzumab ve Fe_2O_3 nano partikül içeren süngerin geliştirilmesi. Alandağ C., Köse S., Öksüz K. E., Göç R. *10. Türk Tibbi Onkoloji Kongresi*, Turkey, 26 April 2023.
[2] Meme kanseri tedavisi için sitratla stabilize edilmiş Fe_3O_4 nanopartiküleri/pertuzumab içeren makro gözenekli süngerin geliştirilmesi. Alandağ C., Aslan S., Öksüz K. E., Yilmaz Göç R. *10. Türk Tibbi Onkoloji Kongresi,* Antalya, Turkey, 26-30 April 2023, pp. 108.
[3] Asetaminofen ile hasar gören böbreklerde eritropoetin ve mezenkimal kök hücre uygulamasinin taramali elektron mikroskobu ile değerlendirilmesi. Göç R., Öksüz K. E., Çelik U., Çoban Z. B., Tandiroğlu A., Yay A. H., Açmaz G., Sezer G., Yakan B. *Karadeniz 9th International Conference on Aplied Sciences, Artvin,* Turkey, 25-26 June 2022, vol. 1, pp. 30-31.
[4] Ardiç yağı katkılı hidrofilik fiber ipliklerin antibakteriyel ve hemolitik aktivitelerinin incelenmesi. Öksüz K. E., Göç R., Dinçer E., Akdeniz E., Şahin F. B., Çelik U. *Karadeniz 9th International Conference on Aplied Sciences,* Artvin, Turkey, 25-26 June 2022, vol. 1, pp. 32.

[5] Biyoaktif nanoseramik katkılı biyopolimer filmlerin üretimi ve karakterizasyonu. Öksüz K. E., Göç R., Beyazit C., Alizadehyegani N., Akbarli N. *Karadeniz 9th International Conference on Aplied Sciences*, Artvin, Turkey, 25-26 June 2022, vol. 1, pp. 29.

[6] Characterization of Poly(vinyl alcohol)/sodium alginate nano fibers produced by electrospinning method. Tuna E. B., Öksüz K. E., Özer A. *6nd International Conference on Material Science and Technology in Cappadocia (IMSTEC'21)*, Nevşehir, Turkey, 26-28 November 2021, vol. 1, no.1, pp. 409-413.

[7] The effect of ball milling speed on the penetration of Cr and Zr into Cu for production of Cu1CrZr Alloy Powder. Kul M., Özer A., Öksüz K. E., Kurt N. *XVI. International Scientific Conference Machines. Technologies.*, Sofija, Bulgaria, 11-14 September 2019, vol. 3, pp. 233-237.

[8] Preparation of highly porous ZrO_2 bone scaffold. Öksüz K. E. 6. *International Conference on Mathematics, Engineering, Natural Medical Sciences, (EJONS 6)*, Adana, Turkey, 8-10 March 2019, vol. 1, pp. 595-600.

[9] Ultrahigh density α-Al_2O_3 ceramics produced by slip casting. Öksüz K. E. 6. *International Conference on Mathematics, Engineering, Natural Medical Sciences, (EJONS 6)*, Adana, Turkey, 8-10 March 2019, vol. 1, pp. 585-594.

[10] İpek serisin–Hap-RGD peptiti içeren nanofibriler liflerin kemik doku mühendisliği çalışmaları için üretilmesi. Bektaş T., Rischka K., Öksüz K. E., Özer A., Özdemir T. 23. *Biyomedikal Bilim ve Teknoloji Sempozyumu, BIOMED2018-TR*, İstanbul, Turkey, 15-16 December 2018, vol. 1, pp. 132.

[11] The sintering properties of Cu1CrZr alloys produced by mechanochemical activation. Öksüz K. E., Sübütay H., Şimşir M., Kumruoğlu L. C., Kul M., Özer A. *4th International Turkic World Conference on Chemical Sciences and Technologies ITWCCST* 2018, Kyyiv, Ukraine, 7-10 September 2018, vol. 1, pp. 0-63.

[12] The effect of ball miling time on the mechanochemical activation properties of Cu1CrZr alloys. Özer A., Öksüz K. E., Sübütay H., Kurt N., Şimşir M., Kumruoğlu L. C., Kul M. *4th International Turkic World Conference on Chemical Sciences and Technologies ITWCCST* 2018, Kiev, Ukraine, 7-10 September 2018, vol.1.

[13] Structural evolution and dielectric properties of $Ba(Ti_{1-x}Zr_x)O_3$ perovskite ceramics. Öksüz K. E., Şen Ş., Şen U. *ICSTR Athens – International Conference on Science TechnologyResearch*, 19-20 July, 2018, Athens, Greece, 19-20 July 2018, vol. 1, pp. 9.

[14] Effect of SiC particle size on the physical and mechanical properties of NiTi alloys. Akkan H., Şimşir M., Öksüz K. E. *10th International Conference on Materials Science Engineering (Bramat 2017)*, Braşov, Romania, 8-11 March 2017, vol. 907, pp. 3-7.

[15] Synthesis and characterization of CeO_2 doped ZrO_2 ceramics by a simple sol-gel route. Özer U. H., Öksüz K. E., Özer A. *10th International Conference on Materials Science Engineering (Bramat 2017)*, Braşov, Romania, 8-11 March 2017, vol. 907, pp. 56-60.

[16] Tribological behavior of distaloy DC components. Öksüz K. E., Akinalp Y. A., Kumruoğlu L. C. *10th International Conference on Materials Science Engineering (Bramat 2017)*, Brasov, Romania, 8-11 March 2017, vol. 907, pp. 121-125.

[17] The influence of composition and calcination temperature on the phases in $(Ba_{1-x}Sr_xZr)O_3$ perovskites. Torman M., Öksüz K. E., Şen Ş., Şen U. *10th International Conference on Materials Science Engineering (Bramat 2017),* Braşov, Romania, 8-11 March 2017, vol. 907, pp. 38-42.

[18] Dielectric properties of Sr-doped $BaZrO_3$ ceramics. Torman M., Öksüz K. E., Şen Ş., Şen U. *10th International Conference on Materials Science Engineering (Bramat 2017),* Braşov, Romania, 8-11 March 2017, vol. 907, pp. 21-24.

[19] Thermal behaviour and structure of $BaZrO_3$ nano particles. Öksüz K. E., Şen Ş., Şen U. *4th Central and Eastern European Commitee for Thermal Analysis and Calorimetry,* Kişinev, Moldova, 28-31 August 2017, vol. 1.

[20] Investigations on microstructural and dielectric properties of $BaTi_{1-x}Zr_xO_3$ ceramics. Öksüz K. E., Şen Ş., Şen U. *6th International Advances in Applied Physics and Materials Science Congress Exhibition,* İstanbul, Turkey, 1-03 June 2016.

[21] TiO_2 grafen nano kompozit partikül takviyeli koruyucu tekstillerin geliştirilmesi. Öksüz K. E., Özer A. *12. Ulusal Kimya Mühendisliği Kongresi (UKMK2016),* İzmir, Turkey, 23-26 August 2016, vol. 1, pp. 1298-1309.

[22] Microstructural and dielectric properties of $BaTi_{1-x}Zr_xO_3$ nano ceramics. Öksüz K. E., Şen Ş., Şen U. *45th Internatioal School Conference on the Physics of Semiconductors*, Katowice-Szczyrk, Poland, 18-24 June 2016.

[23] Effect of sintering temperature on dielectric properties of SiO_2 doped $BaTiO_3$ ceramics. Öksüz K. E., Torman M., Şen Ş., Şen U. *18th International Conference Materials, Methods Technologies*, Burgas, Bulgaria, 26-30 June 2016, vol. 10, pp. 361-366.

[24] Sintering behavior and microstructural evaluation of MgO doped Al_2O_3. Özer A., Kumruoğlu L. C., Çelik M., Emircan Ç., Öksüz K. E. *18th International Conference Materials, Methods Technologies,* Burgas, Bulgaria, 26-30 June 2016, vol. 10, pp. 367-373.

[25] Sintering characteristics and grain growth behavior of SiO_2 doped $BaTiO_3$ ceramics. Öksüz K. E., Şen Ş., Şen U. *International Conference on Material Science and Technology in Cappadocia (IMSTEC'16),* Nevşehir, Turkey, 6-08 April 2016, vol. 1, pp. 401-405.

[26] A Study on Al_2O_3-SiC-B_4C reinforced Cu-Sn matrix composite by warm compaction powder metallurgy. Öksüz K. E. *The 9th International Conference on Materials Science Engineering* – (BraMat 2015), Braşov, Romania, 5-07 March 2015, vol. 1128, pp. 123-126.

[27] Processing and characterization of porous Ti-HA biocomposites. Öksüz K. E., Kumruoğlu L. C., Dinçer C. *International Porous and Powder Materials Symposium and Exhibition, PPM 2015*, İzmir, Turkey, 15-18 September 2015, vol. 1, pp. 602-605.

[28] Effect of SiCp on the microstructure and mechanical properties of sintered distaloy DC composites. Öksüz K. E., Kumruoğlu L. C., Tur O. *5th International Biennial Conference on Ultrafine Grained and Nanostructured Materials, UFGNSM15,* Tehran, Iran, 11-12 November 2015, vol. 11, pp. 49-54.

[29] Phase and morphological evaluation of mechanically activated sintered YAG powders. Öksüz K. E., Apaydin F., Bozdag A. E., Cevik M., Ozer A. *5th*

International Biennial Conference on Ultrafine Grained and Nanostructured Materials, UFGNSM15, Tehran, Iran, 11-12 November 2015, vol. 11, pp. 44-48.

[30] Fabrication and characterization of TiNi/SiCp shape memory composites. Öksüz K. E., Şimşir M., Akkan H. *17th International Metalurgy Materials Congress (IMMC 2014),* İstanbul, Turkey, 11-13 September 2014, vol. 1, pp. 731-737.

[31] Production of V-B reinforced Fe matrix composites. Öksüz K. E., Çevik M., Ahmet Enbiya B., Özer A., Şimşir M. ICMSMM 2014), *XI International Conference on Materials Science, Metal and Manufacturing,* Barcelona, Spain, 18-19 August 2014, vol. 8, pp. 704-708.

[32] Characterization and mechanical properties of PM Fe-Cu/SiCp alloys. Öksüz K. E., Gün T., Şimşir M. *4th International Conference on Ultrafine Grained and Nano-Structured Materials (UFGNSM 2013),* Tehran, Iran, 5-06 November 2013, vol. 829, pp. 73-75.

[33] Fabrication and characterization of porous bioglass ceramic. Öksüz K. E., Bağirov H., Yilmaz H., Buket S., Yildirim V. *International Porous and Powder Materials Symposium and Exhibition,* İzmir, Turkey, 3-06 September 2013, pp. 602-605.

[34] Investigation of mechanical properties and microstructure of AA2024 and AA7075. Öksüz K. E., Bağirov H., Şimşir M., Ceyhun K., Aykut Ö., Yusuf Ziya D., Hayret in Utku B. *4th International Conference on Mechanical and Aerospace Engineering (ICMAE 2013),* Moscow, Russia, 20-21 July 2013, vol. 390, pp. 547-551.

[35] The effect of Al_2O_3-SiC and B_4C additions on mechanical properties of CuSn matrix composites for cutting tools. Öksüz K. E., Bağirov H., Yilmaz H., Buket S., Yildirim V. *The 3rd International Conference on Advanced Materials Research (ICAMR 2013),* Dubai, United Arab Emirates, 19-20, January 2013, vol. 685, pp. 15-18.

[36] S/N ratio analysis of tribological behavior of SiCp reinforced Al2124 based metal matrix composite. Şahin Y., Öksüz K. E. *The 3rd International Conference on Advanced Materials Research (ICAMR 2013)*, Dubai, United Arab Emirates, 19-20 January 2013, vol. 685, pp. 24-29.

[37] The effect of B_4C additions on mechanical properties of Fe-Cu-Co matrix composites for diamond tools. Öksüz K. E., Şahin Y., Yıldırım E. S., Kayir Y. *The 3rd International Conference on Advanced Materials Research (ICAMR 2013),* Dubai, United Arab Emirates, 19-20 January 2013, vol. 685, pp. 30-34.

[38] TM metodu ile üretilen Al_2O_3 takviyeli aluminyum esasli kompozitlerin aşinma davranişi için istatiksel model oluşturulmasi. Şahin Y., Öksüz K. E. *Tasarim-İmalat ve Analiz Kongresi ve CAD-CAM Günleri, (TİMAK)* 2012, Balıkesir, Turkey, 29-30, November 2012, vol. 1, pp. 442-455.

[39] The Microstructure and hardness of Al_2O_3 particle reinforced composite. Şahin Y., Öksüz K. E. *The 3rd International Conference on Mechanical and Aerospace Engineering, ICMAE'2012*, Paris, France, 7-08 July 2012, vol. 232, pp. 39-44.

[40] Caratterizzazione di Materiali Compositi a Matrice Metalica Fe Co e Rinforzo B_4C Impregnati di Diamante. Öksüz K. E., Şimşir M., Şahin Y. *Diamante Applicazioni Tecnologia*, Milan, Italy, 12 June - 12 September 2012, vol. 18, pp. 65-71.

[41] Wear behavior of diamond plus SiCp-reinforced metal matrix composite by Taguchi method. Şahin Y., Öksüz K. E., Şimşir M. *3rd International Conference on*

Manufacturing Science and Engineering (ICMSE 2012), Xiamen, China, 27-29 March 2012, pp. 1309-1312.
[42] Toz metalurjisi ile üretilen ve mermer doğal taş kesmede kulanılan kesici uçlara B_4C ilavesinin etkilerinin incelenmesi. Öksüz K. E., Şimşir M., Şahin Y. *6. Uluslararasi Toz Metalurjisi Konferansi ve Sergisi, (6IPMC)*, Ankara, Turkey, 5-09 October 2011, vol. 1, pp. 588- 591.
[43] Wear studies on diamond reinforced Fe-Co based metal matrix composites. Öksüz K. E., Şimşir M., Şahin Y. *6th International Powder Metalurgy Conference and Exhibition (6IPMC)*, Ankara, Turkey, 5-09 October 2011, vol. 1, pp. 151-158.
[44] The effect of process parameters on mechanical and dry wear behavior of aluminum composite. Şahin Y., Öksüz K. E. *6th International Powder Metalurgy Conference and Exhibition (6IPMC)*, Ankara, Turkey, 5-09 October 2011, vol. 1, pp. 146-150.
[45] Characterization of diamond impregnated B_4C Fe-Co MMC. Öksüz K. E., Şimşir M., Şahin Y. *Euro PM 2011 Congress and Exhibition*, Barcelona, Spain, 9-12 October 2011, vol. 1, pp. 247-253.
[46] Tribological behavior of diamond reinforced Fe-Co composite by Taguchi method. Öksüz K. E., Şimşir M., Şahin Y. *18th International Conference on Composite Materials*, Jeju-Si, South Korea, 21-26 August 2011, vol. 1, pp. 1-6.
[47] Investigation of the wear behavior of B_4C reinforced Fe/Co matrix composites produced by hot press. Şimşir M., Öksüz K. E., Şahin Y. *11th International Conference on the Mechanical Behavior of Materials (ICM)*, Como, Italy, 01 January 2011, vol. 10, pp. 3195-3201.
[48] Soğuma Hızının çelik takviyeli dökme demirin ara yüzey dayancı ve mikroyapiya etkisi. Şimşir M., Öksüz K. E. *3. Isil İşlem Sempozyumu ve Sergisi*, İstanbul, Turkey, 14-15 November 2009, pp. 52-61.

Supported Projects:

Kumruoğlu L. C., Öksüz K. E., Kul M., Tubitak Project, Silisyum Karbür Matrisli-Nitrür, Karbür, Oksit Takviyeli İleri Teknoloji Seramiklerinden Yüksek Sıcaklıklara Dayanıklı Kompozit Ekipman Üretimi, 2018-2019

Şimşir M., Kumruoğlu L. C., Tubitak Project, Grafen Takviyeli CuCr1Zr Matrisli Kompozitlerin Elektriksel, Termal ve Mekanik özelliklerinin İncelenmesi, 2017 - 2019

Öksüz K. E., Project Supported by scientific research projects department, Savunma sanayinde kullanılmak üzere tasarlanan insansız su altı keşif aracı ve uygulamaları, 2021-2022.

Activities in Scientific Journals:

J. Mineral, Metal and Mater., Eng., Commitee Member, 2020 - Continues
SCIREA Journal of Metalurgical Eng., Commitee Member, 2018 - Continues
Composite Materials Research, Commitee Member, 2018 - Continues
Materials Science: Materials Review, Commitee Member, 2018 - Continues

Journal of Biomaterials, Commitee Member, 2018 - Continues
Charact., and App., of Nanomater., Commitee Member, 2018 - Continues
Asian Journal of Applied Science and Techn., Commitee Member, 2016-2017.

Memberships/Tasks in Scientific Organizations:
Scientific & Techn., Res., Assoc., (STRA,USA), Member, 2018 - Continues
American Assoc., for Sci.Techn., (AASCIT,USA), Member, 2015 - Continues
Asia-Pacific Chemical, Biological&Environmental Eng., Soc., (APCBEES, HONG KONK), Member, 2015 -Continues
Turkish Ceramic Society (TSD,TÜRKİYE), Member, 2014 – Continues.

Scientific Refereeing:
Journal of the Australian Ceramic Society, SCI Journal, January 2023
Journal of the Australian Ceramic Society, SCI Journal, January 2023
Journal of the Australian Ceramic Society, SCI Journal, January 2023
Journal of the Australian Ceramic Society, SCI Journal, January 2023
Journal of the Australian Ceramic Society, SCI Journal, December 2022
Journal of the Australian Ceramic Society, SCI Journal, December 2022
Journal of the Australian Ceramic Society, SCI Journal, November 2022
Journal of the Australian Ceramic Society, SCI Journal, November 2022
Journal of the Australian Ceramic Society, SCI Journal, October 2022
Journal of the Australian Ceramic Society, SCI Journal, September 2022
Emerging Materials Research, SCI Journal, March 2022
Journal of the Australian Ceramic Society, SCI Journal, August 2021
Journal of the Australian Ceramic Society, SCI Journal, August 2021
Journal of the Australian Ceramic Society, SCI Journal, August 2021
Journal of the Australian Ceramic Society, SCI Journal, July 2021
Journal of the Australian Ceramic Society, SCI Journal, July 2021
Journal of the Australian Ceramic Society, SCI Journal, July 2021
Journal of the Australian Ceramic Society, SCI Journal, July 2021
Journal of the Australian Ceramic Society, SCI Journal, June 2021
Journal of the Australian Ceramic Society, SCI Journal, June 2021
Journal of the Australian Ceramic Society, SCI Journal, June 2021
Journal of the Australian Ceramic Society, SCI Journal, May 2021
Journal of the Australian Ceramic Society, SCI Journal, May 2021
Journal of the Australian Ceramic Society, SCI Journal, May 2021
Journal of the Australian Ceramic Society, SCI Journal, May 2021
Journal of the Australian Ceramic Society, SCI Journal, March 2021
Journal of the Australian Ceramic Society, SCI Journal, March 2021

Journal of the Australian Ceramic Society, SCI Journal, March 2021
Journal of the Australian Ceramic Society, SCI Journal, March 2021
Journal of the Australian Ceramic Society, SCI Journal, March 2021
Journal of the Australian Ceramic Society, SCI Journal, February 2021
Journal of the Australian Ceramic Society, SCI Journal, February 2021
Journal of the Australian Ceramic Society, SCI Journal, February 2021
Journal of the Australian Ceramic Society, SCI Journal, December 2020
Journal of the Australian Ceramic Society, SCI Journal, December 2020
Journal of the Australian Ceramic Society, SCI Journal, December 2020
Journal of the Australian Ceramic Society, SCI Journal, November 2020
Journal of the Australian Ceramic Society, SCI Journal, November 2020
Journal of the Australian Ceramic Society, SCI Journal, November 2020
Journal of the Australian Ceramic Society, SCI Journal, November 2020
Journal of the Australian Ceramic Society, SCI Journal, December 2020
Eng., Sci., Techn.,An Int., Journal-Jestech, SCI Journal, September 2020
Bioinspired Biomimetic and Nanobiomaterials, SCI Journal, August 2020
Powder Technology, SCI Journal, July 2020
Bioinspired Biomimetic and Nanobiomaterials, SCI Journal, July 2020
ACS Omega, SCI Journal, May 2020
S-N Applied Sciences, SCI Journal, April 2020
Acta Metallurgica Sinica-English Letters, SCI Journal, May 2019
Acta Metallurgica Sinica-English Letters, SCI Journal, November 2018
Acs Nano, SCI Journal, August 2018
Acta Metallurgica Sinica-English Letters, SCI Journal, October 2016
Acta Metallurgica Sinica-English Letters, SCI Journal, February 2016.

Tasks in Event Organizations:

Öksüz K. E., The 4th International Conference on Materials Science and Manufacturing Engineering (MSME 2021), Scientific Congress, Beijing, China, Kasım 2021.

Öksüz K. E., 2021 The 6th International Conference on Smart Materials Technologies (ICSMT 2021), Scientific Congress, Kazan, Russia, Mayıs 2021.

Öksüz K. E., 9th International Conference on Engineering and Innovative Materials (ICEIM 2020), Scientific Congress, Singapore, Singapore, Eylül 2020.

Öksüz K. E., 3rd International Joint Conference on Metallurgical and Materials Engineering (JCMME 2020), Scientific Congress, Seoul, South Korea, Temmuz 2020.

Öksüz K. E., International Conference on Recent Optimization Techniques in Science and Engineering (ROTSE 2020), Scientific Congress, Kolhapur, India, Mayıs 2020.

Öksüz K. E., International Joint Conference on Metallurgical and Materials Engineering (JCMME 2019), Scientific Congress, Auckland, New Zealand, Aralık 2019.

Öksüz K. E., International Conference on Advanced Mechanical Design, Systems and Manufacturing (AMDSA 2019), Scientific Congress, Shenzhen, China, Mart 2019.

Öksüz K. E., International Joint Conference on Metallurgical and Materials Engineering (JCMME 2018), Scientific Congress, Wellington, New Zealand, Aralık 2018.

Öksüz K. E., 3rd International Conference on Smart Materials Technologies (ICSMT 2018), Scientific Congress, Moscow, Russia, Haziran 2018.

Öksüz K. E., International Conference on Advanced Mechanical Design, Systems and Manufacturing (AMDSA 2018), Scientific Congress, Shanghai, China, Nisan 2018.

Öksüz K. E., 5th International Conference on Machine Design and Manufacturing (ICMDME 2018), Scientific Congress, Jakarta, Indonesia, Ocak 2018.

Öksüz K. E., 3rd International Conference on Education and Management Science (ICEMS 2018), Scientific Congress, Shenzhen, China, Ocak 2018.

Öksüz K. E., 2nd International Conference on Smart Materials Technologies (ICSMT 2017), Scientific Congress, SanktPeterburg, Russia, Ağustos 2017.

Metrics:

Publication:	78
Citation (WoS):	57
Citation (Scopus):	92
H-Index (WoS):	4
H-Index (Scopus):	7

Congress and Symposium Activities:

6. International Conference on Mathematics, Engineering, Natural & Medical Sciences, (EJONS 6), Session Moderator, Adana, Turkey, 2019.

Awards

Öksüz K. E., Özer A., 1st Award for Best Poster Presentation, Transilvania University of Brasov, March 2017.

Index

N

O

P

R

S

T

U

V

W

X

Z